Progress in Probability
Volume 24

Seminar on Stochastic Processes, 1990

E. Çınlar
Editor

P.J. Fitzsimmons
R.J. Williams
Managing Editors

Springer Science+Business Media, LLC 1991

E. Çınlar
Department of Civil Engineering and
Operations Research
Princeton University
Princeton, NJ 08544
USA

P.J. Fitzsimmons
R.J. Williams
(Managing Editors)
Department of Mathematics
University of California, San Diego
La Jolla, CA 92093
USA

ISBN 978-0-8176-3488-9 ISBN 978-1-4684-0562-0 (eBook)
DOI 10.1007/978-1-4684-0562-0

Printed on acid-free paper.

Originally published by Birkhäuser Boston in 1991
Softcover reprint of the hardcover 1st edition 1991

ISBN 978-0-8176-3488-9

Camera-ready copy provided by the editors.

9 8 7 6 5 4 3 2 1

FOREWORD

The 1990 Seminar on Stochastic Processes was held at the University of British Columbia from May 10 through May 12, 1990. This was the tenth in a series of annual meetings which provide researchers with the opportunity to discuss current work on stochastic processes in an informal and enjoyable atmosphere. Previous seminars were held at Northwestern University, Princeton University, the University of Florida, the University of Virginia and the University of California, San Diego. Following the successful format of previous years, there were five invited lectures, delivered by M. Marcus, M. Yor, D. Nualart, M. Freidlin and L. C. G. Rogers, with the remainder of the time being devoted to informal communications and workshops on current work and problems. The enthusiasm and interest of the participants created a lively and stimulating atmosphere for the seminar. A sample of the research discussed there is contained in this volume.

The 1990 Seminar was made possible by the support of the Natural Sciences and Engineering Research Council of Canada, the Southwest University Mathematics Society of British Columbia, and the University of British Columbia. To these entities and the organizers of this year's conference, Ed Perkins and John Walsh, we extend our thanks. Finally, we acknowledge the support and assistance of the staff at Birkhäuser Boston.

P. J. Fitzsimmons

R. J. Williams

La Jolla, 1990

LIST OF PARTICIPANTS

A. Al-Hussaini
R. Banuelos
R. Bass
D. Bell
R. Blumenthal
C. Burdzy
R. Dalang
D. Dawson
N. Dinculeanu
P. Doyle
E.B. Dynkin
R. Elliott
S. Evans
N. Falkner
R. Feldman
P. Fitzsimmons
K. Fleischmann
M. Freidlin
R. Getoor
J. Glover
L. Gorostiza
P. Greenwood
J. Hawkes
U. Haussmann
P. Hsu
P. Imkeller
O. Kallenberg
F. Knight
T. McConnell
P. McGill
P. March
M. Marcus
J. Mitro
T. Mountford
D. Nualart
M. Penrose
E. Perkins
M. Perman
J. Pitman
A. Pittenger
Z. Pop-Stojanovic
S. Port
R. Pyke
L.C.G. Rogers
J. Rosen
T. Salisbury
Y.C. Sheu
C.T. Shih
H. Sikic
R. Song
W. Suo
A.S. Sznitman
L. Taylor
E. Toby
R. Tribe
Z. Vondracek
J.B. Walsh
J. Watkins
S. Weinryb
R. Williams
M. Yor
B. Zangeneh
Z. Zhao

CONTENTS

A Note on Trotter's Proof of the Continuity of Local Time for Brownian Motion

A.A. BALKEMA

In his 1939 paper [1], P. Lévy introduced the notion of local time for Brownian motion as the limit of the occupation time of the space interval $(0,\epsilon)$ blown up by a factor $1/\epsilon$:

$$(1) \qquad L_\epsilon(t) = m\{s \in (0,t] \mid 0 < B(s) < \epsilon\}/\epsilon \to L(t) \text{ for } \epsilon \to 0+.$$

Here m is Lebesgue measure on R and B is Brownian motion on R started in 0. In this paper we give a simple proof of the a.s. continuity of local time based on a moment inequality for the occupation time of the Brownian excursion in [2] and the arguments of Trotter's 1958 paper, [3].

In Balkema & Chung [4] the bound $6\sqrt{c}\epsilon^3$ on the second moment of the occupation time of the space interval $(0,\epsilon)$ for the first excursion of duration exceeding $c > 0$ was used to prove relation (1). This bound is based on a general moment inequality in Theorem 9 of [2]. For the proof of the a.s continuity of local time we need the bound $120\sqrt{c}\epsilon^7$ on the fourth moment. The computation is similar and is omitted.

As in [4] let $S(c)$ denote the centered total occupation time divided by ϵ of the space interval $(0,\epsilon)$ for the first $n(c) = [u/\sqrt{2\pi c}]$ positive excursions of duration $> c$. Here $u > 0$ is fixed and we shall let $c > 0$ tend to 0. It was shown in the above paper, see (2.3), that $S(c) \to L_\epsilon(T_u)$ a.s. as $c \to 0$ where $u \mapsto T_u$ denotes the inverse function to local time in zero. Standard computation of the fourth moment of a sum of i.i.d. centered random variables gives a bound $C\epsilon^2(u+u^2)$ on the 4th moment of $S(c)$ for $0<\epsilon<1$. Fatou's lemma then yields as in Lemma 2.2 of the paper above

Lemma 1. $\mathcal{E}(L_\epsilon(T_u)-u)^4 \le \liminf_{c\to 0} \mathcal{E}S(c)^4 \le C\epsilon^2(u+u^2), \quad 0<\epsilon<1.$

The process L in (1) has continuous increasing unbounded sample functions. The inverse process is the Lévy process T which is a pure jump process. Note that $L(t) \stackrel{d}{=} |B(t)|$ and hence for $u, r>0$

$$(2) \qquad P\{T_u<r\} = P\{L(r)>u\} = P\{|B(1)| \ge u/\sqrt{r}\} \le e^{-u^2/2r}.$$

Lemma 2. The process $u \mapsto L_\epsilon(T_u)-u$ is a martingale.

Proof. Observe that $u \mapsto L_\epsilon(T_u)$ is a pure jump increasing Lévy process. This follows from the Itô decomposition, but can also be deduced from the independence of the Brownian motion $B_1(t)=B(T_u+t)$ and the stopped Brownian motion $B(t\wedge T_u)$. The random variable $L_\epsilon(T_u)$ has finite expectation $\mathcal{E}L_\epsilon(T_u)=cu$ and $c=1$ follows by letting $u\to\infty$ in Lemma 1.

The process $t\mapsto L_\epsilon(t)-L(t)$ is no martingale but the submartingale inequality holds at the times $t=T_u$: The jumps in the original process are replaced by continuous increasing functions in the new process. Lemma 1 gives for $\epsilon<1$

$$P\{\max_{t\le T_u}|L_\epsilon(t)-L(t)|>\alpha\} \le C\epsilon^2(u^2+u)/\alpha^4.$$

Let $r\ge 1$. Relation (2) with $u=2r^2$ then gives

$$(3) \qquad P\{\max_{t\le r}|L_\epsilon(t)-L(t)|>\alpha\} \le e^{-2r^3}+6C\epsilon^2r^4/\alpha^4, \quad 0<\epsilon<1.$$

The process L_ϵ defined in (1) and local time L are close if ϵ is small. The remainder of the argument follows Trotter's 1958 paper.

Lévy [1] proved that for almost every realization of Brownian motion the occupation time F defined by

$$F(x,t) = m\{s \le t \mid B(s) \le x\}$$

is a continuous function on $R \times [0,\infty)$. For $x = 0$ the right hand partial derivative $f(x,t) = \partial^+ F(x,t)/\partial x$ exists a.s. as a continuous increasing function in t. (Indeed $f(0,.) = L(.)$ is local time in 0 for Brownian motion.) By spatial homogeneity this holds in each point $x \in R$. Let Δ denote the set of dyadic rationals $k/2^n$. Since the set Δ is countable almost every realization F of occupation time has the property that it is continuous on $R \times [0,\infty)$ and that the function $f(x,.)$ is continuous on $[0,\infty)$ for each $x \in \Delta$. Fix such a realization F and define $f_n : R \times [0,\infty) \to R$ by

$$\begin{aligned} f_n(x,t) &= f(x,t) \text{ if } x = k/2^n \text{ for some integer } k \\ &= 2^n\left(F\left(\frac{k+1}{2^n},t\right) - F\left(\frac{k}{2^n},t\right)\right) \text{ for } k < 2^n x < k+1. \end{aligned}$$

The function f_n is a discrete approximation to $\partial F/\partial x$. Its discontinuities lie on the lines $x = k/2^n$. The function

$$d_n(x,t) = |f_n(x,t) - f_n(x-,t)| + |f_n(x,t) - f_n(x+,t)|$$

measures the size of the discontinuities of f_n.

Proposition 3. Let $t \mapsto f(z,t)$ be a continuous function on $[0,\infty)$ for each dyadic rational $z = k/2^n$. Let $F : R \times [0,\infty) \to R$ be continuous and define f_n and d_n as above. If there exist constants $c_n > 0$ with finite sum $\sum c_n < \infty$ such that

$$d_n(x,t) \le c_n \text{ on } [-n,n] \times [0,n] \text{ for all } n$$

then $\partial F/\partial x$ exists and is continuous on $R \times [0,\infty)$.

Proof. As in Trotter [3] one proves:

a) $f : \Delta \times [0,\infty) \to R$ is uniformly continuous on bounded sets (and hence has a continuous extension f^* on $R \times [0,\infty)$),

b) $f_n \to f^*$ uniformly on bounded sets,

c) $\partial F/\partial x = f^*$ on $R \times [0, \infty)$.

Theorem 4. Occupation time $F(x,t)$ for Brownian motion a.s. has a partial derivative with respect to x which is continuous on $R \times [0, \infty)$.

Proof. With $\epsilon = 2^{-n}$, $\alpha = n^{-2}$ and $r = n$ inequality (3) gives

$$\begin{aligned} p_n &= P\{d_n > 2/n^2 \text{ in some point } (x,t) \in [-n,n] \times [0,n]\} \\ &\leq 2 \cdot (2n2^n + 1) P\{\max_{t \leq n} |L_\epsilon(t) - L(t)| > 1/n^2\} \\ &\leq 2 \cdot (2n2^n + 1) \cdot (e^{-2n^3} + 6C2^{-2n}n^4n^8), \end{aligned}$$

and hence $\sum p_n < \infty$. The first Borel-Cantelli lemma shows that the conditions of Proposition 3 are satisfied a.s. Therefore the conclusion holds a.s.

References

[1] P. LÉVY, Sur certains processus stochastiques homogènes. *Compositio Math.* 7 (1939), 283-339.

[2] K.L. Chung, Excursions in Brownian motion. *Arkiv för Mat.* **14** (1976), 155-177.

[3] H. Trotter, A property of Brownian motion paths. *Illinois J. Math.* **2** (1958), 425-433.

[4] A.A. Balkema & K.L. Chung, Paul Lévy's way to his local time. In this volume.

A.A. Balkema
F.W.I., Universiteit van Amsterdam
Plantage Muidergracht 24
1018 TV Amsterdam, Holland

Paul Lévy's Way to His Local Time

A.A. BALKEMA and K.L. CHUNG

0. Foreword by Chung

In his 1939 paper [1] Lévy introduced the notion of local time for Brownian motion. He gave several equivalent definitions, and towards the end of that long paper he proved the following result. Let $\epsilon > 0$, $t > 0$, $B(0) = 0$,

$$L_\epsilon(t) = m\{s \in [0,t] \mid 0 < B(s) < \epsilon\}/\epsilon \tag{0.1}$$

where $B(t)$ is the Brownian motion in R and m is the Lebesgue measure. Then almost surely the limit below exists for all $t > 0$:

$$\lim_{\epsilon \to 0} L_\epsilon(t) = L(t). \tag{0.2}$$

This process $L(.)$ is Lévy's local time.

As I pointed out in my paper which was dedicated to the memory of Lévy, [2; p.174], there is a mistake in the proof given in [1], in that the moments of occupation time for an excursion were confounded with something else, not specified. Apart from this mistake which I was able to rectify in Theorem 9 of [2], Lévy's arguments can (easily) be made rigorous by standard "bookkeeping". As any serious reader of Lévy's work should know, this is quite usual with his intensely intuitive style of writing. Hence at the time when I wrote [2], I did not deem it necessary to reproduce the details. Nevertheless I scribbled a memorandum for my own file. Later, after I lectured on the subject in Amsterdam in 1975, I sent that memo to Balkema in the expectation that he would render it legible. This valuable sheet of paper has apparently been lost. In my reminiscences of Lévy [3], spoken at the Ecole Polytechnique in June, 1987, I recounted his invention of local time and the original proof of the theorem cited above. It struck me as rather odd that although a supposedly historical account of this topic was given in Volume 4 of Dellacherie-Meyer's encyclopaedic work [4], Lévy's 1939 paper was not even listed in the bibliography. This must be due to the failure of the authors to realize that the contents of that paper were not entirely reproduced in Lévy's 1948 book [5]. Be that as it may, incredible events posterior to the Lévy conference in 1987 (see the Postscript in [3]) have convinced me that very few people have read, much less understood, Lévy's own way to his invention. I have therefore asked Balkema to write a belated exposition based on my 1975 lectures on Brownian motion. Together with the results in my paper [2] on Brownian excursions this forms the basis of the present exposition of Lévy's ideas about local time. Now I wonder who among the latter-day experts on local time will have the curiosity (and humility) to read it?

1. Local time of the zero set of Brownian motion

One of the most striking results on Brownian motion is Lévy's formula:

$$B \stackrel{\mathrm{d}}{=} |B| - L^*$$

where B is Brownian motion and L^* is the local time of $|B|$ in zero defined in terms of the zero set of B. Lévy considered the pair $(M - B, M)$ where M is the max process for Brownian motion:

$$M_t = \max\{B(s) \mid s \leq t\},$$

and proved that the process $Y = M - B$ is distributed like the process $|B|$, using the at that time not yet rigorously established strong Markov property for Brownian motion. In one picture we have the continuous increasing process M and dangling down from it the process Y (distributed like $|B|$). Note that

M increases only on the zero set of Y. Problem: Can one express the sample functions of the increasing process M in terms of the sample functions of the process Y?

Let us define

$$T_u = \inf\{t > 0 \mid M(t) > u\} \qquad u \geq 0.$$

This is the right-continuous inverse process to M. Lévy observed that it is a pure jump process with stationary independent increments. It has Lévy measure $\rho(y, \infty) = \sqrt{(2/\pi y)}$ on $(0, \infty)$. There is a 1-1 correspondence between excursion intervals of Y and jumps of the Lévy process T. Hence the number of excursions of Y in $[0, T_u]$ of duration $> c$ is equal to the number $N = N_c(u)$ of jumps of T of height $> c$ during the interval $[0, u]$. For a Lévy process this number is Poisson distributed with parameter $u\rho(c, \infty) = u\sqrt{(2/\pi c)}$ in our case. In fact if we keep u fixed then $t \to N_{c(t)}$, with $c(t) = 2/\pi t^2$, is the standard cumulative Poisson process on $[0, \infty)$ with intensity u. The strong law of large numbers (for exponential variables) implies

$$N_c(u)/\sqrt{(2/\pi c)} \to u \text{ a.s.} \qquad \text{as } c = c(t) \to 0. \tag{1.1}$$

Now vary u. The counting process $N_c : [0, \infty) \to 0, 1, \ldots$ will satisfy (1.1) for all rational $u \geq 0$ for all ω outside some null set Ω_0 in the underlying probability space. For these realizations we have weak convergence of monotone functions and hence uniform convergence on bounded subsets (since the limit function is continuous). In particular we have convergence for each $u \geq 0$, also if $u = M_t(\omega)$ depends on ω. This proves:

Theorem 1.1 (Lévy). Let B be a Brownian motion and let $N_c^*(t)$ denote the number of excursion intervals of length $> c$ contained in $[0, t]$. Then

$$N_c^*(t)/\sqrt{(2/\pi c)} \to L^*(t) \text{ a.s.} \qquad \text{as } c \to 0$$

for some process L^* with continuous increasing sample paths in the sense of weak convergence. Moreover $(|B|, L^*) \stackrel{\mathrm{d}}{=} (M - B, M)$.

Corollary. L^* is unbounded a.s. and $L^*(0) = 0$.

Note that local time L^* has been defined in terms of the zero set $Z = \{t \geq 0 \mid B(t) = 0\}$. We call this process L^* the local time of the zero set of Brownian motion in order to distinguish it from the process L introduced in (0.2). The process L_ϵ in (0.1) depends on the behaviour of Brownian motion in the ϵ-interval $(0, \epsilon)$. For a discussion of local times for random sets see Kingman [6]. Here we only observe that one can construct another variant of local time in 0 by counting excursions of sup norm $> c$ rather than excursions of duration $> c$. The Lévy measure then is dy/y^2 rather than $dy/\sqrt{(2\pi y^3)}$. This latter procedure has the nice property that it is invariant for time change and hence works for any continuous local martingale.

The next result essentially is an alternative formulation of Theorem 1.1.

Lemma 1.2. Let $u > 0$, and let U_c be the upper endpoint of the $K(c)$th excursion of the Brownian motion B of duration $> c$. Assume that a.s. $K(c) \sim u\sqrt{(2/\pi c)}$ as $c \to 0$. Then $U_c \to T_u^*$ a.s. as $c \to 0$, where

$$(1.2) \qquad T_u^*(\omega) = \inf\{t > 0 \mid L_t^*(\omega) > u\}.$$

Proof. The process $u \mapsto T_u^*$ is a Lévy process since $L^* \stackrel{\mathrm{d}}{=} M$ by Theorem 1.1. Hence it has no fixed discontinuities. Choose a sample point ω in the underlying probability space such that

1) the function $L^*(\omega)$ in Theorem 1.1 is continuous, increasing, unbounded, and vanishes in $t = 0$,

2) the limit relation of Theorem 1.1 holds,

3) $K(c)(\omega) \sim u\sqrt{(2/\pi c)}, \quad c \to 0$,

4) the function $T^*(\omega)$ is continuous at the point u.

We omit the symbol ω in the expressions below. Let $0 < u_1 < u < u_2$ and let $N_i(c)$ denote the number of excursions of length $> c$ in the interval $[0, T_{u_i}^*]$ for $i = 1, 2$. Theorem 1.1 gives the asymptotic relation $N_i(c) \sim u_i\sqrt{(2/\pi c)}$ as $c \to 0$.

Hence for all sufficiently small c we have the inequality $N_1(c) < K(c) < N_2(c)$, and therefore $T^*_{u_1} < U_c < T^*_{u_2}$. The continuity of the sample function T^* at u then implies that $U_c \to T^*_u$.

This innocuous-looking lemma enables us to consider the $S(c)$ in Section 2 with a constant $n(c)$, rather than a random number, which would entail subtle considerations of the dependence between the sequence $\{\psi_n\}$ and the process L^*.

2. Local time as a limit of occupation time

In order to prove Theorem 1.1 using the occupation time of the interval $(0, \epsilon)$, $\epsilon \to 0$, rather than the number of excursions, one needs a bound on the second moment of the occupation time of the interval $(0, \epsilon)$ for the excursions. We begin with a simple but fundamental result.

Theorem 2.1. For fixed $c > 0$ the sequence of excursions of Brownian motion of duration exceeding c is i.i.d. provided the excursions are shifted so as to start in $t = 0$.

Proof. The upper endpoint τ_1 of the first excursion φ_1 of duration $> c$ is optional. By the strong Markov property the process $B_1(t) = B(\tau_1 + t)$, $t \geq 0$, is a Brownian motion and is independent of φ_1. Hence φ_1 is independent of the sequence $(\varphi_2, \varphi_3, \ldots)$ and $\varphi_1 \stackrel{d}{=} \varphi_2$ since φ_2 is the first excursion of B_1 of duration $> c$. Now proceed by induction.

As an aside let us show, as Lévy did, that this theorem by itself gives local time up to a multiplicative constant: Choose a sequence c_n decreasing to zero. We obtain an increasing family of i.i.d. sequences of excursions which contains all the excursions of Brownian motion. Each of these i.i.d. sequences acts as a clock. The large excursions of duration $> c_0$ ring the hours. The next sequence contains all excursions of duration $> c_1$ and ticks off the minutes. The next one the seconds, etc. Note that the number of minutes per hour is random: The sequence of excursions of duration $> c_1$ is i.i.d. and hence the subsequence of excursions of duration $> c_0$ is generated by a selection procedure which gives negative binomial waiting times with expectation $\sqrt{(c_0/c_1)}$. Similarly the number of seconds per hour is negative binomial with expectation $\sqrt{(c_0/c_2)}$. If we standardize the clocks so that the intertick times of the nth clock are $\sqrt{(c_n/c_0)}$ then the clocks become ever more accurate. The limit is local time for Brownian motion. Pursuing this line of thought one can show that the excursions of Brownian motion form a time homogeneous Poisson point process on a product space $[0,\infty)\times E$ where E is the space of continuous excursions and the horizontal axis is parametrized by local time. See Greenwood and Pitman [7] for details.

We now return to our main theme. Let $\psi_1, \psi_2, \ldots$ be the i.i.d. sequence of *positive* excursions of duration $> c$. This is a subsequence of the sequence (φ_n) of theorem 2.1. Given $\epsilon > 0$ let $f_\epsilon(\psi_n)$ denote the occupation time of the space interval $(0,\epsilon)$ for the nth excursion ψ_n:

$$f_\epsilon(\psi_n) = m\{t > 0 \mid 0 < \psi_n(t) < \epsilon\}$$

and set

$$X_n = f_\epsilon(\psi_n)/\epsilon.$$

Section 3 contains the proofs of the following key estimates:

$$\mathcal{E}(X_n) \sim \sqrt{2\pi c} \qquad c \to 0 \tag{2.1}$$

$$\mathcal{E}(X_n^2) \leq 6\epsilon\sqrt{c} \qquad 0 < \epsilon, 0 < c. \tag{2.2}$$

Now define

$$Y_n = X_n - \mathcal{E}X_n$$

$$S(c) = Y_1 + \cdots + Y_{n(c)}$$

where $n(c) = [u/\sqrt{2\pi c}]$ for some fixed $u > 0$. We are interested in the case $c \to 0$. We have by (2.2)

$$\mathcal{E}(Y_n^2) = \sigma^2(X_n) \leq 6\epsilon\sqrt{c}$$

which gives

$$\mathcal{E}(S(c)^2) \leq 6\epsilon u.$$

By (2.1) we have

$$\mathcal{E}(X_1 + \cdots + X_{n(c)}) = n(c)\mathcal{E}X_1 \to u \quad \text{as } c \to 0.$$

Let U_c denote the upper endpoint of the $n(c)$th positive excursion $\psi_{n(c)}$. Note that $\psi_{n(c)} = \varphi_{K(c)}$ is the $K(c)$th excursion of duration exceeding c and that $K(c) \sim 2n(c)$ a.s. by the strong law of large numbers for a fair coin. Lemma 1.2 shows that $U_c \to T_u^*$ a.s. as $c \to 0$ where T_u^* is defined in (1.2). Hence

$$X_1 + \cdots + X_{n(c)} \to L_\epsilon(T_u^*) \text{ a.s.} \quad \text{as } c \to 0. \tag{2.3}$$

Fatou' s lemma then yields

Lemma 2.2. $\mathcal{E}(L_\epsilon(T_u^*) - u)^2 \leq \liminf_{c\to 0} \mathcal{E}(S(c)^2) \leq 6\epsilon u.$

This inequality will enable us to prove (0.2).

Theorem 2.3. Define L_ϵ by (0.1). Then

$$L_\epsilon(t) \to L^*(t) \text{ a.s.} \quad \text{as } \epsilon \to 0 \tag{2.4}$$

in the sense of weak convergence of monotone functions.

Proof. It suffices to show that for each rational $u > 0$ the scaled occupation time

$$L_\epsilon(T_u^*) = m\{t \in [0, T_u^*] \mid 0 < B(t) < \epsilon\}/\epsilon \to u \text{ a.s.} \quad \text{as } \epsilon \to 0.$$

Since occupation time is increasing for fixed $\epsilon > 0$ and local time is continuous this will imply weak convergence. In the definition of $L_\epsilon(t)$ as a ratio both numerator and denominator are increasing in ϵ. Hence it suffices to prove the convergence for $\epsilon_n = n^{-4}$, as $n \to \infty$. We have by Lemma 2.2

$$p_n = P\{|L_{\epsilon_n}(T_u^*) - u| > \frac{1}{n}\} \leq 6n^2\epsilon_n u.$$

Since $\sum p_n$ is finite, the desired result follows from the Borel-Cantelli lemma.

As Chung comments in [3], the preceding proof is in the grand tradition of classical probability. But then, what of the result?

3. The moments of excursionary occupation

In this section we use the results in Chung [2], beginning with a review of the notation. Let

$$\gamma(t) = \sup\{s \mid s \leq t; B(s) = 0\}$$

$$\beta(t) = \inf\{s \mid s \geq t; B(s) = 0\}$$

$$\lambda(t) = \beta(t) - \gamma(t).$$

Then $(\gamma(t), \beta(t))$ is the excursion interval straddling t, and $\lambda(t)$ is its duration. For any Borel set A in $[0, \infty)$:

$$S(t; A) = \int_{\gamma(t)}^{\beta(t)} 1_A(|B(u)|)\, du$$

is the occupation time of A by $|B|$ during the said excursion. Its expectation conditioned on $\gamma(t)$ and $\lambda(t)$ has a density given by

$$\mathcal{E}(S(t; dx) \mid \gamma(t) = s, \lambda(t) = a) = 4xe^{-2x^2/a} dx. \tag{3.1}$$

This result is due to Lévy; a proof is given in [2]. Integration gives

$$\mathcal{E}(S(t; (0, \epsilon)) \mid \gamma(t) = s, \lambda(t) = a) = a(1 - e^{-2\epsilon^2/a}). \tag{3.2}$$

Next it follows from (2.22) and (2.23) in [2] that

$$P\{\lambda(t) \in da\} = \frac{1}{\pi}\sqrt{t/a^3}\, da \quad \text{for } a \geq t. \tag{3.3}$$

In particular if $r > c \geq t > 0$ then $P\{\lambda(t) > c\} > 0$ and

$$P(\lambda(t) \in dr \mid \lambda(t) > c) = \frac{1}{2}\sqrt{c/r^3}\, dr. \tag{3.4}$$

Lévy derived (3.4) from the property of the Lévy process T described in section 1 above. It is a pleasure to secure this fundamental result directly from our excursion theory.

What is the exact relation between the excursion straddling t and the sequence of excursions (φ_n) introduced in Section 2?

Recall that φ_n is the nth excursion of duration exceeding c for given $c > 0$. We claim that φ_1 is distributed like the excursion straddling c conditional on its duration exceeding c. To see this we introduce a new sequence of excursions (η_n) with excursion intervals (γ_n, β_n) of duration $\lambda_n = \beta_n - \gamma_n$. Define η_1 as the excursion straddling $t = c$ with excursion interval (γ_1, β_1); then define η_2 as the excursion straddling $t = \beta_1 + c$ with excursion interval (γ_2, β_2); η_3 as the excursion straddling $t = \beta_2 + c$, etc. Note that the post-β_1 process $B_1(t) = B(\beta_1 + t)$, is a Brownian motion which is independent of the excursion η_1. As in Theorem 2.1 a simple induction argument shows that the sequence (η_n) is i.i.d., at least if we shift the excursions so as to start at $t = 0$. Since for any sample point ω in the underlying probability space $\varphi_1(\omega)$ is the first element of the sequence $(\eta_n(\omega))$ of duration exceeding c, it follows that φ_1 is distributed like the excursion straddling c, conditional on its duration exceeding c.

Now we can compute by (3.2) and (3.4):

$$\frac{1}{\sqrt{c}}\mathcal{E}(S(t;(0,\epsilon)) \mid \lambda(t) > c) = \int_c^\infty r(1 - e^{-2\epsilon^2/r})\frac{dr}{2r^{3/2}}$$

$$\to \int_0^\infty r(1 - e^{-2\epsilon^2/r})\frac{dr}{2r^{3/2}} = \epsilon\sqrt{2\pi} \qquad \text{as } c \to 0.$$

This is (2.1) if we choose $t = c$.

Next Chung proved as a particular case of Theorem 9 in [2]:

$$\mathcal{E}(S(t;(0,\epsilon))^k \mid \gamma(t) = s, \lambda(t) = a) \leq (k+1)!\, \epsilon^{2k} \qquad k \geq 1. \tag{3.5}$$

For $k = 2$ this is the missing estimate mentioned in Section 0. But it is also trivial that

$$S(t;(0,\epsilon)) \leq \lambda(t). \tag{3.6}$$

Using (3.4), (3.5) and (3.6) we have

$$\begin{aligned}\mathcal{E}(S(t;(0,\epsilon))^2 \mid \lambda(t) > c) &= \int_c^\infty \mathcal{E}(S(t;(0,\epsilon))^2 \mid \lambda(t) = r)\frac{\sqrt{c}\,dr}{2r^{3/2}} \\ &\le \int_0^\infty (6\epsilon^4 \wedge r^2)\frac{\sqrt{c}\,dr}{2r^{3/2}} \\ &\le \sqrt{c}\left(6\epsilon^4 \int_{4\epsilon^2}^\infty \frac{dr}{2r^{3/2}} + \int_0^{4\epsilon^2} \frac{\sqrt{r}}{2}dr\right) \\ &\le 6\epsilon^3\sqrt{c}.\end{aligned}$$

Now choose $t = c$. Then $S(c;(0,\epsilon))$ conditional on $\lambda(c) > c$ is distributed like $f_\epsilon(|\varphi_1|)$. Hence

$$\mathcal{E}(X_n^2) \le 6\sqrt{c}\epsilon^3/\epsilon^2 = 6\epsilon\sqrt{c}.$$

This is (2.2).

References

[1] P. LÉVY, Sur certains processus stochastiques homogènes. *Compositio Math.* **7** (1939), 283-339.

[2] K.L. CHUNG, Excursions in Brownian motion. *Arkiv för Mat.* **14** (1976), 155-177.

[3] K.L. CHUNG, Reminiscences of some of Paul Lévy's Ideas in Brownian Motion and in Markov Chains. *Seminar on Stochastic Processes 1988*, 99-108. Birkhäuser, 1989. Also printed with the author's permission but without the Postscript in *Colloque Paul Lévy*, Soc. Math. de France, 1988.

[4] C. DELLACHERIE & P.A. MEYER, *Probabilités et Potentiel.* Chapitres XII à XVI, Hermann, Paris, 1987.

[5] P. LÉVY, *Processus Stochastiques et Mouvement Brownien.* Gauthier-Villars, Paris, 1948 (second edition 1965).

[6] J.F.C. KINGMAN, *Regenerative Phenomena.* Wiley, New York, 1972.

[7] P. GREENWOOD & J. PITMAN, Construction of local time and Poisson point processes from nested arrays. *J. London Math. Soc.* **(2) 22** (1980), 182-192.

A.A. Balkema
F.W.I., Universiteit van Amsterdam
Plantage Muidergracht 24
1018 TV Amsterdam, Holland

K.L. Chung
Department of Mathematics
Stanford University
Stanford, CA 94305

Transformations of Measure on an Infinite Dimensional Vector Space

DENIS BELL

1 Introduction

Let E denote a Banach space equipped with a finite Borel measure ν. For any measurable transformation $T: E \to E$, let ν_T denote the measure defined by $\nu_T(B) = \nu(T^{-1}(B))$ for Borel sets B. A transformation theorem for ν is a result which gives conditions on T under which ν_T is absolutely continuous with respect to ν, and which gives a formula for the corresponding Radon-Nikodym derivative (RND) when these conditions hold.

The study of transformation of a measure defined on a finite dimensional vector space is relatively straightforward. When E is infinite dimensional the situation is much more difficult and in this case treatment of the problem has largely been restricted to cases where ν is a Gaussian measure. In this paper we describe a procedure for deriving a transformation theorem for an arbitrary Borel measure on an infinite dimensional Banach space. Although formal, our argument yields a formula (10) for the RND $d\nu_T/d\nu$ which we believe to be new.

In §2 we give a brief survey of some existing results. In §3 we describe our method, which has also been discussed in [B.2, §5.3]. Finally in §4 we give some applications of our formula, in which we derive the RND's for the theorems described in §2. These applications do not appear in [B.2].

2 Transformation theorems for Gaussian measure

These come in two varieties, classical and abstract. The theory of transformation of the classical Wiener measure was developed by Cameron and Martin, and Girsanov. The Girsanov theorem is as follows:

Let w denote standard real valued Brownian motion and let h be a bounded measurable process adapted to the filtration of w. Let v denote the process

$$v_t = w_t - \int_0^t h_s ds \qquad t \geq 0$$

Define

$$G(w) \equiv \exp\left\{ \int_0^1 h_s dw_s - 1/2\int_0^1 h_s^2 ds \right\} \tag{1}$$

Then $v/[0,1]$ is a standard Brownian motion with respect to the measure $d\mu(w) = G(w)d\nu(w)$.

There has been a series of increasingly more general results concerning the transformation of abstract Gaussian measure. The quintessential paper in this area is due to Ramer [R]. Let (i,H,E) be an abstract Wiener space in the sense of Gross [G] with Gaussian measure ν on E, where the Hilbert space H has inner product $\langle .,. \rangle$ and norm $|\cdot|$.

<u>Theorem</u> (Ramer) Suppose $U \subset E$ is an open set and $T = I + K$ is a homeomorphism from U into E, where I is the identity map on E and K is an $H - C^1$ map from U into H such that its H - derivative is continuous from U into the space of Hilbert-Schmidt maps of H, and $I_H + DK(x) \in GL(H)$ for each x. Then the measure $\nu(T\cdot)$ is absolutely continuous with repect to ν and

$$\frac{d\nu(T\cdot)(x)}{d\nu} = |\delta(DT(x))| \exp\left[-\text{"}\langle K(x),x\rangle - \operatorname{tr} DK(x)\text{"} - 1/2|K(x)|^2 \right] \tag{2}$$

where δ denotes the Carleman-Fredholm determinant, and tr denotes trace, defined with respect to H. (The difference of the random variables contained inside " " is defined as a limit of a certain convergent sequence in L^2; each of the terms may fail to exist by itself.)

The following result is proved in [B.1]:

Theorem (Bell) Let ν be any finite Borel measure on E, differentiable in a direction $r \in E$ in the sense that there exists an L^1 random variable X such that the relations

$$\int_E D_r \theta d\nu = \int_E \theta \cdot X d\nu \qquad (3)$$

hold for all test functions θ defined on E. Suppose X satisfies the conditions: $t \in \mathbb{R} \to X(x + tr)$ is continuous a.a. x and the following random variables are locally integrable

$$\sup_{t \in [0,1]} [X(x + tr]^4, \qquad \sup_{s,t \in [0,1]} \exp\left\{-4\int_s^t X(x + ur)du\right\}$$

Define $T(x) \equiv x - r$, $x \in E$. Then ν_T and ν are equivalent and

$$\frac{d\nu}{d\nu^T}(x) = \exp\left\{-\int_0^1 X(x + ur)du\right\} \qquad (4)$$

3 The scheme

Let ν now denote an arbitrary finite Borel measure on a Banach space. Let $\mathcal{U}$ denote a distinguished subset of the class of functions on E. We make the following

Definition A linear operator $\mathcal{L}$ from $\mathcal{U}$ to $L^2(\nu)$ will be called an integration by parts operator (IPO) for ν if the following relations hold for all C^1 functions Φ: $E \to \mathbb{R}$ and all $h \in \mathcal{U}$ for which either side exists:

$$\int_E D\Phi(x)h(x)d\nu(x) = \int_E \Phi(x)(\mathcal{L}h)(x)d\nu(x)$$

Remark The Malliavin calculus provides a tool for obtaining IPO's for the measures induced by both finite and infinite dimensional-valued stochastic differential equations (the finite dimensional case was established by

Malliavin, Stroock, Bismut et al., see [B.2, Chs.2,3]; this was then extended to infinite dimensions by Bell [B.2 §7.3]).

Suppose $(\mathcal{L}, \mathcal{U})$ is an IPO for ν. The next result is easily verified (see [B.2 §5.3]).

Lemma Let $h \in \mathcal{U} \cap L^2(\nu)$, $\Psi: E \to \mathbb{R} \in L^2(\nu) \cap C^1$. Suppose $\Psi h \in \mathcal{U}$. Then

$$\mathcal{L}(\Psi h)(x) = \Psi(x)\mathcal{L}h(x) - D\Psi(x)h(x) \qquad \text{a.s. } (\nu)$$

Remark The set of functions for which this lemma is valid can be enlarged by a closure argument.

Suppose now that $T: E \to E$ is a map of the form $I + K$, where I is the identity on E and $K \in \mathcal{U}$. The key idea is to construct a homotopy T_t connecting T and the identity map. There are obviously many ways to do this; the simplest is to define

$$T_t(x) = I + tK(x), \qquad t \in [0,1]$$

Suppose that T_t defines a family of invertible maps of E. Note that $\nu_{T_t} \ll \nu$ for each t if and only if there exists a family $\{X_t: t \in [0,1]\}$ of L^1 random variables (i.e. the corresponding RND's) such that for all test functions Φ on E:

(i) $X_0 \equiv 1$

(ii) $$\int_E \Phi(x)d\nu(x) = \int_E \Phi \circ T_t^{-1}(x)X_t(x)d\nu(x) \qquad (5)$$

Note that the RHS in (5) (which we will denote by $f(t)$) is actually independent of t; thus $f'(t) \equiv 0$. This will enable us to derive formulae for X_t, assuming that certain formal manipulations are valid. Formally differentiating the expression for $f(t)$ inside the integral gives

$$0 \equiv \int_E \left\{ D\Phi(T_t^{-1}(x))\frac{d}{dt}T^{-1}(x)\cdot X_t(x) + \Phi\circ T_t^{-1}(x)\frac{d}{dt}X_t(x)\right\} d\nu(x) \tag{6}$$

The first term in the integrand can be simplified by using the easily derived relation

$$D\Phi(T_t^{-1}(x))d/dtT_t^{-1}(x) = -D(\Phi\circ T_t^{-1})(x)K\circ T_t^{-1}(x)$$

Substituting this into (6) gives

$$0 \equiv \int_E \left\{ \Phi\circ T_t^{-1}(x)\frac{d}{dt}X_t(x) - D(\Phi\circ T_t^{-1}(x))K\circ T_t^{-1}(x)X_t(x)\right\} d\nu(x)$$

Assume that for each $t \in [0,1]$, $K\circ T_t\cdot X_t \in \mathcal{U}$. Using the defining property of $\mathcal{L}$ in the last relation gives

$$0 \equiv \int_E \Phi\circ T_t^{-1}(x)\left\{\frac{d}{dt}X_t(x) - \mathcal{L}[K\circ T_t^{-1}\cdot X_t](x)\right\} d\nu(x)$$

Observe that this holds for all test functions Φ if and only if

$$d/dtX_t(x) - \mathcal{L}[K\circ T_t^{-1}\cdot X_t](x) \equiv 0, \qquad \text{a.s. } (\nu) \tag{7}$$

Suppose that $K\circ T_t^{-1}$ and X_t satisfy the respective conditions on h and Ψ in the previous lemma. Then applying the lemma to the second term in (7) yields

$$d/dtX_t(x) - X_t(x)\mathcal{L}[K\circ T_t^{-1}](x) + DX_t(x)K\circ T_t^{-1}(x) = 0 \tag{8}$$

We now write $X_t(x) \equiv X(t,x)$, $X_1 \equiv \partial X/\partial t$, $X_2 \equiv \partial X/\partial x$, and make the substitution $x = T_t(y)$ in (8). We then have

$$X_1(t,T_t(y)) - X(t,T_t(y))\mathcal{L}[K\circ T_t^{1}](T_t(y)) + X_2(t,T_t(y)K(y) = 0$$

Since $K = dT_t/dt$ this reduces to

$$d/dt X(t,T_t(y)) = X(t,T_t(y))\mathcal{L}[K\circ T_t^{-1}](T_t(y))$$

In view of (5,(i)) the above equation has the unique solution

$$X(t,T_t(y)) = \exp\left\{\int_0^t \mathcal{L}[K\circ T_s^{-1}](T_s(y))ds\right\}$$

We thus arrive at the following expression for X:

$$X(t,x) = \exp\left\{\int_0^t \mathcal{L}[K\circ T_s^{-1}](T_s\circ T_t^{-1}(x))ds\right\} \tag{9}$$

In particular

$$\frac{d\nu}{d\nu^T}(x) = \exp\left\{\int_0^1 \mathcal{L}[K\circ T_s^{-1}](T_s\circ T^{-1}(x))ds\right\} \tag{10}$$

Given a measure ν and a map T such that the family of maps T_t are invertible, the scheme is implemented by defining $X(t,x)$ as in (9) and, by reversing the steps in the above argument (note that <u>all</u> the steps are reversible), showing that (5) holds. This will yield a transformation theorem for ν with respect to the maps T_t, with $X(t,x)$ as the corresponding family of RND's. This was done in [B.1] in the special case K = constant; in this case the method yields the non-Gaussian theorem described in §1. One can presumably find a larger class of transformations for which the method is valid.

We now give a simple condition on K under which T_s is invertible for all $s \in [0,1]$. Recall that K is a <u>contraction</u> (on E) if there exists $0 \leqslant c < 1$ such that

$$\|K(x) - K(y)\|_E \leqslant c\|x - y\|_E, \ \forall\ x, y \in E$$

<u>Proposition</u> If K is a contraction then T_s is invertible for all $s \in [0,1]$.

Proof. It clearly suffices to prove that $T = I + K$ is invertible since sK is also a contraction for all $s \in [0,1]$. To show T is surjective, suppose y is any element of E and define $K'(x) = y - K(x)$, $x \in E$. Then K' is also a contraction. It therefore follows from the contraction mapping theorem that K' has a fixed point $x_0 \in E$. We then have $y - K(x_0) = x_0$, i.e. x_0 satisfies $T(x_0) = y$. To see that T is injective, suppose that $T(x_1) = T(x_2)$. This implies $\|x_1 - x_2\|_E = \|K(x_1) - K(x_2)\|_E \leq c\|x_1 - x_2\|_E$. Since $c < 1$ this implies $x_1 = x_2$.

4 Applications

(A) Suppose ν is the standard Wiener measure on the space of paths $C_0[0,1]$. Then the Ito integral

$$\mathcal{L}k = \int_0^1 k_s' \, dw_s$$

defines an IPO for ν. The domain $\mathcal{U}$ of $\mathcal{L}$ consists of the set of adapted paths k (which we think of as being functions of w) with square integrable time derivatives k'. This property of the Ito integral was first observed by Gaveau and Trauber [G-T]. (One can actually use functional techniques to define an IPO for ν with an extended domain containing non-adapted paths, and this gives rise to the Skorohod integral.)

Suppose h $(= h(w))$ satisfies the conditions in the Girsanov theorem, and define $K: C_0[0,1] \to \mathcal{U}$ by $K(w) \equiv -\int h_u du$ and $T \equiv I + K$. Suppose T is invertible and let S denote the inverse of T. The Girsanov theorem states that $\nu_S \ll \nu$ and $d\nu_S/d\nu = G$. Since G is positive this is equivalent to saying that $\nu_T \ll \nu$ and $(d\nu_T/d\nu)\circ T = 1/G$. We will use (10) to derive this formula.

Note that (10) gives

$$\frac{d\nu}{d\nu^T} \circ T(w) = \exp\left\{ \int_0^1 \mathcal{L}[K \circ T_s^{-1}](T_s(w))ds \right\}$$

Using the Ito integral form of $\mathcal{L}$ we have

$$\frac{d\nu}{d\nu^T} \circ T(w) = \exp\left\{ \int_0^1 \int_0^1 -h_u d\left[w_u - s\int_0^u h_v dv\right] ds \right\}$$

$$= \exp-\left\{ \int_0^1 \left[\int_0^1 h_u dW_u - s\int_0^1 h_u^2 du \right] ds\right\}$$

$$= \exp-\left\{ \int_0^1 h_u dW_u - 1/2\int_0^1 h_u^2 du\right\}$$

The last expression is equal to $1/G(W)$ as required.

(B) Let ν denote a Gaussian measure corresponding to an abstract Wiener space (i,H,E). Then ν has an IPO $\mathcal{L}$ defined by the divergence operator

$$\mathcal{L}K(x) = \langle K(x),x\rangle - \operatorname{tr} DK(x)$$

where $\langle .,. \rangle$ denotes the inner product on H and tr the trace with respect to H. An initial domain for $\mathcal{L}$ can be taken to be the set of C^1 functions from E into E^* (where E^* is identifed with a subset of E under the inclusions defined by the map i). However this domain can be extended in Ramer's sense and the extended domain $\mathcal{U}$ consists of precisely the class of functions $K: E \to H$ defined in the statement of Ramer's theorem. For $K \in \mathcal{U}$, one then has

$$\mathcal{L}K(x) = \text{"}\langle K(x),x\rangle - \operatorname{tr} DK(x)\text{"}$$

Thus

$$\mathcal{L}[K \circ T_s^{-1}](T_s(x)) = \text{"}\langle K(x),x\rangle + s|K(x)|^2 - \operatorname{tr} D[K \circ T_s^{-1}](T_s(x))\text{"}$$

In order to obtain (2) from this it will be necessary to do some manipulations on trace term above. Under the present assumptions these will necessarily be of a formal nature since, as we remarked earlier, the trace might fail to exist. One could overcome this difficulty by working with the approximations used to define " ", then passing to the limit. However in order to avoid having to this we will assume that K is a C^1 map into E^*. Under this assumption all the terms in " " exist separately. We have

$$\int_0^1 \mathcal{L}[K\circ T_s^{-1}](T_s\ (x))ds =$$

$$\langle K(x),x\rangle - \text{tr}\int_0^1 D[K\circ T_s^{-1}](T_s(x))ds + 1/2|K(x)|^2$$

Substituting this into (10) gives

$$\frac{d\nu}{d\nu^T}\circ T(x) = \exp\left\{ \langle K(x),x\rangle + 1/2|K(x)|^2 - \text{tr}\int_0^1 D[K\circ T_s^{-1}](T\ (x)ds\right\}$$

$$= |\text{Det } DT(x)|^{-1}\exp\left\{\langle K(x),x\rangle + 1/2|K(x)|^2\right\} \tag{11}$$

where (11) follows from the identity:

$$\exp\left\{-\text{tr}\int_0^1 D[K\circ T_s^{-1}](T_s(x))ds\right\} = |\text{Det } DT(x)|^{-1} \tag{12}$$

It is particularly easy to verify (12) under the assumption that K is a contraction from E into H, for in this case $\|DK(x)\|_{L(H)} < 1$. We then have

$$\int_0^1 D[K\circ T_s^{-1}](T_s(x))ds$$

$$= \int_0^1 DK(x)[DT_s(x)]^{-1}ds$$

$$= \int_0^1 DK(x)[I + sDK(x)]^{-1} ds$$

$$= \int_0^1 d/ds \, Log[I + sDK(x)] ds$$

$$= Log[I + DK(x)]$$

where Log is defined by a power series in the algebra of operators on H. This implies (12). It follows from (11) that

$$\frac{d\nu(T\cdot)}{d\nu}(x) = |Det\ DT(x)| \exp-\left\{<K(x),x> + 1/2|K(x)|^2\right\}$$

Thus we obtain the formula given by the transformation theorem of Kuo [K]. Under Ramer's assumptions it is necessary to introduce the term tr DK(x) into the exponential in order to obtain convergence, and the corresponding adjustment required outside the exponential converts the standard determinant into the Carleman-Fredholm determinant which appears in (2).

(C) Suppose that ν a finite Borel measure on E for which (3) holds. Define $\mathcal{U} \equiv \{cr: c \in \mathbb{R}\}$ and $\mathcal{L}$ on $\mathcal{U}$ by $\mathcal{L}(cr) \equiv c \cdot X$. Note that for $T_s \equiv I - sr$, $T_s^{-1} = I + sr$. Thus 10) gives

$$\frac{d\nu}{d\nu^T}(x) = \exp\left\{-\int_0^1 X(x + (1 - s)r) ds\right\}$$

$$= \exp\left\{-\int_0^1 X(x + sr) ds\right\}$$

Hence we obtain the formula in (4).

REFERENCES

[B.1] Bell D 1985 A quasi-invariance theorem for measues on Banach spaces, Trans Amer Math Soc **290** no.2: 841-845

[B.2] Bell D 1987 The Malliavin Calculus Pitman Monographs and Surveys in Pure and Applied Mathematics # 34, Wiley/New York

[G-T] Gaveau B and Trauber P 1982 L'integrale stochastique comme operateur de divergence dans L'espace fonctionnel, J Funct Anal **46**: 230-238

[G] Gross L Abstract Wiener Spaces 1965 Proc Fifth Berkeley Sympos Math Statist and Probability Vol 2, part 1: 31-42

[K] Kuo H-H 1971 Integration on infinite dimensional manifolds, Trans Amer Math Soc **159**: 57-78

[R] Ramer R 1974 On Nonlinear transformations of Gaussian measures, J Funct Anal **15**: 166-187

Denis Bell
Department of Mathematics
University of North Florida
Jacksonville
Florida 32216

Stochastic Integration in Banach Spaces

J. K. BROOKS and N. DINCULEANU

Introduction

The purpose of this paper is twofold: first, to extend the definition of the stochastic integral for processes with values in Banach spaces; and second, to define the stochastic integral as a genuine integral, with respect to a measure, that is, to provide a general integration theory for vector measures, which, when applied to stochastic processes, yields the stochastic integral along with all its properties. For the reader interested only in scalar stochastic integration, our approach should still be of interest, since it sheds new light on the stochastic integral, enlarges the class of integrable processes and presents new convergence theorems involving the stochastic integral.

The classical theory of stochastic integration for real valued processes, as it is presented, for example, by Dellacherie and Meyer in [D–M], reduces, essentially, to integration with respect to a square integrable martingale; and this is done by defining the stochastic integral, first for simple processes, and then extending it to a larger class of processes, by means of an isometry between certain L^2–spaces of processes. This method has been used also by Kunita in [K] for processes with values in Hilbert spaces, by using the existence of the inner product to prove the isometry mentioned above. But this approach cannot be used for Banach spaces, which lack an inner product. A number of technical difficulties emerge for Banach valued processes, and one truly appreciates the geometry that the Hilbert space setting provides in stochastic integration, after considering the general case. A new approach is needed for Banach valued processes.

On the other hand, the classical stochastic integral, as described above, is not a genuine integral, with respect to some measure. It would be desirable, as in classical Measure Theory, to have a space of "integrable" processes, with a norm on it, for which it is a Banach space, and an integral for the integrable processes, which would coincide with the stochastic integral. Also desirable would be to have Vitali and Lebesgue convergence theorems for the integrable processes. Such a goal is legitimate and many attempts have been made to fulfill it.

Any measure theoretic approach to stochastic integration has to use an integration theory with respect to a vector measure. Pellaumail [P] was the first to attempt such an approach; but due to the lack of a satisfactory integration theory, this goal was not achieved—even the establishment of a cadlag modification of the stochastic integral could not be obtained. Kussmaul [Ku.1] used the idea of Pellaumail and was able to define a measure theoretic stochastic integral, but only for real valued processes. He used in [Ku.2] the same method for Hilbert valued processes, but the goal was only partially fulfilled, again due to the lack of a satisfactory general integration theory.

The integration theory used in this paper is a general bilinear vector integration, with respect to a Banach valued measure with finite semivariation, developed by the authors in [B–D.2]. This theory seems to be tailor–made for application to the stochastic integral. For the convenience of the reader, we give a short presentation in section 1, and a more complete presentation in Appendix I. The technical difficulties encountered in applying this theory to stochastic integration have required us to extend and modify the integration theory given in [B–D.2] and to add a series of new results. We mention in this respect the extension theorem of vector measures (Theorem AI.3) which is an improvement over the existing extension theorems.

In order to apply this theory to define a stochastic integral with respect to a Banach valued process X, we construct a stochastic measure I_X on the ring $\mathcal{R}$ generated by the predictable rectangles. The process X is called *summable* if I_X can be extended to a σ–additive measure with finite semivariation on the σ–algebra $\mathcal{P}$ of predictable sets. Roughly speaking, the stochastic integral

$H \cdot X$ is the process $(\int_{[0,t]} H dI_X)_{t \geq 0}$ of integrals with respect to I_X.

The summable processes play in this theory the role played by the square integrable martingales in the classical theory. It turns out that every Hilbert valued square integrable martingale is summable; but we show by an example that for any infinite dimensional Banach space E, there is an E–valued summable process which is not even a semimartingale.

Not only does our theory allows to consider stochastic integration for a larger class of processes than the semimartingales, but even in the classical case our theory provides a larger space of integrable processes. Our space of integrable processes with respect to a given summable process X is a Lebesgue–type space, endowed with a seminorm; but, unlike the classical Lebesgue spaces, the simple processes are not necessarily dense. This creates considerable difficulty, since usually most properties in integration theory are proved first for simple functions and then are extended by continuity to the whole space. To overcome this difficulty, we proved a Lebesgue–type theorem (Theorem 3.1) which insures the convergence of the integrals (rather than the convergence in the Lebesgue space itself). We are able then to prove that our Lebesgue–type space is complete, that Vitali and Lebesgue convergence theorems are valid in this space, as well as weak compactness criteria and weak convergence theorems for the stochastic integral, which are new even in the scalar case.

The stochastic integral is extended then in the usual manner to processes that are "locally integrable" with respect to "locally summable" processes. It turns out that any caglad adapted process is locally integrable with respect to any locally summable process. This allows the definition of the quadratic variation which, in turns, is used in a separate paper [B–D.7] to prove the Itô formula for Banach valued processes, for use in the theory of stochastic differential equations in Banach spaces.

When is X summable? This crucial problem is treated in section 2. The answer to this problem, which constitutes one of the main results of this paper, can be stated, roughly, as follows: X is summable if and only if I_X is bounded on the ring $\mathcal{R}$ (Theorems 2.3 and 2.5). It is quite unexpected that the mere boundedness of I_X of $\mathcal{R}$ implies not only that I_X is σ–additive on $\mathcal{R}$, but that I_X has a σ–additive extension to $\mathcal{P}$. The proof of this result is quite involved and uses the above mentioned new extension theorem for vector measures as well as the theory of quasimartingales. The reader is referred to Appendix II for pertinent results concerning quasimartingales used in connection with the summability theory.

We mention that various definitions of a stochastic integral have been given in a Banach space setting (Pellaumail [P], Yor [Y_1], [Y_2], Gravereaux and Pellaumail [G–P], Métivier [M.1], Métivier and Pellaumail [M–P], Kussmaul [Ku.2] and Pratelli [Pr]). However, either the Banach spaces were too restrictive, or the construction did not yield the convergence theorems necessary for a full development of the stochastic integration theory.

Contents

1. Preliminaries

In this section we shall present some of the notation used throughout this paper. In addition, for the reader's convenience we shall quickly sketch, in a few paragraphs, the vector integration used in defining the stochastic integral. A full treatment is presented in Appendix AI. Finally, we present here the stochastic integral (that is the pathwise Stieltjes integral) with respect to processes with finite variation. The stochastic integral proper, with respect to summable processes, will be presented in section 3.

Notations

Throughout the paper, E, F, G will be Banach spaces. The norm of a Banach space will be denoted by $|\cdot|$. The dual of any Banach space M is denoted by M^*, and the unit ball of M by M_1. The space of bounded linear operators from F to G is denoted by $L(F, G)$. We write $E \subset L(F, G)$ to mean that E is isometrically embedded in $L(F, G)$. Examples of such embeddings are: $E = L(\mathbb{R}, E)$; $E \subset L(E^*, \mathbb{R}) = E^{**}$; $E \subset L(F, E \hat{\otimes}_\pi F)$; if E is a Hilbert space over the reals, $E = L(E, \mathbb{R})$.

We write $c_0 \not\subset G$ to mean that G does not contain a copy of c_0, that is, G does not contain a subspace which is isomorphic to the Banach space c_0.

A subspace $Z \subset D^*$ is said to be *norming* for D if for every $x \in D$ we have

$$|x| = \sup\{|\langle x, z\rangle| : z \in Z_1\}.$$

Obviously, D^* is norming for D, and $D \subset D^{**}$ is norming for D^*. Useful examples of a norming space are the following.

Let $(\Omega, \mathcal{F}, P)$ be a probability space, and $1 \leq p \leq \infty$. If $p < \infty$, then $L^p_E \equiv L^p_E(\Omega, \mathcal{F}, P)$ is the space of $\mathcal{F}$–measurable, E–valued functions such that $\|f\|^p_p = \int |f|^p dP < \infty$. If $p = \infty$, then L^∞_E denotes the space of E–valued, essentially bounded, $\mathcal{F}$–measurable functions. Note that $L^q_{E^*}$ is contained in $(L^p_E)^*$, where $\frac{1}{p} + \frac{1}{q} = 1$; if E^* has the Radon–Nikodym property, then $L^q_{E^*} = (L^p_E)^*$. One can show that $L^q_{E^*}$ is a norming space for L^p_E; if $\mathcal{F}$ is generated by a ring $\mathcal{R}$, then even the E^*–valued, simple functions over $\mathcal{R}$ form a norming space for L^p_E.

Vector integration

Let S be a non–empty set, Σ a σ–algebra of subsets of S and let $m : \Sigma \to E \subset L(F, G)$ be a σ–additive measure with finite semivariation $\tilde{m}_{F,G}$ (see AI for the definition of $\tilde{m}_{F,G}$).

For $z \in G^*$, let $m_z : \Sigma \to F^*$ be the σ–additive measure, with finite variation $|m_z|$, defined by $\langle x, m_z(A)\rangle = \langle m(A)x, z\rangle$, for $A \in \Sigma$ and $x \in F$. We denote by $m_{F,G}$ the set of all measures $|m_z|$ with $z \in G^*_1$.

If D is any Banach space, we denote by $\mathcal{F}_D(m_{F,G})$, the vector space of functions $f : S \to D$ belonging to the intersection

$$\cap\{L^1_D(|m_z|) : z \in G^*_1\},$$

and such that

$$\tilde{m}_{F,G}(f) := \sup\{\int |f| d|m_z| : z \in G^*_1\} < \infty.$$

Then $\tilde{m}_{F,G}(\cdot)$ is a seminorm and $\mathcal{F}_D(m_{F,G})$ is complete for this seminorm. We note that $\mathcal{F}_D(m_{F,G})$ contains all bounded measurable functions. But, unlike the classical integration theory, the step functions are not necessarily dense in $\mathcal{F}_D(m_{F,G})$.

The most important case is when $D = F$, for then we can define an integral $\int f dm \in G^{**}$, for $f \in \mathcal{F}_F(m_{F,G})$ as follows: since $f \in L^1_F(|m_z|)$ for every $z \in G^*$, the mapping $z \to \int f dm_z$ is an element of G^{**}, which we denote by $\int f dm$:

$$\langle z, \int f dm\rangle = \int f dm_z, \text{ for } z \in G^*.$$

Under certain conditions, we have $\int f dm \in G$, for example, if f is the limit in $\mathcal{F}_F(m_{F,G})$ of simple functions. If the set of measures $m_{F,G}$ is uniformly σ–additive, for example if $F = I\!R$, then $\int f dm \in F$, for any f in the closure, in $\mathcal{F}_F(m_{F,G})$, of the bounded measurable functions. Without this added hypothesis, this need not be true in general—a fact which causes many complications in vector integration theory.

Processes with finite variation

Let $(\Omega, \mathcal{F}, P)$ be a probability space and $(\mathcal{F}_t)_{t\geq 0}$ a filtration satisfying the usual conditions. Let $X : I\!R_+ \times \Omega \to E$ be a process.

We say that X has *finite variation* if for every $\omega \in \Omega$ and $t \geq 0$, the function $s \to X_s(\omega)$ has finite variation $\mathrm{Var}_{[0,t]} X_{(\cdot)}(\omega)$ on $[0,t]$. For every $t \geq 0$, we denote

$$|X|_t(\omega) = |X_0(\omega)| + \mathrm{Var}_{[0,t]} X_{(\cdot)}(\omega).$$

The process $|X| = (|X|_t)_{t\geq 0}$ is called the *variation process* of X. We note that $|X|_0 = |X_0|$. We say that X has *bounded variation* if $|X|_\infty(\omega) := |X|^*(\omega) = \sup_t |X|_t(\omega) < \infty$, for every $\omega \in \Omega$. The process X is said to have *integrable variation* if $|X|^* \in L^1(P)$.

For the remainder of this section we shall assume that $X : I\!R_+ \times \Omega \to E \subset L(F,G)$ is a cadlag process with finite variation $|X|$. Then $|X|$ is also cadlag. If X is adapted, then $|X|$ is also adapted (see [D.3]).

We say that a process $H : I\!R_+ \times \Omega \to F$ is *locally integrable* with respect to X, if for each $\omega \in \Omega$ and $t \geq 0$, the function $s \mapsto H_s(\omega)$ is Stieltjes integrable with respect to $s \mapsto |X|_s(\omega)$ on $[0,t]$; then we can define the Stieltjes integral $\int_{[0,t]} H_s(\omega) dX_s(\omega)$. The function $t \mapsto \int_{[0,t]} H_s(\omega) dX_s(\omega)$ is cadlag and has finite variation $\leq \int_{[0,t]} |H_s(\omega)| d|X|_s(\omega)$.

We say that H is *integrable* with respect to X if for each $\omega \in \Omega$ the Stieltjes integral $\int_{[0,\infty)} H_s(\omega) dX_s(\omega)$ is defined. Then, evidently, H is locally integrable with respect to X. If H is jointly measurable, to say that H is locally integrable with respect to X means that $\int_{[0,t]} |H_s(\omega)| d|X|_s(\omega) < \infty$ for every $\omega \in \Omega$ and $t \geq 0$.

If H is jointly measurable and locally integrable with respect to X, then we can consider the G–valued process $\left(\int_{[0,t]} H_s(\omega)dX_s(\omega)\right)_{t\geq 0}$. This process is cadlag and has finite variation; it is adapted if both X and H are adapted.

Assume X is *cadlag, adapted, with finite variation* and H is *jointly measurable, adapted* and *locally integrable* with respect to X. Then the *cadlag, adapted* process $\left(\int_{[0,t]} H_s(\omega)dX_s(\omega)\right)_{t\geq 0}$ is called the *stochastic integral* of H with respect to X and is denoted $H \cdot X$ or $\int HdX$:

$$(H \cdot X)_t(\omega) = \int_{[0,t]} H_s(\omega)dX_s(\omega), \text{ for } \omega \in \Omega \text{ and } t \geq 0.$$

We list now some properties of the stochastic integral:

1) The stochastic integral $H \cdot X$ has finite variation $|H \cdot X|$ satisfying

$$|H \cdot X|_t(\omega) \leq (|H| \cdot |X|)_t(\omega) < \infty,$$

where $|H| = (|H_t|)_{t\geq 0}$ and $|X| = (|X|_t)_{t\geq 0}$. If both H and X are real valued, then $|H \cdot X| = |H| \cdot |X|$.

2) If T is a stopping time, then X^T has finite variation and

$$X^T = 1_{[0,T]} \cdot X \text{ and } X^{T-} = 1_{[0,T)} \cdot X$$

3) Let T be a stopping time. Then H is locally integrable with respect to X^T (respectively X^{T-}) iff $1_{[0,T]}H$ (respectively $1_{[0,T)}H$) is locally integrable with respect to X. In this case we have

$$H \cdot X^T = (1_{[0,T]}H) \cdot X = (H \cdot X)^T$$

and

$$H \cdot X^{T-} = (1_{[0,T)}H) \cdot X = (H \cdot X)^{T-}.$$

4) If H is real valued and K is F–valued, then K is locally integrable with respect to $H \cdot X$ iff KH is locally integrable with respect to X. In this case we have

$$K \cdot (H \cdot X) = (KH) \cdot X.$$

4′) If H is F–valued and K is a real valued process such that KH is locally integrable with respect to X, then K is locally integrable with respect to $H \cdot X$ and we have

$$K \cdot (H \cdot X) = (KH) \cdot X.$$

5) $$\Delta(H \cdot X) = H \Delta X$$

where $\Delta X_t = X_t - X_{t-}$ is the jump of X at t.

In sections 3 and 4 we shall define the stochastic integral for processes X which are summable or locally summable, and we shall prove that the stochastic integral still has all these properties. A locally summable process is not necessarily with finite variation; and a process with finite variation is not necessarily locally summable. If X has (locally) integrable variation, then it is (locally) summable (Theorem 3.32 *infra*). The processes with integrable variation will be studied in section 3.

2. Summable processes

In this section, we shall introduce the notion of summability of a process X. This concept replaces, in some sense, in the Banach space setting, the classic assumption of X being a square integrable martingale, and allows us to define the stochastic integral $\int H dX$ for a larger class of predictable processes H than has been previously considered. For Hilbert valued processes X, we recover the classical stochastic integral. As we mentioned in the introduction, it turns out, surprisingly, that a mere boundedness condition on the stochastic measure I_X, induced by X, implies the summability of X.

Throughout this paper, $(\Omega, \mathcal{F}, P)$ is a probability space, $(\mathcal{F}_t)_{t \geq 0}$ is a filtration satisfying the usual conditions; $1 \leq p < \infty$; and $X : \mathbb{R}_+ \times \Omega \to E \subset L(F, G)$ is a cadlag, adapted process, with $X_t \in L^p_E(P) \equiv L^p_E$ for every $t \in \mathbb{R}_+$ (the terminology of Dellacherie and Meyer, [D–M], will be used).

We shall denote $\mathcal{R} = \mathcal{A}[0, \infty)$, the ring of subsets of $\mathbb{R}_+ \times \Omega$ generated by the predictable rectangles $[0_A]$, with $A \in \mathcal{F}_0$ and $(s, t] \times A$, with $0 \leq s < t < \infty$ and $A \in \mathcal{F}_s$. The σ–algebra of predictable sets is generated by $\mathcal{R}$.

There is a close connection between summability and quasimartingales (Theorem 2.5 *infra*). Facts concerning quasimartingales, taken from [B–D.5] and [Ku.1], are presented in Appendix AII.

Definition of summable processes

We define the finitely additive *stochastic measure* $I_X : \mathcal{R} \to L^p_E$, first for predictable rectangles by

$$I_X([0_A]) = 1_A X_0 \text{ and } I_X((s,t] \times A) = 1_A(X_t - X_s),$$

and then we extend it in an additive fashion to $\mathcal{R}$. We note that $I_X([0,t]\times\Omega) = X_t$, for $t \geq 0$. Frequently we shall write I in place of I_X. Since $E \subset L(F,G)$, we consider $L^p_E \subset L(F, L^p_G)$, and therefore the semivariation of I_X can be computed relative to the pair (F, L^p_G). The reader is referred to Appendix AI for relevant information concerning vector measures, such as semivariation, strong additivity, etc. Explicity, $\tilde{I}_{F,G}$, which denotes the semivariation of I_X relative to (F, L^p_G) is defined by

$$\tilde{I}_{F,G}(A) = \sup \|\Sigma I_X(A_i)x_i\|_{L^p_G}, \text{ for } A \in \mathcal{R},$$

where the supremum is extended over all finite families of vectors $x_i \in F_1$ and disjoint sets A_i from $\mathcal{R}$ contained in A. If I_X can be extended to $\mathcal{P}$, the semivariation of the extension is defined on sets belonging to $\mathcal{P}$ in an analogous fashion. We say that I_X has finite semivariation relative to (F, L^p_G) if $\tilde{I}_{F,G}(A) < \infty$, for every $A \in \mathcal{R}$.

2.1 DEFINITION. *We say that X is p–summable relative to (F,G) if I_X has a σ–additive L^p_G–valued extension (which will be unique), still denoted by I_X, to the σ–algebra $\mathcal{P}$ of predicatable sets and, in addition, I_X has finite semivariation on $\mathcal{P}$ relative to (F, L^p_G).*

If $p = 1$, we say, simply, that X is *summable relative to* (F,G).

If we consider $E = L(\mathbb{R}, E)$, and if X is p–summable relative to $(\mathbb{R}, E)$, we say that X is *p–summable*, without specifying the pair $(\mathbb{R}, E)$.

Remarks. (a) X is p–summable relative to $(\mathbb{R}, E)$ if and only if I_X has a σ–additive extension to $\mathcal{P}$, since in this case I_X is bounded in L^p_E on $\mathcal{P}$ and automatically has finite semivariation relative to $(\mathbb{R}, L^p_E)$.

(b) If $1 \leq p' < p < \infty$, and if X is p–summable relative to (F,G), then X is p'–summable relative to (F,G). In particular p–summable relative to (F,G) implies summable relative to (F,G). For this reason, most theorems stated and proved for summable processes remain valid for p–summable processes.

(c) If X is p–summable relative to (F, G), then X is p–summable relative to $(\mathbb{R}, E)$.

(d) If X is p–summable relative to (F, G), then for any $t \geq 0$ we have $X_{t-} \in L^p_E$ and $I_X([0, t) \times \Omega) = X_{t-}$. In fact, if $t_n \nearrow t$ then $X_{t_n} = I_X([0, t_n] \times \Omega) \to I_X([0, t) \times \Omega)$ in L^p_E and $X_{t_n} \to X_{t-}$ pointwise.

(e) We shall prove in the next sections that the following classes of processes are summable.

1) If $X : \mathbb{R}_+ \times \Omega \to E$ is a process with integrable variation then X is p–summable relative to any pair (F, G) such that $E \subset L(F, G)$ (Theorem 3.32 *infra*).

2) If E and G are Hilbert spaces, then any square integrable martingale $X : \mathbb{R} \to E \subset L(F, G)$ is 2–summable relative to (F, G) (Theorem 3.24 *infra*).

(f) By proposition AI.5, X is p–summable relative to (F, G) iff I_X has a σ–additive extension to $\mathcal{P}$ and I_X has bounded semivariation on $\mathcal{R}$ (rather than on $\mathcal{P}$) with respect to (F, L^p_G). It follows that the problem of summability reduces to a great extent to that of the σ–additive extension of I_X from $\mathcal{R}$ to $\mathcal{P}$.

(g) Once the summability of X is assured, we can apply Appendix AI to the measure I_X and define an integral with respect to I_X. This will lead to the stochastic integral which will be studied in section 3.

Extension of I_X to stochastic intervals

The σ–algebra $\mathcal{P}$ of predictable subsets of $\mathbb{R}$ contains stochastic intervals of the form

$$(S, T] = \{(t, \omega) \in \mathbb{R} \times \Omega : S(\omega) < t \leq T(\omega)\},$$

where $S \leq T$ are stopping times (possibly infinite). Other stochastic intervals are similarly defined. If I_X is extended to $\mathcal{P}$, it is convenient to extend it further to sets of the form $\{\infty\} \times A$, with $A \in \mathcal{F}_\infty := \vee_{t \geq 0} \mathcal{F}_t$, by setting $I_X(\{\infty\} \times A) = 0$. Then $\mathcal{P} \cup (\{\infty\} \times \mathcal{F}_\infty)$ is the σ–algebra $\mathcal{P}[0, \infty]$ of predictable subsets of $\overline{\mathbb{R}}_+ \times \Omega$, where $\overline{\mathbb{R}}_+ = [0, \infty]$, and the above extension is still σ–additive. Then $I_X\ ((S, T])$ has the same value whether $(S, T]$ is regarded as a subset of $\mathbb{R}$, or as a subset of $\overline{\mathbb{R}}_+ \times \Omega$ defined by $(S,T] = \{(t,\omega) \in \overline{\mathbb{R}}_+ \times \Omega : S(\omega) < t \leq T(\omega)\}$. Similar considerations hold for other types of predictable stochastic intervals, and in particular for $I_X([T])$ if T is a predictable stopping time.

The following theorem extends the computation of I_X from predictable rectangles to stochastic intervals.

2.2 THEOREM. *Assume that X is p–summable relative to (F,G) and regard I_X as the unique extension of I_X to $\mathcal{P}$. Then*

(a) There is a random variable, denoted by X_∞, belonging to L^p_E, such that $\lim_{t\to\infty} X_t = X_\infty$ in L^p_E, and $I_X((t,\infty)\times A) = 1_A(X_\infty - X_t)$, for $A \in \mathcal{F}_t$. If X has a pointwise left limit $X_{\infty-}$, then $X_{\infty-} = X_\infty$ a.s.

Consider now X extended at ∞, by a representative of X_∞, and define $X_{\infty-}$ to be X_∞.

(b) For any stopping time T, we have $X_T \in L^p_E$ and $I_X([0,T]) = X_T$.

(c) If T is a predictable stopping time, then $X_{T-} \in L^p_E$ and $I_X([0,T)) = X_{T-}$ and $I_X([T]) = \Delta X_T$.

(d) If $S \leq T$ are stopping times, then $I_X((S,T]) = X_T - X_S$. If S is predictable, then $I_X([S,T]) = X_T - X_{S-}$. If T is predictable, then $I_X((S,T)) = X_{T-} - X_S$. If both S and T and predictable, then $I_X([S,T)) = X_{T-} - X_{S-}$.

Proof. Let $t_n \nearrow \infty$. Since I_X is σ–additive on $\mathcal{P}$, we have $I_X([0,\infty)\times\Omega) = \lim_n I_X([0,t_n]\times\Omega) = \lim X_{t_n}$ in L^p_E. Set $X_{\infty-} = X_\infty = I_X([0,\infty)\times\Omega)$. The rest of (a) easily follows.

To prove (b), assume first that T is a simple stopping time; it follows that $I_X((T,\infty)) = X_\infty - X_T$. For the general case, when T is an arbitrary stopping time, let $T_n \downarrow T$, where the T_n are simple stopping times. Since I_X is σ–additive, we have $I_X((T,\infty)) = \lim_n I_X((T_n,\infty)) = \lim_n(X_\infty - X_{T_n})$ in L^p_E. By right continuity of X, we have $X_\infty - X_T = \lim_n(X_\infty - X_{T_n})$ a.s., hence $X_T \in L^p_E$ and (b) follows.

To prove (c), let T be predictable and let $T_n \nearrow T$, where each T_n is a stopping time. Hence $I_X([0,T)) = \lim_n I_X([0,T_n]) = \lim_n X_{T_n} = \lim_n(X_{T_n}1_{\{T<\infty\}} + X_{T_n}1_{\{T=\infty\}}) = X_{T-}1_{\{T<\infty\}} + X_\infty 1_{\{T=\infty\}} = X_{T-}$, in L^p_E, and the rest of (c) follows, as well as (d).

Summability criteria

The following theorems give necessary and sufficient conditions for a process X to be p–summable. It is interesting to note that, if $E \not\supset c_0$, the mere boundedness of I_X on $\mathcal{R}$ implies that X is p–summable relative to $(\mathbb{R}, E)$; and bounded semivariation on $\mathcal{R}$, relative to (F, L^p_G) implies that X is p–summable relative to (F, G).

Summability of X reduces to σ–additivity of I_X which will be studied in the next subsection.

One of the main results of this section is the following.

2.3 THEOREM. *Assume that E does not contain a copy of c_0. If I_X is bounded in L^p_E on $\mathcal{R}$, then X is p–summable relative to $(\mathbb{R}, E)$. If I_X has bounded semivariation on $\mathcal{R}$, relative to (F, L^p_G), then X is p-summable relative to (F, G).*

The above theorem will follow from our fundamental σ–additive extension Theorem 2.5 *infra*, and the fact that if a vector measure has finite semivariation on $\mathcal{R}$, relative to a pair (F, G), then its extension to $\mathcal{P}$, if it exists, has finite semivariation on $\mathcal{P}$ relative to (F, G) (Theorem AI.5 *infra*).

We state a corollary of the above theorem.

2.4 COROLLARY. *Assume X is real valued and regard $\mathbb{R} \subset L(F, F)$. Then X is summable relative to (F, F) if and only if I_X has bounded semivariation on $\mathcal{R}$ relative to (F, L^p_F).*

σ–additivity and the extension of I_X

For every $g \in L^q_{E^*}$, we denote by $G = (G_t)_{t \geq 0}$ the martingale defined by $G_t = E(g|\mathcal{F}_t)$, and by XG the real valued process $((\langle X_t, G_t\rangle))_{t\geq 0}$, where $(x, x^*) \mapsto \langle x, x^*\rangle$ is the "duality mapping" on $G \times G^*$. We also denote $\langle f, g\rangle = E(\langle f(\cdot), g(\cdot)\rangle)$ the duality mapping in $L^p_G \times L^q_{G^*}$.

The following theorem gives a characterization of a process X to have a σ–additive extension of I_X to $\mathcal{P}$. Note that just requiring boundedness of I_X on $\mathcal{R}$ implies that $\langle I_X, z\rangle$ is σ–additive for any z belonging to a norming space $Z \subset L^q_{E^*}$, and in the case $E \not\supset c_0$, this is sufficient for I_X to have a σ–additive extension from $\mathcal{P}$ into L^p_E. The proof of this theorem relies heavily on the general extension Theorems AI.1, AI.2, AI.3 in the appendix AI for vector measures.

The main part of the theorem is the equivalence of (1) and (2). This is done by proving the equivalence of the first 6 assertions. The equivalence with the rest of assertions is done for the sake of completeness.

2.5 THE EXTENSION THEOREM. *If E does not contain a copy of c_0, then the following assertions (1) — (10) are equivalent. If E is any general Banach space, then assertions (2) — (10) are equivalent and (1) implies (2).*

(1) I_X can be extended to a σ–additive measure on $\mathcal{P}$.

(2) I_X is bounded on $\mathcal{R}$, the ring generated by the predictable rectangles in $\mathbb{R}$.

Let $Z \subset L^q_{E^*}$ be any norming subspace for L^p_E.

(3) For each $g \in Z$, the real measure $\langle I_X, g\rangle$ is bounded on $\mathcal{R}$.

(4) For each $g \in Z$, XG is a quasimartingale on $(0,\infty)$.

(5) For each $g \in Z$, XG is a quasimartingale on $(0,\infty]$.

(6) For each $g \in Z$, the measure $\langle I_X, g\rangle$ is σ–additive and bounded on $\mathcal{R}$.

(7) For each $x^* \in E^*$, the measure $I_{x^*X} : \mathcal{R} \to L^p$ is bounded in L^p on $\mathcal{R}$.

(8) For each $x^* \in E^*$, the measure $I_{x^*X} : \mathcal{R} \to L^p$ is bounded in L^p and is σ–additive on $\mathcal{R}$.

(9) For each $g \in Z$, XG is a quasimartingale on $(0,\infty)$ (or on $(0,\infty]$) and $(XG)^* := \sup_t |(XG)_t|$ is integrable.

(10) For each $g \in Z$, XG is a quasimartingale on $(0,\infty)$ (or on $(0,\infty]$) of class (D).

Proof. The proof will be done in the following way: $1 \Longrightarrow 2 \Longleftrightarrow 3 \Longleftrightarrow 4 \Longleftrightarrow 5 \Longleftrightarrow 6 \Longrightarrow 1$; $2 \Longrightarrow 6 \Longrightarrow 7 \Longrightarrow 2$; $7 \Longleftrightarrow 8$ and $5 \Longleftrightarrow 9 \Longrightarrow 10 \Longrightarrow 6 \Longleftrightarrow 5$. The only implication that requires E not to contain a copy of c_0 is $6 \Longrightarrow 1$. All other implications are valid for any Banach space E.

The implication $1 \Longrightarrow 2$ is evident (since any σ–additive measure on a σ–algebra is bounded). The implication $2 \Longrightarrow 3$ is also evident. To prove $3 \Longrightarrow 2$ we remark that for each set $A \in \mathcal{R}$, the linear functional $g \mapsto \langle I_X(A), g\rangle$ on Z is continuous. Since Z is norming for L^p_E, we can embed $L^p_E \subset Z^*$ isometrically.

If we assume 3, then

$$\sup\{|\langle I_X(A), g\rangle| : A \in \mathcal{R}\} < \infty \text{ for each } g \in Z;$$

by the Banach–Steinhaus theorem we deduce that

$$\sup\{\|I_X(A)\|_p; A \in \mathcal{R}\} < \infty,$$

that is (2).

Let us prove 3 $\Longleftrightarrow$ 4. Let $g \in L^q_{E^*}$ and consider the real measure $\langle I_X, g\rangle$ on $\mathcal{R}$ defined as follows:

$$\langle I_X, g\rangle(A) = \int \langle I_X(A), g\rangle dP, \text{ for } A \in \mathcal{R}.$$

We shall use the results concerning quasimartingales given in Appendix AII. We shall show that $\langle I_X, g\rangle$ is bounded on $\mathcal{R}$ if and only if XG is a quasimartingale on $(0, \infty)$. To prove this, we first show that

$$\langle I_X, g\rangle(A) = \mu_{XG}(A), \text{ for } A \in \mathcal{R},$$

where μ_{XG} is the Doléans function of the process XG. In fact, for $B \in \mathcal{F}_0$ we have

$$\langle I_X, g\rangle([0_B]) = \int 1_B \langle X_0, g\rangle dP = \int 1_B X_0 G_0 dP = \mu_{XG}([0_B]).$$

For $(s,t] \times B$, with $B \in \mathcal{F}_s$, we have

$$\begin{aligned} \langle I_X, g\rangle((s,t] \times B) &= \int \langle 1_B(X_t - X_s), g\rangle dP \\ &= \int_B \langle X_t, G_t\rangle dP - \int_B \langle X_s, G_s\rangle dP = \mu_{XG}((s,t] \times B). \end{aligned}$$

Hence, $\langle I_X, g\rangle$ is bounded on $\mathcal{A}(0,\infty)$ if and only if μ_{XG} is bounded on $\mathcal{A}(0,\infty)$, which is true if and only if XG is a quasimartingale on $(0, \infty)$.

It follows that 3 $\Longleftrightarrow$ 4, since $\mathcal{R} = \mathcal{A}[0] \cup \mathcal{A}(0,\infty)$, $\langle I_X, g\rangle = \mu_{XG}$ on $\mathcal{A}(0,\infty)$ and I_X is always bounded on $\mathcal{A}[0]$.

We now show 4 $\Longleftrightarrow$ 5. Obviously 5 $\Longrightarrow$ 4. If (4) holds, then from 2 $\Longleftrightarrow$ 3 $\Longleftrightarrow$ 4 proved above, we deduce that I_X is bounded on $\mathcal{R}$. Thus for $g \in Z$, we have

$$\|X_t G_t\|_1 = \int |X_t G_t| dP \le \|g\|_q \|X_t\|_p = \|g\|_q \|I_X([0,t] \times \Omega)\|_p,$$

hence

$$\sup_t \|X_tG_t\|_1 \leq \|g\|_q \sup\{\|I_X(A)\|_p : A \in \mathcal{R}\| < \infty.$$

Thus XG is a quasimartingale on $[0,\infty]$, that is (5).

Next we prove $5 \iff 6$. The implication $6 \implies 5$ is evident. Assume (5) and let $g \in Z$. Then XG is a quasimartingale on $(0,\infty]$, where $(XG)_\infty = 0$ by definition. For each n, define the stopping time $T_n = \inf\{t : |X_t| > n\}$. Then $T_n \nearrow \infty$ and $|X_t| \leq n$ on $[0, T_n)$. At this stage we do not know if X_{T_n} belongs to L^p_E, but since XG is a quasimartingale on $(0,\infty]$, we know that $X_{T_n}G_{T_n} \in L^1$, and

$$|X_tG_t| \leq n|G_t|1_{\{t<T_n\}} + |X_{T_n}G_{T_n}|1_{\{t\geq T_n\}}.$$

Since G is a uniformly integrable martingale, it follows that $X^{T_n}G^{T_n}$ is a quasimartingale of class (D) on $(0,\infty]$, hence the corresponding measure $\mu_{(XG)^{T_n}}$ is σ–additive with bounded variation on $\mathcal{A}(0,\infty]$, therefore it can be extended to a σ–additive measure with bounded variation on the σ–algebra $\mathcal{P}(0,\infty]$, the class of predictable subsets of $(0,\infty] \times \Omega$. Now for each predictable rectangle $(s,t] \times A$, with $s < t \leq \infty$ and $A \in \mathcal{F}_s$ we have

$$\mu_{(XG)^{T_n}}((s,t] \times A) = \mu_{XG}((s,t] \times A) \cap [0,T_n]),$$

therefore

$$\mu_{(XG)^{T_n}}(B) = \mu_{XG}(B \cap [0,T_n]), \text{ for } B \in \mathcal{P}(0,\infty].$$

It follows that μ_{XG} is σ–additive on the σ–ring $\mathcal{P}(0,\infty] \cap [0,T_n]$; consequently, μ_{XG} is σ–additive on the ring $\mathcal{B} = \cup_{1\leq n<\infty}\mathcal{P}(0,\infty] \cap [0,T_n]$. On the other hand, μ_{XG} is bounded on $\mathcal{R}(0,\infty)$ since XG is a quasimartingale on $(0,\infty]$, hence μ_{XG} has bounded variation on $\mathcal{A}(0,\infty]$. It follows that μ_{XG} is σ–additive and has bounded variation on the ring $\mathcal{B} \cap \mathcal{A}(0,\infty]$, which generates $\mathcal{P}(0,\infty]$; hence μ_{XG} can be extended to a σ–additive measure with bounded variation on $\mathcal{P}(0,\infty]$. Since $\langle I_X, g\rangle = \mu_{XG}$ on $\mathcal{A}(0,\infty)$ it follows that $\langle I_X, g\rangle$ is bounded and σ–additive on $\mathcal{A}(0,\infty)$. Since $\langle I_X, g\rangle$ is bounded and σ–additive on $\mathcal{A}[0]$, it follows that $\langle I_X, g\rangle$ is bounded and σ–additive on $\mathcal{R} = \mathcal{A}[0,\infty)$; hence (6) holds.

To prove $6 \Longrightarrow 1$, we assume that E does not contain c_0. If we assume (6) then $\langle I_X, g\rangle$ is bounded and σ-additive on $\mathcal{R}$, for $g \in Z$. By Theorem AI.3, I_X can be extended to a σ-additive measure on $\mathcal{P} = \sigma(\mathcal{R})$, that is (1).

We show now that $6 \Longrightarrow 7$. If we assume (6), then by the equivalence $2 \Longleftrightarrow 6$ proved above, I_X is bounded on $\mathcal{R}$. Then for each $x^* \in E^*$, the measure $I_{x^*X} = x^* I_X$ is bounded on $\mathcal{R}$, which is (7).

Next we show $7 \Longrightarrow 2$. Assume (7), and let $x^* \in E^*$ and $\varphi \in L^q$. Then $g = x^*\varphi \in L^q_{E^*}$. For $A \in \mathcal{R}$ we have

$$|\langle I_X, g\rangle(A)| = |E(I_{x^*X}(A)\varphi| \leq \|I_{x^*X}(A)\|_p \|\varphi\|_q,$$

hence the measure $\langle I_X, g\rangle$ is bounded on $\mathcal{R}$. It follows that $\langle I_X, g\rangle$ is bounded on $\mathcal{R}$ for every step function $g \in L^q_{E^*}$. Since the step functions of $L^q_{E^*}$ form a norming space for L^p_E, we proved (3) for this particular norming space. Now, since $2 \Longleftrightarrow 3$ for any norming space, assertion (2) follows.

Now we prove $7 \Longleftrightarrow 8$. Obviously $8 \Longrightarrow 7$. Assume (7) and prove (8). By the implication $7 \Longrightarrow 2$ proved above, I_X is bounded in L^p on $\mathcal{R}$. By the equivalence $2 \Longleftrightarrow 5$ applied to I_{x^*X}, we deduce that $\langle I_{x^*X}, \varphi\rangle$ is σ-addtive and bounded for every $\varphi \in L^q$. Since L^p does not contain a copy of c_0, by applying Theorem AI.3, it follows that I_{x^*X} can be extended to a σ-additive measure on $\mathcal{P}$ with values in L^p. In particular, I_{x^*X} is σ-additive and bounded on $\mathcal{R}$, which is (8).

Finally we prove the implications $5 \Longleftrightarrow 9 \Longrightarrow 10 \Longrightarrow 6 \Longleftrightarrow 5$.

Let us prove that $5 \Longleftrightarrow 9$. Obviously $9 \Longrightarrow 5$. Assume (5) and let $g \in Z$. Then XG is a quasimartingale on $(0, \infty]$. We have to prove that $(XG)^*$ is integrable. The proof will be carried out in several steps.

(a) By Theorem AII.9, XG has a decomposition $XG = M + V$, where M is a real valued local martingale and V is a real valued predictable process with integrable variation $|V|$. For each t, since $X_t G_t$ and V_t are integrable, we deduce that M_t is integrable. Then $M = XG - V$ is a quasimartingale on $(0, \infty]$, thus the stochastic measure I_M is bounded in L^1 on $\mathcal{R}$. As a quasimartingale, we define $M_\infty = 0$; thus, for any stopping time T, we have $M_T \in L^1$. I_M can be extended to the algebra $\mathcal{A}$ generated by the stochastic intervals $[0_A]$, with $A \in \mathcal{F}_0$ and $(S, T]$, with $S \leq T$, by $I_M((S, T]) = M_T - M_S$.

(b) I_M is bounded on $\mathcal{A}(0,\infty]$. To see this let

$$a = \sup\{\|I_M(A)\|_1 : A \in \mathcal{R}\} < \infty.$$

If T is a simple stopping time, then $[0,T] \in \mathcal{R}$, hence $\|M_T\|_1 = \|I_M([0,T])\|_1 \leq a$. If T is an arbitrary stopping time, then there is a decreasing sequence (T_n) of simple stopping times converging to T. Then $M_{T_n} \to M_T$ in L^1, hence $\|M_T\|_1 = \lim \|M_{T_n}\|_1 \leq a$. Thus $\|I_M((S,T])\|_1 \leq 2a$, if $S \leq T$ are stopping times. Hence I_M is bounded on $\mathcal{A}(0,\infty]$.

(c) There exists an increasing sequence $T_n \nearrow \infty$ of stopping times such that, for each n, M^{T_n} is a uniformly integrable martingale and $(M^{T_n})^* \in L^1$. In fact, define the stopping times $U_n = \inf\{t : |M_t| \geq n\}$. Let (V_n) be an increasing sequence of stopping times, with $V_n \nearrow \infty$, such that each M^{V_n} is a uniformly integrable martingale. The $T_n = U_n \wedge V_n$ is the required sequence, since for each n we have

$$(M^{T_n})^* = \sup_{t\geq 0} |M_t^{T_n}| \leq n + |M_{T_n}| \in L^1.$$

(d) The sequence $(M^{T_n})^*$ is increasing and bounded in L^1. In fact, by the corollary of Theorem 12.12 in [Ku.1] we have

$$\|(M^{T_n})^*\|_1 \leq 40\widetilde{I}_{M^{T_n}}(\mathbb{R}) = 40\widetilde{I}_M([0,T_n]) \leq 40\widetilde{I}_M(\mathbb{R}),$$

where $\widetilde{I}_M$ is the semivariation of I_M relative to $(\mathbb{R}, L^1)$.

(e) $M^* \in L^1$, since

$$M^* = \sup_t |M_t| = \sup_t \sup_n |M_t^{T_n}|$$

thus

$$\|M^*\|_1 = \sup_n \|(M^{T_n})^*\|_1 \leq 40\widetilde{I}_M(\mathbb{R}) < \infty.$$

(f) Since $(XG)^* \leq M^* + V^*$, we deduce that $(XG)^*$ is integrable, which proves (9).

Obviously, $9 \Longrightarrow 10$. Now we shall assume (10) and prove (6). Let $g \in Z$. Then XG is a quasimartingale of class (D) on $(0,\infty]$. The corresponding measure μ_{XG} is σ-additive with bounded variation on $\mathcal{A}(0,\infty]$. From the equality $\langle I_X, g\rangle = \mu_{XG}$ on $\mathcal{A}(0,\infty)$, it follows that $\langle I_X, g\rangle$ is σ–additive and bounded on $\mathcal{A}(0,\infty)$; hence it is also bounded on $\mathcal{R} = \mathcal{A}[0,\infty)$, which is (6), which in turn is equivalent to (5). This concludes the proof of the theorem.

3. The stochastic integral

In this section we shall define the stochastic integral with respect to a p–summable process X and study various properties of this integral, including various types of convergence theorems, some of them derived from the study of the weak topology of the Lebesgue space constructed in Appendix AI.

Definition of the integral $\int H dI_X$.

The setting for this section is the same as that of section 2. *We shall always assume in this section that $X : \mathbb{R} \to E \subset L(F,G)$ is a p-summable process relative to the pair (F,G)*; hence the stochastic measure I_X is a σ–additive measure on $\mathcal{P}$ with values in $L^p_E \subset L(F, L^p_G)$. As in the previous section, we can extend I_X to $\mathcal{P}[0,\infty]$, with $I_X(\{\infty\}\times A) = 0$ for $A \in \mathcal{F} = \vee_{t\geq 0}\mathcal{F}_t$. As usual we identify functions with their equivalence classes in L^p_E or L^p_G.

Since I_X has bounded semivariation relative to the pair (F, L^p_G), we can apply the integration theory of section 1 and Appendix I, with $\Sigma = \mathcal{P}$ or $\Sigma = \mathcal{P}[0,\infty]$, $m = I_X$, E replaced by L^p_E, G replaced by L^p_G and $Z \subset (L^p_G)^*$ a norming subspace for L^p_G, (for example, we can take Z to be the space of simple functions in $L^q_{G^*}$, where $\frac{1}{p} + \frac{1}{q} = 1$). For the reader's convenience, we shall translate some of the general theory in AI to our particular setting.

For $z \in Z$, consider the measure

$$m_z = (I_X)_z : \mathcal{P}[0,\infty] \to F^*$$

defined, for $A \in \mathcal{P}[0,\infty]$ and $y \in F$ as follows:

$$\langle y, m_z(A)\rangle = \langle m(A)y, z\rangle = \int \langle I_X(A)(\omega)y, z(\omega)\rangle dP(\omega).$$

Then we have

$$(\widetilde{I}_X)_{F,L^p_G} = \widetilde{m}_{F,L^p_G} = \sup\{|m_z| : z \in Z, \|z\|_q \leq 1\}.$$

We note that $\{\infty\} \times \Omega$ is $|m_z|$–negligible for every z. If p is fixed, to simplify notation, we shall write $I = I_X$ and $\tilde{I}_{F,G} = \tilde{I}_{F,L^p_G}$. We shall also write $I_{F,G} = (I_X)_{F,G} = (I_X)_{F,L^p_G}$ for the set of positive σ–additive measures $|(I_X)_z| = |m_z|$ with $z \in Z$ and $|z| \leq 1$.

For any Banach space D, we denote by $\mathcal{F}_D(I_{F,G}) = \mathcal{F}_D(I_{F,L^p_G})$ the space of all predictable processes $H : \mathbb{R} \to D$ such that

$$\tilde{I}_{F,G}(H) = \tilde{m}_{F,L^p_G}(H) = \sup\{\int |H| d|m_z| : |z| \leq 1\} < \infty.$$

The definition of $\mathcal{F}_D(I_{F,G})$ and $\tilde{I}_{F,G}(H)$ is independent on the norming space Z. For any extension of H to $\overline{\mathbb{R}}_+ \times \Omega$, the value of $\tilde{I}_{F,G}(H)$ is the same. We know that $\mathcal{F}_D(I_{F,G})$ is a vector space with seminorm $\tilde{I}_{F,G}$, and $\mathcal{F}_D(I_{F,G})$ is complete for this seminorm. For any set $\mathcal{C} \subset \mathcal{F}_D(I_{F,G})$, we denote by $\mathcal{F}_D(\mathcal{C}, I_{F,G})$ the closure of $\mathcal{C}$ in $\mathcal{F}_D(I_{F,G})$.

If $D = F$, we can define the integral $\int H dI_X \in Z^*$, for $H \in \mathcal{F}_F(I_{F,G}) = \mathcal{F}_F(I_{F,L^p_G})$, and the mapping $H \to \int H dI_X$ is a continuous linear mapping from $\mathcal{F}_F(I_{F,G})$ into Z^*. We have

$$\langle z, \int H dI_X \rangle = \int H d(I_X)_z, \text{ for } z \in Z,$$

and

$$\| \int H dI_X \|_{Z^*} \leq \tilde{I}_{F,G}(H).$$

The integral $\int H dI_X$ depends on the norming space Z. But the integral corresponding to Z is the restriction to Z of the integral corresponding to $(L^p_G)^*$.

To further simplify notation, we write

$$\mathcal{F}_{F,G}(X) = \mathcal{F}_{F,L^p_G}(X) = \mathcal{F}_F((I_X)_{F,L^p_G}).$$

If $H \in \mathcal{F}_{F,G}(X)$, then for every $t \geq 0$ we have $1_{[0,t]}H \in \mathcal{F}_{F,G}(X)$. We denote

$$\int_{[0,t]} H dI_X = \int 1_{[0,t]} H dI_X.$$

Also we define

$$\int_{[0,\infty]} H dI_X := \int_{[0,\infty)} H dI_X := \int H dI_X.$$

Thus for each $H \in \mathcal{F}_{F,G}(X)$, we obtain a family $(\int_{[0,t]} HdI_X)_{t\in[0,\infty]}$ of elements in Z^*. We are interested in the subspace of $\mathcal{F}_{F,G}(X)$ which consists of processes H such that for every $t \in [0,\infty]$, the integral $\int_{[0,t]} HdI_X$ belongs to the subspace L^p_G of Z^*. In this case we denote by the same symbol, the equivalence class $\int_{[0,t]} HdI_X$ as well as any representative of this class. If in each equivalence class $\int_{[0,t]} HdI_X$ we choose a representative, we obtain a process $(\int_{[0,t]} HdI_X)_{t\in[0,\infty]}$ with values in G, such that $\int_{[0,t]} HdI_X \in L^p_G$ for each t. This process does not necessarily have a cadlag modification. This situation is discussed in detail in the following subsections. Before this, we shall discuss some general convergence theorems.

The Vitali and Lebesgue theorems can now be stated for sequences (H^n) in $\mathcal{F}_{F,G}(X)$ which converge *in measure* to a process H (and satisfy additional conditions), and the conclusion is that $H^n \to H$ in $\mathcal{F}_{F,G}(X)$, hence $\int H^n dI_X \to \int HdI_X$ in $(L^p_G)^{**}$. Pointwise convergence of the H^n to H will not suffice for this conclusion unless the family of measures $I_{F,G}$ is uniformly σ-additive. We will postpone the statements of these theorems until we will be able to add an important property to the conclusion, namely, that the integrals belong to L^p_G, and there exists a subsequence (n_k) such that $\int_{[0,t]} H^{n_k} dI_X \to \int_{[0,t]} HdI_X$, uniformly on compact time intervals (Theorems 3.14 and 3.15 *infra*).

At this time we shall state a very useful version of the Lebesgue theorem for pointwise convergence in which the conclusion involves $\int H^n dI_X \to \int HdI_X$ weakly in L^p_G—but not necessarily the convergence of H^n to H in $\mathcal{F}_{F,G}(X)$.

3.1 THEOREM. *Let $(H^n)_{0\leq n<\infty}$ be a sequence of elements from $\mathcal{F}_{F,G}(X)$ such that $|H^n| \leq |H^0|$ for each n and assume that $H^n \to H$ pointwise.*

If $\int H^n dI_X \in L^p_G$ for each $n \geq 1$ and if the sequence $(\int H^n dI_X)_n$ converges pointwise on Ω, weakly in G, then $\int HdI_X \in L^p_G$, and $\int H^n dI_X \to \int HdI_X$ in the $\sigma(L^p_G, L^q_{G^*})$ topology of L^p_G, as well as pointwise, weakly in G. If $(\int H^n dI_X)_n$ converges pointwise, strongly in G, then $\int H^n dI_X \to \int HdI_X$ strongly in L^1_G.

Proof. Since $|H| \leq |H^0|$, we deduce that $H \in \mathcal{F}_{F,G}(X)$. Let $z \in L^q_{G^*}$. We can apply Lebesgue's theorem to (H^n) in the space $L^1_F(|m_z|)$, and deduce that

$H^n \to H$ in $L^1(|m_z|)$ and thus $\int H^n dm_z \to \int H dm_z$, that is

$$E(\langle(\int H^n dI_X)(\cdot), z(\cdot)\rangle) \to \langle \int H dI_X, z\rangle.$$

If $h \in L^\infty(P)$, then $hz \in L^q_{G^*}$, hence, replacing z with hz; we obtain

$$E(h(\cdot)\langle(\int H^n dI_X)(\cdot), z(\cdot)\rangle) \to \langle \int dI_X, hz\rangle.$$

Thus the sequence $(\langle(\int H^n dI_X)(\cdot), z(\cdot)\rangle)$ is weakly Cauchy in $L^1(P)$, hence the indefinite integrals of the above sequence are uniformly absolutely continuous with respect to P. If we let $\phi(\omega) := \lim_n(\int H^n dI_X)(\omega)$ weakly in G, then the Vitali convergence theorem implies that $\langle\phi(\cdot), z(\cdot)\rangle \in L^1(P)$ and $\langle(\int H^n dI_X)(\cdot), z(\cdot)\rangle$ converges in $L^1(P)$ to $\langle\phi(\cdot), z(\cdot)\rangle$, hence the expectations $E(\langle(\int H^n dI_X)(\cdot), z(\cdot)\rangle)$ converge to $E(\langle\phi, z\rangle)$. Since $z \in L^q_{G^*}$ was arbitrary, we deduce that $\phi \in L^p_G$ (by Corollary 2, p. 236 in [D.1]). We then deduce that $\langle\phi, z\rangle = E(\langle\phi(\cdot), z(\cdot)\rangle) = \langle\int H dI_X, z\rangle$, hence $\int H dI_X = \phi \in L^p_G$ and $\int H^n dI_X \to \int H dI_X$ pointwise, weakly in G. From the above, it follows that $\int H^n dI_X \to \int H dI_X$ in the $\sigma(L^p_G, L^q_{G^*})$ topology of L^p_G. In particular, the above sequence converges in the $\sigma(L^1_G, L^\infty_{G^*})$ topology of L^1_G, hence by Theorem 4.4 in [B–D.3], the indefinite integrals $\int |\int H^n dI_X| dP$ are uniformly σ–additive on $\mathcal{F}$. If $\phi(\omega) = \lim_n(\int H^n dI_X)(\omega)$, strongly in G, we can apply the Vitali theorem for L^1_G and deduce that $\int H^n dI_X \to \int H dI_X$ in L^1_G.

The stochastic integral

We shall be interested in the subspace of $\mathcal{F}_{F,G}(X)$ of processes H that in addition to the property that $\int_{[0,t]} H dI_X \in L^p_G$, for each t, also have the property that the process $(\int_{[0,t]} H dI_X)_{t\in[0,\infty]}$ has a cadlag modification. Note that, since X is cadlag, this holds for simple processes of the form

$$H = 1_{\{0\}\times A_0} x_0 + \Sigma_{1\le i\le n} 1_{(s_i,t_i]\times A_i} x_i,$$

where the sets in the definition of H are predictable. We have

$$\int_{[0,t]} H dI_X = 1_{A_0} X_0 x_0 + \Sigma_{1\le i\le n} 1_{A_i} x_i (X_{t_i\wedge t} - X_{s_i\wedge t})$$

and the right–hand side is cadlag.

We now define our Lebesgue space of processes.

3.2 DEFINITION. *We denote by* $L^1_{F,G}(X)$ *the space of processes* $H \in \mathcal{F}_{F,G}(X)$ *satisfying the following two conditions:*

(1) $\int_{[0,t]} H dI_X \in L^p_G$ *for every* $t \in [0,\infty]$;

(2) The process $(\int_{[0,t]} H dI_X)_{t\in[0,\infty]}$ *has a cadlag modification.*

The processes $H \in L^1_{F,G}(X)$ *are said to be integrable with respect to* X.

If $H \in L^1_{F,G}(X)$, *then any cadlag modification of* $(\int_{[0,t]} H dI_X)_{t\in[0,\infty]}$ *is called the stochastic integral of* H *with respect to* X *and is denoted by* $\int H dX$ *or* $H \cdot X$:

$$(H \cdot X)_t = (\int H dX)_t = \int_{[0,t]} H dI_X \text{ a.s.}$$

We note that if X is real valued, we regard $\mathbb{R}$ as being embedded in $L(F, F)$, and thus the space of F–valued integrable processes is denoted by $L^1_{F,F}(X)$.

We shall see later (Corollary 3.11 *infra*) that $L^1_{F,G}(X)$ is complete relative to the seminorm $\tilde{I}_{F,G}$, and $L^1_{F,G}(X) \supset \mathcal{E}$, the class of predictable "elementary processes" (see Corollary 3.6 *infra*). If $I_{F,G}$ if uniformly σ–additive, then $\mathcal{F}_{F,G}(\mathcal{B}, X) \subset L^1_{F,G}(X)$ (Corollary 3.12 *infra*), where $\mathcal{B}$ is the set of bounded processes.

We note that the stochastic integral is uniquely defined up to an evanescent set. For $t = \infty$, we have

$$\begin{aligned}(H \cdot X)_\infty &= \int_{[0,\infty]} H dX = \int_{[0,\infty)} H dI_X + \int_{\{\infty\}\times\Omega} H dI_X \\ &= \int_{[0,\infty)} H dI_X = \int H dI_X.\end{aligned}$$

For simple, $\mathcal{R}$–measurable processes H, the stochastic integral can be computed pathwise, as a Stieltjes integral:

$$(H \cdot X)_t(\omega) = (\int_{[0,t]} H dI_X)(\omega) = \int_{[0,t]} H_s(\omega) dX_s(\omega).$$

This property remains valid whenever both the stochastic integral and the pathwise Stieltjes integral appearing above are defined. Moreover, we prove below that if $H \in \mathcal{F}_{F,G}(X)$ and if the Stieltjes integral $\int_{[0,t]} H_s(\omega) dX_s(w)$ is defined for every $t \geq 0$, then necessarily $H \in L^1_{F,G}(X)$.

3.3 THEOREM. *Assume that X has finite variation $|X|$ and that X is p–summable relative to (F,G). If $H \in \mathcal{F}_{F,G}(X)$ and if $\int_{[0,t]} |H_s(\omega)| d|X|_s(\omega) < \infty$, for every $t \in \mathbb{R}_+$ and $\omega \in \Omega$, then $H \in L^1_{F,G}(X)$ and*

$$(H \cdot X)_t(\omega) = \int_{[0,t]} H_s(\omega) dX_s(\omega).$$

Proof. As we mentioned above, if $H = 1_A \cdot x$, for $x \in F$ and $A \in \mathcal{R}$, then the theorem is true. By a monotone class argument, this also holds if $A \in \mathcal{P}$, hence for H any simple predictable process.

Now suppose that H satisfies the hypotheses of the above theorem. Let (H_n) be a sequence of simple, predictable processes such that $H^n \to H$ pointwise and $|H^n| \le |H|$ for each n. Let $t > 0$ and $\omega \in \Omega$. Using the Lebesgue theorem in $L^1_F(dX_{(\cdot)}(\omega))$, we deduce that

$$\int_{[0,t]} |H^n_s(\omega) - H_s(\omega)| d|X|_s(\omega) \to 0,$$

and

$$\int_{[0,t]} H^n_s(\omega) dX_s(\omega) \to \int_{[0,t]} H_s(\omega) dX_s(\omega).$$

Now we use the Lebesgue theorem 3.1 to conclude that $\int_{[0,t]} H dI_X \in L^p_G$ and $\int_{[0,t]} \int H^n dI_X \to \int_{[0,t]} H dI_X$ pointwise. Hence $(\int_{[0,t]} H dI_X)(\omega) = \int_{[0,t]} H_s(\omega) dX_s(\omega)$ a.s. Since the Stieltjes integral is cadlag, as a function of t, we have $H \in L^1_{F,G}(X)$ and $(H \cdot X)_t(\omega) = \int_{[0,t]} H_s(\omega) dX_s(\omega)$.

Remark. This equality will remain valid for locally integrable processes (Theorem 4.4 *infra*).

3.4 PROPOSITION. *If $H \in L^1_{F,G}(X)$, then for every $t \in [0,\infty]$ we have $(H \cdot X)_{t-} \in L^p_G$ and*

$$(H \cdot X)_{t-} = \int_{[0,t)} H dI_X.$$

In particular,

$$(H \cdot X)_{\infty-} = (H \cdot X)_\infty = \int H dI_X.$$

The mapping $t \to (H \cdot X)_t$ is cadlag in L^1_G.

Proof. Let $t_n \nearrow t$. The $1_{[0,t_n]} H \to 1_{[0,t)} H$ pointwise, $|1_{[0,t_n]} H| \le |H|$ for each n, and $\int 1_{[0,t_n]} H dI_X = (H \cdot X)_{t_n} \in L^p_G$ and $(H \cdot X)_{t_n} \to (H \cdot X)_{t-}$. By

Theorem 3.1, we have $\int 1_{[0,t)} H dI_X \in L^p_G$ and $\int 1_{[0,t_n]} H dI_X \to \int 1_{[0,t)} H dI_X$ pointwise. Hence $(H \cdot X)_{t-} = \int 1_{[0,t)} H dI_X$. The final conclusion follows from Theorem 3.1.

Notation and remarks

If $\mathcal{C} \subset \mathcal{F}_{F,G}(X)$, we denote the closure of $\mathcal{C}$ in $\mathcal{F}_{F,G}(X)$ by $\mathcal{F}_{F,G}(\mathcal{C}, X)$.

If $\mathcal{C}$ consists of processes H such that $\int H dI_X \in L^p_G$, for every $H \in \mathcal{C}$, then by continuity of the integral we still have $\int H dI_X \in L^p_G$ for every $H \in \mathcal{F}_{F,G}(\mathcal{C}, X)$. We shall see later (Corollary 3.11) that if $\mathcal{C} \subset L^1_{F,G}(X)$, then $\mathcal{F}_{F,G}(\mathcal{C}, X) \subset L^1_{F,G}(X)$. In this case we write $L^1_{F,G}(\mathcal{C}, X) = \mathcal{F}_{F,G}(\mathcal{C}, X)$.

Particular spaces $\mathcal{C}$ of interest are:

(1) The space $\mathcal{B}_F$ of bounded, predictable processes with values in F. We write $\mathcal{F}_{F,G}(\mathcal{B}, X)$ for $\mathcal{F}_{F,G}(\mathcal{B}_F, X)$;

(2) The space $\mathcal{S}_F(\mathcal{R})$ (respectively, $\mathcal{S}_F(\mathcal{P})$) of simple, F–valued processes over $\mathcal{R} = \mathcal{A}[0,\infty)$ (respectively, over $\mathcal{P}$). The closures of these sets in $\mathcal{F}_{F,G}(X)$ will be denoted by $\mathcal{F}_{F,G}(\mathcal{S}(\mathcal{R}), X)$ (respectively, $\mathcal{F}_{F,G}(\mathcal{S}(\mathcal{P}), X)$);

(3) The space $\mathcal{E}_F$ of predictable, elementary, F–valued processes of the form

$$H = H_0 1_{\{0\}} + \Sigma_{1 \le i \le n} H_i 1_{(T_i, T_{i+1}]}$$

where $(T_i)_{0 \le i \le n+1}$ is an increasing family of stopping times with $T_0 = 0$, and H_i is bounded and $\mathcal{F}_{T_i}$–measurable for each i. We let $\mathcal{F}_{F,G}(\mathcal{E}, X)$ denote the closure of this set.

We shall see (Corollary 3.6 *infra*) that $\mathcal{S}_F(\mathcal{R})$ and $\mathcal{E}_F$ are contained in $L^1_{F,G}(X)$, hence $L^1_{F,G}(\mathcal{S}(\mathcal{R}), X) = \mathcal{F}_{F,G}(\mathcal{S}(\mathcal{R}), X)$ and $L^1_{F,G}(\mathcal{E}, X) = \mathcal{F}_{F,G}(\mathcal{E}, X)$.

By Proposition AI.11, we have

$$\mathcal{F}_{\mathbb{R},E}(\mathcal{S}(\mathcal{R}), X) = \mathcal{F}_{\mathbb{R},E}(\mathcal{B}, X) = L^1_{\mathbb{R},E}(\mathcal{B}, X).$$

More generally, if the set of measures $I_{F,G}$ is uniformly σ–additive, then

$$\mathcal{F}_{F,G}(\mathcal{S}(\mathcal{R}), X) = \mathcal{F}_{F,G}(\mathcal{B}, S) = L^1_{F,G}(\mathcal{B}, X).$$

Moreover, if X has integrable variation, or if X is a square integrable martingale with values in a Hilbert space E, we have $\mathcal{F}_{F,G}(\mathcal{S}(\mathcal{R}), X) = L^1_{F,G}(X) = \mathcal{F}_{F,G}(X)$ (see Theorems 3.27 and 3.32 *infra*).

The stochastic integral of elementary processes

For simple predictable processes defined on $\overline{\mathbb{R}}_+ \times \Omega$, of the form

$$H = \Sigma_{1 \leq i \leq n} 1_{A_i} y_i \text{ with } A_i \in \mathcal{P}[0,\infty] \text{ and } y_i \in F,$$

we have

$$\int H dI_X = \Sigma_{1 \leq i \leq n} I_X(A_i) x_i \in L^p_G.$$

If H' is the restriction of H to $\mathbb{R}$, then H' is predictable and $\int H' dI_X = \int H dI_X$. However, it is not certain that H is integrable with respect to X, because of the cadlag requirement. We shall see that if $I_{F,G}$ is uniformly σ–additive, then these processes are integrable with respect to X (see Theorem 3.12 *infra*). In particular, the real valued, simple, predictable processes are integrable with respect to X since $I_{\mathbb{R},E}$ is uniformly σ–additive.

The simplest class of integrable processes with respect to X is that of the simple processes over the algebra $\mathcal{A}[0,\infty]$ of the form

$$H = y_0 1_{A_0 \times \{0\}} + \Sigma_{1 \leq i \leq n} y_i 1_{A_i} 1_{(t_i, t_i+1]},$$

where $0 = t_0 \leq t_1 < \cdots < t_n < t_{n+1} \leq \infty$, $y_i \in F$ and $A_i \in \mathcal{F}_{t_i}$. According to the definition of the integral for simple processes, for each $t \in [0,\infty]$, the integral $\int_{[0,t]} H dI_X$ can be computed pathwise:

$$\int_{[0,t]} H dI_X = y_0 1_{A_0} X_0 + \Sigma_{1 \leq i \leq n} 1_{A_i} y_i (X_{t_{i+1} \wedge t} - X_{t_i \wedge t}).$$

This integral belongs to L^p_G and is cadlag, hence H is integrable with respect to X and the stochastic integral $(H \cdot X)_t = \int_{[0,t]} H dI_X$ can be computed pathwise by the above sum. In particular, this is the case of simple processes H over $\mathcal{R} = \mathcal{A}[0,\infty)$, having the above form but with $t_{n+1} < \infty$.

A more general class is that of the simple processes of the form

$$H = 1_{[0_A]} y_0 + \Sigma_{1 \leq i \leq n} y_i 1_{(T_i, T_{i+1}]},$$

where $A \in \mathcal{F}_0$, $(T_i)_{1 \leq i \leq n+1}$ is an increasing family of stopping times, and $y_i \in F$. From Corollary 3.6 *infra* it will follow that any such process is integrable with respect to X and the stochastic integral can be computed pathwise:

$$(H \cdot X)_t = 1_A X_0 y_0 + \Sigma_{1 \leq i \leq n} y_i (X_{T_{i+1} \wedge t} - X_{T_i \wedge t}).$$

A still larger class of integrable processes is that of the *elementary processes* of the form

$$H = H_0 1_{\{0\}} + \Sigma_{1 \le i \le n} H_i 1_{(T_i, T_{i+1}]},$$

where $(T_i)_{0 \le i \le n+1}$ is an increasing family of stopping times with $T_0 = 0$ and for $0 \le i \le n$, H_i is an F–valued, *bounded*, random variable which is $\mathcal{F}_{T_i}$–measurable. We shall prove below (Corollary 3.6) that the stochastic integral of such a process can be computed pathwise:

$$(H \cdot X)_t = H_0 X_0 + \Sigma_{1 \le i \le n} H_i (X_{T_{i+1} \wedge t} - X_{T_i \wedge t}).$$

This will follow from the following result.

3.5 PROPOSITION. *Let $S \le T$ be stopping times and let $h : \Omega \to F$ be an $\mathcal{F}_S$–measurable, bounded, random variable. Then*

$$\int h 1_{(S,T]} dI_X = h(X_T - X_S).$$

If S is predictable and h is $\mathcal{F}_{S-}$–measurable, then

$$\int h 1_{[S,T]} dI_X = h(X_T - X_{S-})$$

and

$$\int h 1_{[S]} dI_X = h \Delta X_S.$$

Proof. If $h = 1_A y$, with $A \in \mathcal{F}_S$ and $y \in F$, then

$$\int h 1_{(S,T]} dI_X = I_X(1_{(S_A, T_A]}) y = (X_{T_A} - X_{S_A}) y = h(X_T - X_S).$$

Thus the equality holds when h is a simple function. For the general case, let h_n be simple functions converging pointwise to h with $|h_n| \le |h|$ for each n. By applying the Lebesgue Theorem 3.1, we obtain the desired result.

Assume now that S is predictable and h is $\mathcal{F}_{S-}$–measurable. If $h = 1_A y$, with $A \in \mathcal{F}_{S-}$ and $y \in F$, then S_A is a predictable stopping time and

$$\begin{aligned} \int 1_A y 1_{[S]} dI_X &= \int 1_{[S_A]} y \, dI_X = \Delta X_{S_A} y \\ &= 1_A y I_X([S]); \end{aligned}$$

thus

$$\int h1_{[S,T]}dI_X = \int 1_A y 1_{[S]}dI_X + \int 1_A y 1_{(S,T]}dI_X$$
$$= 1_A y\left(\int 1_{[S]}dI_X + \int 1_{(S,T]}dI_X\right)$$
$$= 1_A y \int 1_{[S,T]}dI_X = h \int 1_{[S,T]}dI_X.$$

As before, the conclusion holds for simple functions, and using the Lebesgue Theorem 3.1, we obtain the general case.

3.6 COROLLARY. *Every elementary process*

$$H = H_0 1_{\{0\}} + \Sigma_{1\leq i\leq n} H_i 1_{(T_i,T_{i+1}]}$$

is integrable with respect to X and its stochastic integral can be computed pathwise, as a Stiltjes integral:

$$(H\cdot X) = H_0X_0 + \Sigma_{1\leq i\leq n} H_i(X^{T_{i+1}} - X^{T_i}).$$

Stochastic integrals and stopping times

In this subsection we continue to assume that X is p–summable relative to (F,G). We shall examine the relationship between stochastic integrals and stopping times. First we extend Proposition 3.5 to a more general situation.

3.7 THEOREM. *Let $S \leq T$ be stopping times and assume either*

(a) $h : \Omega \to I\!R$ is bounded, $\mathcal{F}_S$–measurable, and $H \in \mathcal{F}_{F,G}(X)$;

or

(b) $h : \Omega \to F$ is bounded, $\mathcal{F}_S$–measurable, and $H \in \mathcal{F}_{I\!R}((I_X)_{F,G})$.

(1) If $\int 1_{(S,T]} H dI_X \in L^p_G$, in case (a), and $\int 1_{(S,T]} H dI_X \in L^p_E$ in case (b), then

$$\int h1_{(S,T]} H dI_X = h \int 1_{(S,T]} H dI_X.$$

(1′) If S is predictable, h is $\mathcal{F}_{S-}$–measurable and $\int 1_{[S,T]} H dI_X \in L^p_G$ in case (a), and $\int 1_{[S,T]} H dI_X \in L^p_E$ in case (b), then

$$\int h1_{[S,T]} H dI_X = h \int 1_{[S,T]} H dI_X.$$

(2) If H is integrable with respect to X, then $1_{(S,T]}H$ and $h1_{(S,T]}H$ are integrable with respect to X and

$$(h1_{(S,T]}H) \cdot X = h[(1_{(S,T]}H) \cdot X].$$

(2') If S is predictable, h is $\mathcal{F}_{S-}$-measurable, and H is integrable with respect to X, then $1_{[S,T]}H$ and $h1_{[S,T]}H$ are integrable with respect to X and

$$(h1_{[S,T]}H) \cdot X = h[(1_{[S,T]}H) \cdot X].$$

Proof. We shall only prove (1) and (2). The case when S is predictable is similar.

Assume first hypothesis (a). Let H be of the form $H = 1_{(s,t]\times A}y$, where $A \in \mathcal{F}_s$ and $y \in F$. By Proposition 3.5, we have

$$\begin{aligned}\int h1_{(S,T]}H\,dI_X &= \int h1_A y(1_{(S\vee s,T\wedge t]}dI_X \\ &= h1_A y(X_{T\wedge t} - X_{S\vee s}) = h\int 1_{(S,T]}H\,dI_X \in L^p_G.\end{aligned}$$

It follows that for $B \in \mathcal{R}$, we have

$$\int h1_{(S,T]}1_B y\,dI_X = h\int 1_{(S,T]}1_B y\,dI_X \in L^p_G.$$

For any $z \in L^q_{G^*}$, we have then

$$\int h1_{(S,T]}1_B y\,d(I_X)_z = \int 1_{(S,T]}1_B y\,d(I_X)_{hz}.$$

The class of sets B for which the above equality holds for all $z \in L^q_{G^*}$ is a monotone class which contains $\mathcal{R}$, hence the above equality holds for all $B \in \mathcal{P}$, and $z \in L^q_{G^*}$.

Hence, for any predictable, simple process H, we have

$$\int h1_{(S,T]}H\,d(I_X)_z = \int 1_{(S,T]}H\,d(I_X)_{hz}.$$

If $H \in \mathcal{F}_{F,G}(X)$, Lebesgue's theorem implies that the above equality holds for H. Assume now that $\int 1_{(S,T]}H\,dI_X \in L^p_G$. Then $h\int 1_{(S,T]}H\,dI_X \in L^p_G$ and

$$\begin{aligned}\langle h\int 1_{(S,T]}H\,dI_X, z\rangle &= \langle \int 1_{(S,T]}H\,dI_X, hz\rangle \\ &= \int 1_{(S,T]}H\,d(I_X)_{hz} = \int h1_{(S,T]}H\,d(I_X)_z \\ &= \langle \int h1_{(S,T]}H\,dI_X, z\rangle.\end{aligned}$$

Since $L^q_{G^*}$ is norming for both L^p_G and $(L^q_{G^*})^*$, we deduce that $\int h1_{(S,T]}HdI_X = h\int 1_{(X,T]}HdI_X \in L^p_G$, and this proves the theorem under hypothesis (a).

Assume (b), and let $H : \mathbb{R} \to \mathbb{R}$ be predictable with $\widetilde{I}_{F,G}(H) < \infty$, that is

$$H \in \mathcal{F}_{\mathbb{R}}(I_{F,G}) \subset \mathcal{F}_{\mathbb{R}}(I_{\mathbb{R},E}) = \mathcal{F}_{\mathbb{R},E}(X).$$

Also assume that $\int 1_{(S,T]}HdI_X \in L^p_E$. Consider first the case $h = h'y$ where $y \in F$ and h' is real valued, bounded, and $\mathcal{F}_S$–measurable. The $1_{(S,T]}Hy \in L^p_G$, and by Theorem AI.14, $\int 1_{(S,T]}HydI_X = y\int 1_{(S,T]}HdI_X$. By the first part of the proof, we have

$$\int h1_{(S,T]}HdI_X = h'\int 1_{(S,T]}HydI_X = h\int 1_{(S,T]}HdI_X.$$

This equality then holds for any $\mathcal{F}_S$–measurable simple function. By approximating the general h with a dominated sequence of simple functions, and using the Lebesgue Theorem 3.1, we obtain the desired conclusion.

We now establish a theorem which is essential for the proof of the main convergence theorem. This theorem will be completed with additional properties in Theorem 3.9 *infra*.

3.8 THEOREM. *Let $H \in L^1_{F,G}(X)$ and let T be any stopping time. Then $1_{[0,T]}H \in L^1_{F,G}(X)$ and*

$$(H \cdot X)^T = (1_{[0,T]}H) \cdot X.$$

If T is predictable, then $1_{[0,T)}H \in L^1_{F,G}(X)$ and

$$(H \cdot X)^{T-} = (1_{[0,T)}H) \cdot X.$$

Proof. Suppose that T is a simple stopping time of the form $T = \Sigma_{1\le i\le n}1_{A_i}t_i$, with $0 \le t_1 < \cdots < t_n \le +\infty$, $A_i \in \mathcal{F}_{t_i}$ mutually disjoint, and $\cup_{1\le i\le n}A_i = \Omega$. For each $\omega \in \Omega$, there is a unique i such that $\omega \in A_i$ and hence $T(\omega) = t_i$. Then

$$(H \cdot X)_T(\omega) = (H \cdot X)_{t_i}(\omega) = (\int_{[0,t_i]} HdI_X)(\omega)$$

hence

$$\begin{aligned}(H\cdot X)_T &= \Sigma_{1\le i\le n}1_{A_i}\int_{[0,t_i]} H\,dI_X\\ &= \int_{[0,\infty]} H\,dI_X - \Sigma_{1\le i\le n}1_{A_i}\int_{(t_i,\infty]} H\,dI_X\\ &= \int_{[0,\infty]} H\,dI_X - \Sigma_{1\le i\le n}\int_{(t_i,\infty]} 1_{A_i}H\,dI_X,\end{aligned}$$

by Theorem 3.7, since $A_i \in \mathcal{F}_{t_i}$; and hence

$$(H\cdot X)_T = \int_{[0,\infty]} H\,dI_X - \int 1_{(T,\infty]}H\,dI_X = \int 1_{[0,T]}H\,dI_X.$$

We can establish the above equality for a general stopping time T by approximating it by $T_n \searrow T$, where the T_n are simple stopping times, and then applying the Lebesgue Theorem 3.1; we note that $\int 1_{[0,T]}H\,dI_X \in L^p_G$.

Replacing T by $T\wedge t$, we have

$$(H\cdot X)^T_t = \int_{[0,t]} 1_{[0,T]}H\,dI_X.$$

Thus the process $(\int 1_{[0,t]}1_{[0,T]}H\,dI_X)_{t\ge 0}$ has values in G, and is cadlag, hence $1_{[0,T]}H \in L^1_{F,G}(X)$ and $(1_{[0,T]}H\cdot X) = (H\cdot X)^T$.

For the predictable case, we approximate the predictable stopping time T by an increasing sequence of stopping times $T_n \nearrow T$, and use the Lebesgue Theorem 3.1 to obtain the conclusion.

The next theorem gives a more complete description of the properties of X^T. The proofs follow from our previous results and the definitions.

3.9 THEOREM. *Let T be a stopping time.*

(a) X^T is p–summable relative to (F,G) and we have

$$X^T = 1_{[0,T]}\cdot X \text{ and } I_{X^T}(A) = I_X([0,T]\cap A), \text{ for } A\in\mathcal{P}[0,\infty].$$

(a′) If T is predictable, then X^{T-} is p–summable relative to (F,G) and we have

$$X^{T-} = 1_{[0,T)}\cdot X \text{ and } I_{X^{T-}}(A) = I_X([0,T)\cap A) \text{ for } A\in\mathcal{P}[0,\infty].$$

(b) For every predictable F–valued process H, we have

$$\operatorname{svar}_{F,L^p_G} I_{X^T}(H) = \operatorname{svar}_{F,L^p_G} I_X(1_{[0,T]}H).$$

(b′) If T is predictable, then

$$\mathrm{svar}_{F,L^p_G} I_{X^{T-}}(H) = \mathrm{svar}_{F,L^p_G} I_X(1_{[0,T)}H).$$

(c) We have $H \in \mathcal{F}_{F,G}(X^T)$ if and only if $1_{[0,T]}H \in \mathcal{F}_{F,G}(X)$, and in this case we have

$$\int H dI_{X^T} = \int 1_{[0,T]} H dI_X.$$

(c′) If T is predictable, then $H \in \mathcal{F}_{F,G}(X^{T-})$ if and only if $1_{[0,T)}H \in \mathcal{F}_{F,G}(X)$, and in this case we have

$$\int H dI_{X^{T-}} = \int 1_{[0,T)} H dI_X.$$

(d) If $H \in L^1_{F,G}(X)$, then $1_{[0,T]}H \in L^1_{F,G}(X)$ and $H \in L^1_{F,G}(X^T)$. In this case

$$(H \cdot X)^T = H \cdot X^T = (1_{[0,T]}H) \cdot X.$$

(d′) If T is predictable and $H \in L^1_{F,G}(X)$, then $1_{[0,T)}H \in L^1_{F,G}(X)$ and $H \in L^1_{F,G}(X^{T-})$. In this case we have

$$(H \cdot X)^{T-} = H \cdot X^{T-} = (1_{[0,T)}H) \cdot X.$$

(e) If the set of measures $(I_X)_{F,G}$ is uniformly σ–additive, then so is $(I_{X^T})_{F,G}$; if T is predictable, then $(I_{X^{T-}})_{F,G}$ is also uniformly σ–additive.

Convergence theorems

We maintain the assumption that X is p-summable relative to (F, G). We have already proved a Lebesgue–type convergence theorem (Theorem 3.1) for processes in $\mathcal{F}_{F,G}(X)$ concerning the convergence of the integrals. In this section we shall consider the Lebesgue and Vitali theorem for convergence in $L^1_{F,G}(X)$, as well as pointwise uniform convergence of the integrals on compact time intervals for a suitable subsequence.

The key result needed for the uniform convergence property is the following theorem, which will imply that the space $L^1_{F,G}(X)$ is complete.

3.10 THEOREM. *Let (H^n) be a sequence in $L^1_{F,G}(X)$ and assume that $H^n \to H$ in $\mathcal{F}_{F,G}(X)$. Then $H \in L^1_{F,G}(X)$. Moreover, for every t, we have*

$(H^n \cdot X)_t \to (H \cdot X)_t$ *in* L^p_G, *and there exists a subsequence* (n_r) *such that* $(H^{n_r} \cdot X)_t \to (H \cdot X)_t$ a.s., *as* $r \to \infty$, *uniformly on every compact time interval.*

Proof. (H^n) is a Cauchy sequence in $L^1_{F,G}(X)$, converging in $\mathcal{F}_{F,G}(X)$ to H. By passing to a subsequence, if necessary, we can assume that $\widetilde{I}_{F,G}(H^n - H^{n+1}) \leq \frac{1}{4^n}$ for each n. Let $t_0 > 0$. For each n, let $Z^n = H^n \cdot X$, and define the stopping time

$$u_n = \inf\{t : |Z^n_t - Z^{n+1}_t| > \frac{1}{2^n}\} \wedge t_0.$$

Let $G_n = \{u_n < t_0\}$. For each stopping time v, we have, by Theorem 3.8, $Z^n_v = \int_{[0,v]} H^n dI_X$, hence

$$\begin{aligned} E(|Z^n_v - Z^{n+1}_v|) &= E(|\int_{[0,v]} (H^n - H^{n+1}) dI_X|) \\ &= \|\int_{[0,v]} (H^n - H^{n+1}) dI_X\|_{L^1_G} \leq \|\int_{[0,v]} (H^n - H^{n+1}) dI_X\|_{L^p_G} \\ &\leq \widetilde{I}_{F,G}(H^n - H^{n+1}) \leq \frac{1}{4^n}. \end{aligned}$$

In particular, for $v = u_n$, we have

$$E(|Z^n_{u_n} - Z^{n+1}_{u_n}|) \leq \frac{1}{4^n}.$$

On the other hand,

$$P(G_n) \leq 2^n E(|Z^n_{u_n} - Z^{n+1}_{u_n}|) \leq \frac{1}{2^n}.$$

To see this, we note that if $\omega \in G_n$, then $u_n(\omega) < t_0$; we take a sequence $t_k \searrow u_n(\omega)$, with $t_k < t_0$ such that $|Z^n_{t_k}(\omega) - Z^{n+1}_{t_k}(\omega)| > \frac{1}{2^n}$, for each k. Then we use the right continuity of Z^n and Z^{n+1} to conclude that $|Z^n_{u_n}(\omega) - Z^{n+1}_{u_n}(\omega)| \geq \frac{1}{2^n}$. Thus

$$E(|Z^n_{u_n} - Z^{n+1}_{u_n}|) \geq \frac{1}{2^n} P(G_n),$$

and the desired inequality follows.

Let $G_0 = \limsup_n G_n$. Then $P(G_0) = 0$. For $\omega \notin G_0$, there is a k such that if $n \geq k$, we have $\omega \notin G_n$, hence $u_n(\omega) = t_0$. Thus

$$\sup_{t < t_0} |Z^n_t(\omega) - Z^{n+1}_t(\omega)| \leq \frac{1}{2^n}.$$

Hence for $\omega \notin G_0$, the sequence $(Z_t^n(\omega))$ is Cauchy in G uniformly for $t < t_0$.

The process $Z_t(\omega) := \lim_n Z_t^n(\omega)$, defined for $t < t_0$ and $\omega \notin G_0$, with values in G, is cadlag, adapted, and $|Z_t^n(\omega) - Z_t(\omega)|_G \leq \frac{1}{2^{n-1}}$, hence $\|Z_t^n - Z_t\|_{L_G^p} \leq \frac{1}{2^{n-1}}$. It follows that for $t < t_0$, we have $Z_t \in L_G^p$ and $Z_t^n \to Z_t$ in L_G^p. On the other hand, $1_{[0,t]}H^n \to 1_{[0,t]}H$ in $\mathcal{F}_{F,G}(X)$, hence $Z_t^n \to \int_{[0,t]} H dI_X$ in $(L_{G^*}^q)^*$. It follows that

$$\int_{[0,t]} H dI_X = Z_t \in \mathrm{L}_G^p.$$

Since Z is cadlag, we deduce that $H \in L_{F,G}^1(X)$, where we extend Z_t consistently, for $t \in [0,\infty)$, and we have also $(H \cdot X)_t = Z_t$, for each t. Thus $L_{F,G}^1(X)$ is complete. Since t_0 was arbitrary, it follows that $(H^{n_r} \cdot X)_t \to (H \cdot X)_t$ a.s., uniformly on every compact time interval, for a suitable subsequence (n_r).

3.11 COROLLARY. *$L_{F,G}^1(X)$ is complete.*

3.12 COROLLARY. *If $I_{F,G}$ is uniformly σ–additive, the $L_{F,G}^1(X)$ contains all the F–valued, bounded, predictable processes (in particular, this is the case if $F = \mathbb{R}$).*

In fact, in this case $\mathcal{E}_F$, the space of elementary processes is dense in $\mathcal{F}_{F,G}(\mathcal{B}, X)$. Since $\mathcal{E}_F \subset L_{F,G}^1(X)$, we have $\mathcal{F}_{F,G}(\mathcal{B}, X) \subset L_{F,G}^1(X)$.

Remark. We shall see that if X has integrable variation, or if E, G are Hilbert spaces and X is a square integrable martingale, then $L_{F,G}^1(X) = \mathcal{F}_{F,G}(\mathcal{B}, X) = \mathcal{F}_{F,G}(X)$ (see Theorems 3.27 and 3.32 *infra*).

Uniform convergence of processes yields convergence in $L_{F,G}^1(X)$, as the next theorem shows.

3.13 THEOREM. *Let (H^n) be a sequence from $\mathcal{F}_{F,G}(X)$ which converges uniformly pointwise to a process H. Then*

(a) $H \in \mathcal{F}_{F,G}(X)$ and $H^n \to H$ in $\mathcal{F}_{F,G}(X)$.

Assume, in addition, that $H^n \in L_{F,G}^1(X)$, for each n. Then

(b) $H \in L_{F,G}^1(X)$, and $H^n \to H$ in $L_{F,G}^1(X)$.

(c) For every $t \in [0,\infty]$, we have $(H^n \cdot X)_t \to (H \cdot X)_t$ in L_G^p.

(d) There is a subsequence (n_r) such that $(H^{n_r} \cdot X)_t \to (H \cdot X)_t$ a.s., uniformly on compact time intervals.

Proof. Assertion (a) is immediate. Assertions (b) and (d) follow from Theorem 3.10. Assertion (c) follows from the continuity of the integral.

For the Vitali and Lebesgue theorems, pointwise convergence does not ensure convergence in $L^1_{F,G}(X)$, unless $I_{F,G}$ is uniformly σ-additive. The following two theorems follow from the preceding two theorems and the general Vitali and Lebesgue convergence theorems AI.9 and AI.10 in Appendix I.

3.14 THEOREM. *(Vitali). Let H^n be a sequence from $\mathcal{F}_{F,G}(X)$ and let H be an F-valued predictable process. Assume that*

(1) $\tilde{I}_{F,G}(H^n 1_A) \to 0$ as $\tilde{I}_{F,G}(A) \to 0$, uniformly for n;

and either one of the conditions (2), (2′) below:

(2) $H^n \to H$ in $\tilde{I}_{F,G}$-measure;

(2′) $H^n \to H$ pointwise and $I_{F,G}$ is uniformly σ-additive (this is the case, for example, if the H^n are real valued, i.e. $F = \mathbb{R}$).

Then

(a) $H \in \mathcal{F}_{F,G}(X)$ and $H^n \to H$ in $\mathcal{F}_{F,G}(X)$.

Conversely, if $H^n, H \in \mathcal{F}_{F,G}(\mathcal{B}, X)$ and if $H^n \to H$ in $\mathcal{F}_{F,G}(X)$, then conditions (1) and (2) are satisfied.

Under the hypotheses (1) and (2) or (2′), assume in addition that $H^n \in L^1_{F,G}(X)$ for each n.

Then

(b) $H \in L^1_{F,G}(X)$ and $H^n \to H$ in $L^1_{F,G}(X)$;

(c) For every $t \in [0, \infty]$, we have $(H^n \cdot X)_t \to (H \cdot X)_t$ in L^p_G;

(d) There is a subsequence (n_r) such that $(H^{n_r} \cdot X)_t \to (H \cdot X)_t$, as $r \to \infty$, a.s. uniformly on compact time intervals.

3.15 THEOREM. *(Lebesgue). Let (H^n) be a sequence from $\mathcal{F}_{F,G}(X)$ and let H be an F-valued predictable process. Assume that*

(1) There is a process $\phi \in \mathcal{F}_{\mathbb{R}}(\mathcal{B}, I_{F,G})$ such that $|H^n| \le \phi$ for every n,

and either one of the conditions (2), (2′) below:

(2) $H^n \to H$ in $\tilde{I}_{F,G}$-measure;

(2′) $H^n \to H$ pointwise and $I_{F,G}$ is uniformly σ–additive (this is the case if the H^n are real valued, i.e. $F = \mathbb{R}$).

Then

(a) $H^n \in \mathcal{F}_{F,G}(\mathcal{B}, X)$ and $H^n \to H$ in $\mathcal{F}_{F,G}(X)$.

Assume in addition that $H^n \in L^1_{F,G}(X)$ for each n. Then

(b) $H \in L^1_{F,G}(X)$ and $H^n \to H$ in $L^1_{F,G}(X)$;

(c) For every $t \in [0, \infty]$, we have $(H^n \cdot X)_t \to (H \cdot X)_t$ in L^p_G;

(d) There is a subsequence (n_r) such that $(H^{n_r} \cdot X)_t \to (H \cdot X)_t$, as $r \to \infty$, uniformly on compact time intervals.

The stochastic integral of caglad and bounded processes

The stochastic integral $H \cdot X$ can be computed pathwise for the class of *σ–elementary processes* $H \in \mathcal{F}_{F,G}(X)$ of the form

$$H = H_0 1_{\{0\}} + \Sigma_{1 \le i < \infty} H_i 1_{(T_i, T_{i+1}]},$$

where (T_i) is a sequence of stopping times with $T_i \nearrow \infty$, H_0 is bounded and $\mathcal{F}_0$–measurable, and for each i, H_i is bounded and $\mathcal{F}_{T_i}$–measurable.

This result will follow from the following general theorem.

3.16 THEOREM. *Let $H \in \mathcal{F}_{F,G}(X)$ and assume that there is a sequence $T_n \nearrow \infty$ of stopping times such that $1_{[0,T_n]}H \in L^1_{F,G}(X^{T_n})$ for each n. Then $H \in L^1_{F,G}(X)$ and $H \cdot X = \lim_n (1_{[0,T_n]}H) \cdot X$ pointwise.*

Proof. Let $t \in [0, \infty]$. Note that, by Theorem 3.9 we have $1_{[0,T_n]}H \in L^1_{F,G}(X)$ for each n. Then, for $t \ge 0$ we have $1_{[0,t]}1_{[0,T_n]}H \to 1_{[0,t]}H$ pointwise, $|1_{[0,t]}1_{[0,T_n]}H| \le |H|$,

$$\int_{[0,t]} 1_{[0,T_n]} H \, dI_{X^{T_n}} \in L^p_G,$$

and

$$\int_{[0,t]} 1_{[0,T_n]} H \, dI_X = ((1_{[0,T_n]}H) \cdot X^{T_n})_t.$$

We shall show that this sequence converges pointwise, as $n \to \infty$. For $m \le n$, we have $(1_{[0,T_n]}H \cdot X)^{T_m}_t = (1_{[0,T_m]}H \cdot X)_t$; for a given $\omega \in \Omega$, we choose $m = m_\omega$, such that $t < T_m(\omega)$. Then, for $n \ge m$, we have

$$\begin{aligned}\lim_n \Big(\int_{[0,t]} 1_{[0,T_n]} H \, dI_X\Big)(\omega) &= \lim_n (1_{[0,T_n]}H \cdot X)^{T_m}_t(\omega) \\ &= (1_{[0,T_m]}H \cdot X)_t(\omega).\end{aligned}$$

This proves the pointwise convergence as asserted. Applying the Lebesgue theorem 3.1, we have $\int_{[0,t]} H dI_X \in L^p_G$ and $\int_{[0,t]} 1_{[0,T_n]} H dI_X \to \int_{[0,t]} H dI_X$ pointwise.

For each ω, and $m = m_\omega$ as above, we have

$$(\int_{[0,t]} H dI_X)(\omega) = (1_{[0,T_m]} H \cdot X)_t(\omega),$$

hence the process $(\int_{[0,t]} H dI_X)_{t \geq 0}$ is cadlag; thus $H \in L^1_{F,G}(X)$ and

$$(H \cdot X)_t = \int_{[0,t]} H dI_X = \lim_n (1_{[0,T_n]} H \cdot X)_t.$$

3.17 COROLLARY. *$L^1_{F,G}(X)$ contains all the σ–elementary processes of $\mathcal{F}_{F,G}(X)$. If we put such a process H in the standard form:*

$$H = H_0 1_{\{0\}} + \Sigma_{1 \leq n < \infty} H_n 1_{(T_n, T_{n+1}]},$$

then the stochastic integral $H \cdot X$ can be computed pathwise:

$$(H \cdot X)_t = H_0 X_0 + \Sigma_{1 \leq n < \infty} H_n (X_{T_{n+1} \wedge t} - X_{T_n \wedge t}).$$

Remark. There are σ–elementary processes which do not belong to $\mathcal{F}_{F,G}(X)$; such processes are not integrable with respect to X. However, we shall see in section 4 that such processes are "locally integrable" with respect to any "locally summable process," even if the random variables H_n are not bounded (Theorem 4.5 *infra*).

The next theorem considers all caglad processes of $\mathcal{F}_{F,G}(X)$ — not just the σ–elementary processes.

3.18 THEOREM. *$L^1_{F,G}(X)$ contains all caglad processes of $\mathcal{F}_{F,G}(X)$. In particular, $L^1_{F,G}(X)$ contains all bounded, caglad, adapted, F–valued processes.*

Proof. Let H be first a bounded, caglad, adapted process. Then H_+ is cadlag and adapted. For each n, define the stopping times $T(n,0) = 0$, and for $k \geq 1$,

$$T(n, k+1) = \inf\{t > T(n,k) : |H_{t+} - H_{T(n,k)+}| > \frac{1}{n}\} \wedge (T(n,k) + \frac{1}{n}).$$

Now define the σ–elementary processes

$$H^n = \Sigma_{k\geq 0} H_{T(n,k)+} 1_{(T(n,k),T(n,k+1)]}.$$

We note that if $|H| \leq M$, then $|H^n| \leq M$ for each n, hence $H^n \in \mathcal{F}_{F,G}(X)$. By the preceding Corollary 3.17, we have $H^n \in L^1_{F,G}(X)$. Since H is caglad, from the definition of the above family of stopping time, we deduce that $H^n \to H$ uniformly. Then $H \in L^1_{F,G}(X)$ by Theorem 3.13.

Now assume $H \in \mathcal{F}_{F,G}(X)$ and that H is caglad, hence H is locally bounded. Let $S_n \nearrow \infty$ be a sequence of stopping times, such that each $1_{[0,S_n]}H$ is bounded. Since each such process is caglad, we have $1_{[0,S_n]}H \in L^1_{F,G}(X^{S_n})$ for each n; hence, by Theorem 3.16, $H \in L^1_{F,G}(X)$.

Summability of the stochastic integral

The following theorem states that under certain conditions, the stochastic integral $H \cdot X$ is itself summable, and $K \cdot (H \cdot X) = (KH) \cdot X$. This property follows from the associativity property established in Appendix I for the general vector integrals (Theorem AI.15).

3.19 THEOREM. *I. Let $H \in \mathcal{F}_{\mathbb{R}}((I_X)_{F,G}) \subset \mathcal{F}_{\mathbb{R},E}(X)$. Assume that $H \in L^1_{\mathbb{R},E}(X)$ and $\int_A H dI_X \in L^p_E$ for every $A \in \mathcal{P}$. Then:*

(a) $H \cdot X$ is p–summable relative to (F,G) and

$$dI_{H\cdot X} = d(HI_X).$$

where HI_X is the measure defined by $(HI_X)(A) = \int_A H dI_X$ for $A \in \mathcal{P}$.

(b) For any predictable process $K \geq 0$, we have

$$(\tilde{I}_{H\cdot X})_{F,G}(K) = (\tilde{I}_X)_{F,G}(KH).$$

(c) $K \in L^1_{F,G}(H \cdot X)$ if and only if $KH \in L^1_{F,G}(X)$ and in this case, we have

$$K \cdot (H \cdot X) = (KH) \cdot X).$$

(d) Assume $(I_X)_{F,G}$ is uniformly σ–additive. Then $(I_{H\cdot X})_{F,G}$ is uniformly σ–additive if and only if $H \in \mathcal{F}_{\mathbb{R}}(\mathcal{B}, (I_X)_{F,G})$.

II. Let $H \in L^1_{F,G}(X)$ and assume that $\int_A H dI_X \in L^p_G$ for $A \in \mathcal{P}$. Then:

(a) $H \cdot X$ is p–summable relative to $(I\!R, G)$ and

$$dI_{H \cdot X} = d(HI_X).$$

(b) For any predictable process $K \geq 0$, we have

$$(\tilde{I}_{H \cdot X})_{I\!R,G}(K) \leq (\tilde{I}_X)_{F,G}(KH).$$

(c) If K is a real valued predictable process such that $KH \in L^1_{F,G}(X)$, then $K \in L^1_{I\!R,G}(H \cdot X)$, and in this case we have

$$K \cdot (H \cdot X) = (KH) \cdot X.$$

(d) Assume that $(I_X)_{F,G}$ is uniformly σ–additive and that $H \in \mathcal{F}_{F,G}(\mathcal{B}, X)$. Then $(I_{H \cdot X})_{I\!R,G}$ is uniformly σ–additive.

Proof. We only need to prove assertion I(a), and then apply Theorem AI.15. We notice first that by Proposition AI.12(a), $d(HI_X)$ is σ–additive on $\mathcal{P}$. Next we prove the equalities

$$I_{H \cdot X}(A) = \int_A H dI_X$$

and

$$(\tilde{I}_{H \cdot X})_{F,G}(A) = (\tilde{I}_X)_{F,G}(1_A H)$$

first for predictable rectangles A and then for every $A \in \mathcal{R}$.

From the first equality we deduce that $I_{H \cdot X}$ can be extended to a σ–additive measure on $\mathcal{P}$ with values in L^p_E. From the second equality it follows that $I_{H \cdot X}$ has bounded semivariation on $\mathcal{R}$ relative to (F, G):

$$\sup\{\tilde{I}_{H \cdot X})_{F,G}(A); A \in \mathcal{R}\} \leq (\tilde{I}_X)_{F,G}(H) < \infty.$$

By remark (f) following Definition 2.1, $H \cdot X$ is summable relative to (F, G). From the first of the above equalities we deduce that the σ–additive measures $dI_{H \cdot X}$ and $d(HI_X)$ are equal on $\mathcal{R}$; therefore they are equal on $\mathcal{P}$.

Assertion II(a) is proved in the same way, using the inequality

$$(\tilde{I}_{H \cdot X})_{I\!R,G}(A) \leq (\tilde{I}_X)_{F,G}(1_A H), \text{ for } A \in \mathcal{R}.$$

The jumps of the stochastic integral

The following theorem yields the jumps of the stochastic integral.

3.20 THEOREM. *For any process $H \in L^1_{F,G}(X)$, we have*

$$\Delta(H \cdot X) = H\Delta X.$$

Proof. Assume H is bounded. By Theorem 3.8 we have $\Delta X_t = X_t - X_{t-} \in L^p_E$ and

$$\Delta(H \cdot X)_t = (H \cdot X)_t - (H \cdot X)_{t-} = \int_{[t]} H dI_X$$
$$= \int_{[t]} H_t dI_X = H_t \int_{[t]} dI_X = H_t \Delta X_t,$$

by Proposition 3.5, since H_t is $\mathcal{F}_{t-}$–measurable.

Assume now that $H \in L^1_{F,G}(X)$. For each n, the stopping time $T_n = \inf\{t : |H_t| \geq n\}$ is predictable and $1_{[0,T_n)}|H| \leq n$. By the above case,

$$\Delta(1_{[0,T_n)} H \cdot X) = 1_{[0,T_n)} H\Delta X.$$

On the other hand,

$$\Delta(1_{[0,T_n)} H \cdot X)_t = \int 1_{[t]} 1_{[0,T_n)} H dI_X$$
$$= \int 1_{[t]} 1_{\{t<T_n\}} H dI_X = 1_{\{t<T_n\}} \int 1_{[t]} H dI_X$$
$$= 1_{\{t<T_n\}} \Delta(H \cdot X)_t.$$

Thus

$$1_{\{t<T_n\}} \Delta(H \cdot X)_t = 1_{[0,T_n)}(T) H_t \Delta X_t,$$

and the desired equality follows by letting $n \to \infty$.

The stochastic integral with respect to a martingale

3.21 THEOREM. *Let X be p–summable relative to (F, G) and let $H \in \mathcal{F}_{F,L^p_G}(X)$. If X is a martingale and if $\int_{[0,t]} H dI_X \in L^p_G$ for every $t \in [0, \infty]$, then $H \in L^1_{F,L^p_G}(X)$ and $H \cdot X$ is a uniformly integrable martingale, bounded in L^p_G. In particular, for $p = 2$, if X is a 2–summable, square integrable martingale, if*

$H \in \mathcal{F}_{F,L^2_G}(X)$ and if $\int_{[0,t]} H dI_X \in L^2_G$ for $t \in [0,\infty]$, then $H \in L^2_{F,L^2_G}(X)$ and $H \cdot X$ is a square integrable martingale.

Proof. Let $t \in [0,\infty)$ and $A \in \mathcal{F}_t$ and prove that $E(1_A(\int_{[0,\infty]} H dI_X - \int_{[0,t]} H dI_X)) = 0$ that is

$$(*) \qquad E[1_A(\int 1_{(t,\infty]} H dI_X] = 0.$$

If $H = 1_{\{0\}\times B}x$, with $B \in \mathcal{F}_0$ and $x \in F$, then $(*)$ holds. Assume $H = 1_{(u,v]\times B}x$, with $B \in \mathcal{F}_u$ and $x \in F$. If $v \le t \vee u$, then $(*)$ holds. Assume $t \vee u < v$; then $\int 1_{(t,\infty]} H dI_X = 1_B x(X_v - X_{t\vee u})$, thus $1_A \int 1_{(t,\infty]} H dI_X = 1_{A\cap B} x(X_v - X_{t\vee u})$. By taking expectations of both sides, and noting that $A \cap B \in \mathcal{F}_{t\vee u}$, we obtain $(*)$. Thus $(*)$ holds for $\mathcal{R}$–measurable simple processes H.

Assume now H is predictable, and let $y^* \in G^*$. The $\mathcal{R}$–measurable, simple processes are dense in $L^1_F((I_X)_z)$, where $z = 1_A y^* \in L^q_{G^*}$. Let (H^n) be a sequence of such processes converging to H in $L^1_F((I_X)_z)$.

Then $\int 1_{(t,\infty]} H^n d(I_X)_z \to \int 1_{(t,\infty]} H d(I_X)_z$, that is $\langle \int_{(t,\infty]} H^n dI_X, z\rangle \to \langle \int_{(t,\infty]} H dI_X, z\rangle$. Thus

$$E(\langle 1_A \int_{(t,\infty]} H^n dI_X, y^*\rangle) \to E(\langle 1_A \int_{(t,\infty]} H dI_X, y^*\rangle)$$

that is

$$\langle E(1_A \int_{(t,\infty]} H^n dI_X), y^*\rangle \to \langle E(1_A \int_{(t,\infty]} H dI_X), y^*\rangle.$$

By the previous case, the left–hand side is 0, for each $y^* \in G^*$ and every n, hence $E(1_A \int_{(t,\infty]} H dI_X) = 0$. It follows that $(\int_{[0,t]} H dI_X)_{t\ge 0}$ is a uniformly integrable martingale. Since every martingale has a cadlag modification, ([B–D.4]), we deduce that $H \in L^1_{F,L^p_G}(X)$ and the theorem is proved.

3.22 Corollary. *If L^p_G is reflexive and if X is a p–summable martingale, relative to (F,G), then $L^1_{F,L^p_G}(X) = \mathcal{F}_{F,L^p_G}(X)$.*

Remarks. (1) We shall see in the next section that if X is a local martingale and is locally summable, and if H is locally integrable with respect to X, then $H \cdot X$ is a local martingale (Theorem 4.14 *infra*). The case when $H \cdot X$ is a martingale, but not necessarily uniformly integrable is also considered.

(2) A martingale, or a square integrable martingale is not necessarily summable. But if E and G are Hilbert spaces and if $X : \mathbb{R} \to E \subset L(F, G)$ is a square integrable martingale, then X is 2–summable (see Theorem 3.24 *infra*).

Square integrable martingales

In this subsection, E and G are Hilbert spaces over the reals and F is a Banach space such that $E \subset L(F, G)$. For example, $E = L(\mathbb{R}, E)$, $E = L(E, \mathbb{R})$; $E \subset L(G, E \hat{\otimes}_{HS} G)$, where HS indicates that the Hilbert–Schmidt norm is used on $E \otimes G$. The inner product in any Hilbert space is denoted by $\langle \cdot, \cdot \rangle$.

The main result of this section is that any E–valued square integrable martingale M is 2–summable relative to any embedding $E \subset L(F, G)$, and that the semivariation of I_M is independent of this embedding.

We say that a martingale $M : \mathbb{R} \to E$ is square integrable if $M_t \in L^2_E$ for every $t \in [0, \infty)$ and $\sup_t \|M_t\|_2 < \infty$. This is equivalent to the existence of a random variable $M_\infty \in L^2_E$ such that for every t we have $M_t = E(M_\infty | \mathcal{F}_t)$.

We shall make a slight departure from our usual notation. We shall write $L^1_{F, L^2_G}(X)$, $(\widetilde{I}_M)_{F, L^2_G}$, etc., in place of $L^1_{F,G}(X)$, $(\widetilde{I}_M)_{F,G}$, respectively. This notational change will only be made in this subsection.

3.22 PROPOSITION. *(1) If $M : \mathbb{R} \to E$ is a square integrable martingale, then I_M can be extended to a σ–additive measure on $\mathcal{P}$ with values in L^2_E.*

(2) If M and N are E–valued square integrable martingales, then for any pair of disjoint sets A, B from $\mathcal{P}$, and for any $x, y \in F$, we have

$$I_M(A) \perp I_N(B) \text{ in } L^2_E \text{ and } I_M(A)x \perp I_N(B)y \text{ in } L^2_G.$$

Proof. (a) Assume first that M and N are E–valued square integrable martingales. Suppose A and B are disjoint sets from $\mathcal{R}$. By expressing A and B as a finite union of disjoint predictable rectangles, it is easy to show that $E(\langle I_M(A), I_N(B) \rangle_E) = 0$.

(b) Now we shall prove assertion (1). If $A \in \mathcal{R}$ then A is a disjoint union of predictable rectangles $[0_{A_0}]$ and $((s_i, t_i] \times A_i)_{1 \le i \le n}$. Let $T = \max\{t_i : 1 \le i \le n\} < \infty$ and let $B = [0, T] \times \Omega$. Then $\|I_M(A)\|_2^2 + \|I_M(B - A)\|_2^2 =$

$\|I_M(B)\|_2^2 \leq q$, where $q = \sup_t \|M_t\|_2^2$. Thus I_M is L_E^2–bounded on $\mathcal{R}$. Since L_E^2 is reflexive, L_E^2 does not contain c_0; by Theorem 2.5, I_M can be extended to a σ–additive measure on $\mathcal{P}$.

(c) We now prove assertion (2). By (b), we can consider I_M and I_N as having been extended to σ–additive measures on $\mathcal{P}$. If $A \in \mathcal{R}$, let Σ_A be the class of sets $B \in \mathcal{P}$ such that $I_M(A) \perp I_N(B - A)$. Since Σ_A is a monotone class containing $\mathcal{R}$, we have $\Sigma_A = \mathcal{P}$. If $B \in \mathcal{P}$, let Σ'_B be the class of sets $A \in \mathcal{P}$ such that $I_M(A) \perp I_N(B - A)$. Again, $\Sigma'_B = \mathcal{P}$. Hence if A and B are disjoint subsets of $\mathcal{P}$, we have $I_M(A) \perp I_N(B)$. The second assertion of (2) follows by considering the G–valued square integrable martingales Mx and Ny.

3.23 Theorem. *Let M be an E–valued square integrable martingale. Then*

(1) M is 2–summable relative to (F, G);

(2) The semivariation $(\widetilde{I}_M)_{F,L_G^2}$ is independent of the embedding $E \subset L(F, G)$ and satisfies

$$(\widetilde{I}_M)_{F,L_G^2}(A) = \|I_M(A)\|_{L_E^2}, \quad \text{for } A \in \mathcal{P};$$

(3) The set of measures $(I_M)_{F,L_G^2}$ is uniformly σ–additive.

Proof. Assertions (1) and (3) follows from Proposition 3.22 and assertion (2). To prove assertion (2), let $A \in \mathcal{P}$ and let (A_i) be a finite family of disjoint sets form $\mathcal{P}$, with union A; let (x_i) be a finite family of elements from F_1. Using the orthogonality properties in assertion (2) of Proposition 3.22, we deduce that

$$\begin{aligned} \|\Sigma_i I_M(A_i)x_i\|^2 &= \Sigma_i \|I_M(A_i)x_i\|^2 \\ &\leq \Sigma_i \|I_M(A_i)\|^2 = \|\Sigma I_M(A_i)\|^2 = \|I_M(A)\|^2, \end{aligned}$$

hence $(\widetilde{I}_M)_{F,L_G^2}(A) \leq \|I_M(A)\|_{L_E^2}^2$. The reverse inequality obviously holds.

3.24 Corollary. *An E–valued, square integrable martingale M is summable relative to (F, G) and*

$$(\widetilde{I}_M)_{F,L_G^1}(A) \leq (\widetilde{I}_M)_{F,L_G^2}(A) = \|I_M(A)\|_{L_G^2}, \quad \text{for } A \in \mathcal{P}.$$

The set of measures $(I_M)_{F,L_G^1}$ is uniformly σ–additive.

3.25 COROLLARY. *If M is a real valued square integrable martingale, then M is 2–summable relative to (E,E), for any Hilbert space E, and*

$$(\tilde{I}_M)_{\mathbb{R},L^1_{\mathbb{R}}} \leq (\tilde{I}_M)_{E,L^1_E} \leq (\tilde{I}_M)_{E,L^2_E} = (\tilde{I}_M)_{\mathbb{R},L^2_{\mathbb{R}}}.$$

Remark. In the proof of the 2–summability of M relative to (F,G), it was essential that both E and G are Hilbert spaces. If G is not a Hilbert space we may have $(\tilde{I}_M)_{F,G} = \infty$, as it is shown by an example given by Yor [Y.2]: Let M be the real Brownian motion on $[0,1]$. We can embed $\mathbb{R} \subset L(\ell^1,\ell^1)$ and then $L^1_{\mathbb{R}} \subset L(\ell^1, L^1_{\ell^1})$. Since M is a square integrable martingale, I_M has a σ–additive extension to $\mathcal{P}$, with values in $L^2_{\mathbb{R}} \subset L^1_{\mathbb{R}}$, hence M is summable relative to $(\mathbb{R},\mathbb{R})$. But I_M has infinite semivariation relative to $(\ell^1, L^1_{\ell^1})$. In fact, if I_M had finite semivariation relative to $(\ell^1, L^1_{\ell^1})$, then M would be summable relative to (ℓ^1,ℓ^1), therefore, by Corollary 3.12, every bounded σ–elementary process with values in ℓ^1 would be integrable with respect to M. However it is proved in [Y.2] that for the following process

$$H = \Sigma_{1 \leq n < \infty} e_n 1_{(\frac{n-1}{n},\frac{n}{n+1}]},$$

where $e_n = (\delta_{in})_{i\in\mathbb{N}} \in \ell^1$, we have $E(\| \int H dI_M \|_{\ell^1}) = \infty$, therefore $\int H dI_M$ does not belong to $L^1_{\ell^1}$. It follows that I_M does not have finite semivariation relative to $(\ell^1, L^1_{\ell^1})$.

We proved in Corollary 3.12 that if $(I_X)_{F,G}$ is uniformly σ–additive, then the space $L^1_{F,G}(X)$ contains all the bounded predictable processes; however, we do not know if, in general, the bounded predictable processes are dense in $L^1_{F,G}(X)$. This is true, as the next theorem shows, if X is a square integrable Hilbert–valued martingale.

3.26 THEOREM. *If M is an E–valued, square integrable martingale, then*

$$L^1_{F,L^2_G}(M) = \mathcal{F}_{F,L^2_G}(M) = \mathcal{F}_{F,L^2_G}(\mathcal{B},M).$$

Proof. The first equality follows from Remark (1) following Theorem 3.21.

Now suppose that $H \in L^1_{F,L^2_G}(M)$. We shall show that $H \in \mathcal{F}_{F,L^2_G}(\mathcal{B}, M)$. We note that $|H| \in \mathcal{F}_{\mathbb{R},L^2_E}(M)$, hence $|H| \in L^1_{\mathbb{R},L^2_E}(M)$ and by Theorem 3.21, $|H| \cdot M$ is an E–valued square integrable martingale; thus

$$(\widetilde{I}_{|H|\cdot M})_{\mathbb{R},L^2_E}(A) = (\widetilde{I}_{|H|\cdot M})_{F,L^2_G}(A) = \|I_{|H|\cdot M}(A)\|_{L^2_E}, \text{ for } A \in \mathcal{P}.$$

By Theorem 3.19, for $A \in \mathcal{P}$, we have

$$(\widetilde{I}_{|H|\cdot M})_{F,L^2_G}(A) = (\widetilde{I}_M)_{F,L^2_G}(1_A|H|).$$

It follows that

$$\begin{aligned}(\widetilde{I}_M)_{F,L^2_G}(1_A H) &= (\widetilde{I}_M)_{F,L^2_G}(1_A|H|).\\ &= \|I_{|H|\cdot M}(A)\|_{L^2_E} = (\widetilde{I}_{|H|\cdot M})_{\mathbb{R},L^2_E}(A).\end{aligned}$$

Since $|H|\cdot M$ is a square integrable martingale, the set of measures $(I_{|H|\cdot M})_{\mathbb{R},L^2_E}$ is uniformly σ–additive, hence $(\widetilde{I}_M)_{F,L^2_G}(1_{A_n}H) = (\widetilde{I}_{|H|\cdot M})_{\mathbb{R},L^2_E}(A_n) \to 0$, if $A_n \searrow \phi$. By Proposition AI.8(b), we have $H \in \mathcal{F}_{F,G}(\mathcal{B}, M)$.

We recall that if M is an E–valued, square integrable martingale, then $|M|^2$ is a submartingale of class (D) and has a Doob–Meyer decomposition $|M|^2 = N + \langle M, M\rangle$, where N is a martingale of class (D) and $\langle M, M\rangle$ is a predictable, integrable, increasing process called the sharp bracket of M. Then $\mu_{|M|^2} = \mu_{\langle M,M\rangle}$ on $\mathcal{P}$, where

$$\mu_{|M|^2}(A) = E(I_{|M|^2}(A)), \text{ for } A \in \mathcal{R}$$

and

$$\mu_{\langle M,M\rangle}(A) = E(I_{\langle M,M\rangle}(A)) = E(\int 1_A d\langle M, M\rangle),$$

for $A \in \mathcal{B}([0,\infty)) \times \mathcal{F}$.

If we set $z = M_\infty \in L^2_E$, we can consider the scalar measure $\langle I_M, z\rangle$ on $\mathcal{P}$, which is positive; in fact $\langle I_M, M_\infty\rangle = \mu_{\langle M,M\rangle}$.

The relationship between all these measures and the seminorm $(\widetilde{I}_M)_{F,G}$ is given by the following theorem. This theorem also shows that the mapping $H \to \int H dI_M$, from $L^1_{F,G}(M)$ into L^2_G, which is continuous in general, is an isometry in the case of a square integrable martingale with values in $\mathbb{R}$, or in the case the martingale is Hilbert–valued, but $F = \mathbb{R}$.

3.27 THEOREM. *Let M be an E–valued, square integrable martingale, and $H \in L^1_{F,L^2_G}(M)$. If either M is scalar valued or H is scalar valued, then*

$$(\widetilde{I}_M)_{F,L^2_G}(H) = \| \int H dI_M \|_{L^2_G}, = \|H\|_{L^2_F(\mu_{\langle M,M\rangle})} = \|H\|_{L^2_F(\langle I_M, M_\infty\rangle)}.$$

Proof. For $A \in \mathcal{F}_0$ and $x \in F$, we have (since either M is real, or $F = I\!R$),

$$\begin{aligned}\|I_M([0_A])x\|^2_{L^2_G} &= \|1_A M_0 x\|^2_{L^2_G} = E(1_A |M_0|^2 |x|^2)\\ &= E(\int 1_{[0_A]} |x|^2 dI_{|M|^2}),\end{aligned}$$

and for stopping times $S \leq T$, and $x \in F$, we have

$$\begin{aligned}&\|I_M(1_{(S,T]})x\|^2_{L^2_G} = \|(M_T - M_S)x\|^2_{L^2_G}\\ &= E(|M_T - M_S|^2 |x|^2) = E((|M_T|^2 - |M_S|^2)|x|^2)\\ &= E(\int 1_{(S,T]} |x|^2 dI_{|M|^2}).\end{aligned}$$

Let H be a simple process of the form

$$H = 1_{[0_A]} x_0 + \Sigma_{1 \leq i \leq n} 1_{(T_i, T_{i+1}]} x_i,$$

where $A \in \mathcal{F}_0$, and $(T_i)_{1 \leq i \leq n+1}$ is an increasing family of stopping times, and $x_i \in F$, for $0 \leq i \leq n$. Since the sets $[0_A]$ and $(T_i, T_{i+1}]$ are mutually disjoint, we have

$$\begin{aligned}&\| \int H dI_M \|^2_{L^2_G} = \|I_M([0_A])x_0\|^2_{L^2_G} + \Sigma_{1 \leq i \leq n} \|I_M((T_i, T_{i+1}])x_i\|^2_{L^2_G}\\ &= E(\int |H|^2 dI_{|M|^2}) = \int |H|^2 d\mu_{|M|^2}\\ &= \int |H|^2 d\mu_{\langle M,M\rangle} = E(\int |H|^2 d\langle M, M\rangle) = \int |H|^2 d\langle I_M, M_\infty\rangle.\end{aligned}$$

Since $(I_M)_{F,L^2_G}$ is uniformly σ–additive, the $\mathcal{R}$–simple processes are dense in $L^1_{F,L^2_G}(\mathcal{B}, M)$, hence by Theorem 3.26, they are dense in $L^1_{F,L^2_G}(M)$.

Let $H \in L^1_{F,L^2_G}(M)$, and let (H^n) be a sequence of $\mathcal{R}$–simple processes such that $H^n \to H$ in $L^1_{F,L^2_G}(M)$. By taking a subsequence if necesary, we can assume that $H^n \to H$ pointwise I_M–a.e. The continuity of the integral implies that $\int H^n dI_M \to \int H dI_M$ in L^2_G.

Since the measure $\langle I_M, M_\infty\rangle$ is dominated by $(\widetilde{I}_M)_{I\!R,L^2_E}$, we deduce that $H^n \to H$, $\langle I_M, M_\infty\rangle$–a.e. At the same time, (H^n) is Cauchy in $L^1_F(\langle I_M, M_\infty\rangle)$, using the isometry proved above. It follows that $H^n \to H$ in $L^2_F(\langle I_M, M_\infty\rangle)$, and from the above mentioned isometry, we deduce

$$\| \int H dI_M \|^2_{L^2_G} = \int |H|^2 d\langle I_M, M_\infty\rangle.$$

Finally,

$$(\widetilde{I}_M)_{F,L^2_G}(H) = \sup \| \int s dI_M \|_{L^2_G} = \sup(\int |s|^2 d\langle I_M, M_\infty\rangle)^{\frac{1}{2}}$$
$$= (\int |H|^2 d\langle I_M, M_\infty\rangle)^{\frac{1}{2}} = \| \int H dI_M \|_{L^2_G},$$

where the supremum is taken over all simple, predictable, F–valued processes s such that $|s| \leq |H|$.

3.28 COROLLARY. *The spaces $L^1_{F,L^2_G}(M)$ and $L^2_F(\langle M, M\rangle)$ contain the same predictable processes and are isometrically isomorphic.*

Remark. The classical approach to scalar stochastic integrals with respect to a real valued, square integrable martingale M is to prove the isometry $H \to \int H dI_M$, for the $\mathcal{R}$–simple processes H from $L^2(\mu_{\langle M,M\rangle})$ into L^2, and to extend this isometry to all of $L^2(\mu_{\langle M,M\rangle})$.

In our approach, we obtain this isometry directly from the space $L^1_{F,L^2_G}(M)$ into L^2_G.

Processes with integrable variation

Let $X : \mathbb{R} \to E$ be a *cadlag, adapted* process with *integrable variation* $|X|$; that is, $|X|_\infty \in L^1(P)$. Then $X_t \in L^1_E(P)$ and $|X|_t \in L^1(P)$ for every $t \in [0, \infty]$.

Then, there is a σ–additive measure $\mu_X : \mathcal{B}[0,\infty] \times \mathcal{F} \to E$ with *bounded variation* $|\mu_X|$ satisfying

$$\mu_X(B) = E(\int 1_B(s,\omega) dX_s(\omega))$$

and

$$|\mu_X|(B) = E(\int 1_B(s,\omega) d|X|_s(\omega))$$

for every $B \in \mathcal{B}[0,\infty] \times \mathcal{F}$. It follows that

$$|\mu_X| = \mu_{|X|}.$$

Moreover, if $E \subset L(F,G)$ and if $H : \mathbb{R} \to F$ is jointly measurable, then

$$H \in L^1_F(\mu_X) \text{ iff } E(\int |H_s(\omega)| d|X|_s(\omega)) < \infty.$$

In this case we have (see [D.2]):

$$\int H d\mu_X = E(\int H_s(\omega) dX_s(\omega)).$$

3.30 PROPOSITION. *If X has integrable variation, then the measure I_X is σ-additive and has bounded variation on $\mathcal{R}$, and for every $B \in \mathcal{R}$ we have*

$$\mu_X(B) = E(I_X(B))$$

and

$$|I_X|(B) = |\mu_X|(B) = \mu_{|X|}(B) = E(I_{|X|}(B)).$$

Proof. From the definition of I_X, we deduce that for every $B \in \mathcal{R}$ we have

$$I_X(B)(\omega) = \int 1_B(s,\omega) dX_s(\omega)$$

and

$$I_{|X|}(B) = \int 1_B(s,\omega) d|X|_s(\omega);$$

hence

$$\mu_X(B) = E(I_X(B)) \text{ and } \mu_{|X|}(B) = E(I_{|X|}(B).$$

Then

$$\begin{aligned} \|I_X(B)\|_{L^1_E} &= E(|\int 1_B(s,\omega) dX_s(\omega)|) \\ &\leq E(\int 1_B(s,\omega) d|X|_s(\omega)) \\ &= \mu_{|X|}(B). \end{aligned}$$

Since $\mu_{|X|}$ is σ-additive, it follows that $I_X : \mathcal{R} \to L^1_E$ is σ-additive and has bounded variation $|I_X|$ on $\mathcal{R}$, satisfying $|I_X| \leq \mu_{|X|} = |\mu_X|$.

Conversely, for $B \in \mathcal{R}$ we have

$$\begin{aligned} |\mu_X(B)| &= E(I_X(B))| \leq E(|I_X(B)|) \\ &= \|I_X(B)\|_{L^1_E} \leq |I_X|(B); \end{aligned}$$

therefore

$$|\mu_X| \le |I_X| \text{ on } \mathcal{R},$$

and the conclusion follows.

3.32 THEOREM. *A cadlag, adapted process $X : I\!R \to E$ with integrable variation $|X|$ is summable relative to any pair (F,G) such that $E \subset L(F,G)$. In this case, the set of measures $(I_X)_{F,L^1_G}$ is uniformly σ-additive and we have*

$$L^1_F(\mathcal{P},\mu_X) = L^1_F(\mathcal{P},I_X) \subset L^1_{F,L^1_G}(X)$$

and

$$L^1_{F,L^1_G}(\mathcal{S}(\mathcal{R}),X) = L^1_{F,L^1_G}(X) = \mathcal{F}_{F,L^1_G}(X).$$

Proof. The first equality follows from $|\mu_X| = |I_X|$. The inclusion follows from the inequality $(\widetilde{I}_X)_{F,L^1_G} \le |I_X|$: since the step processes over $\mathcal{P}$ are dense in $L^1_F(\mathcal{P},I_X)$, from Theorem 3.12 we deduce that $L^1_F(\mathcal{P},I_X) \subset \mathcal{F}_{F,L^1_G}(\mathcal{B},X) \subset L^1_{F,L^1_G}(X)$. On the other hand, by Theorem AI.8 we have $\mathcal{F}_{F,L^1_G}(\mathcal{B},X) = \mathcal{F}_{F,L^1_G}(X)$.

Remark. We can define I_X for every (not necessarily predictable) rectangle $\{0\} \times A$ or $(s,t] \times A$ with $A \in \mathcal{F}$, by

$$I_X(\{0\} \times A) = 1_A X_0 \text{ and } I_X((s,t] \times A) = 1_A(X_t - X_s).$$

and we still have

$$\|I_X(B)\|_{L^1_G} \le |\mu_X|(B)$$

for B in the algebra generated by the above rectangles. Since this algebra generates the σ-algebra $\mathcal{B}(I\!R_+) \times \mathcal{F}$, it follows that I_X can be extended as a σ-additive measure with finite variation on the whole algebra $\mathcal{B}(I\!R_+) \times \mathcal{F}$, not only on $\mathcal{P}$, and we still have $|I_X| = |\mu_X|$ on $\mathcal{B}(I\!R_+) \times \mathcal{F}$. We can then apply the integration theory of Appendix I, with $\Sigma = \mathcal{B}(I\!R_+) \times \mathcal{F}$ and obtain the space $\mathcal{F}_{F,L^1_G}(\Sigma,X)$. Then we can define a "stochastic integral" $(H \cdot X)_t = \int_{[0,t]} H dI_X$ in the case the integral belongs to L^1_G. This integral is still cadlag, but is not necessarily adapted.

Weak completeness of $L^1_{F,G}(\mathcal{B}, X)$

The following theorem gives sufficient conditions for $L^1_{F,G}(\mathcal{B}, X)$ to be weakly sequentially complete. It is a corollary of the general theorem AI.19 in Appendix I.

3.33 THEOREM. *Assume that F is reflexive, that $(I_X)_{F,G}$ is uniformly σ-additive and $c_0 \not\subset G$. Then $L^1_{F,G}(\mathcal{B}, X)$ is weakly sequentially complete.*

In fact, L^p_G does not contain c_0 (see [Kw]), and we can apply Theorem AI.19.

Weak compactness in $L^1_{F,G}(\mathcal{B}, X)$

We shall apply the general theory of weak compactness in Appendix I to $L^1_{F,G}(\mathcal{B}, X)$. Recall that a subset K in a Banach space is said to be *conditionally weakly compact* if every sequence of elements from K contains a subsequence which is weakly Cauchy.

The next theorem follows from Theorem AI.20.

3.34 THEOREM. *Let X be p-summable relative to (F, G). Assume F is reflexive and $(I_X)_{F,G}$ is uniformly σ-additive.*

Let $K \subset L^1_{F,G}(\mathcal{B}, X)$ be a set satisfying the following conditions:

(1) K is bounded in $L^1_{F,G}(\mathcal{B}, X)$;

(2) $H1_{A_n} \to 0$ in $L^1_{F,G}(\mathcal{B}, X)$, uniformly for $H \in K$, whenever $A_n \in \mathcal{P}$ and $A_n \searrow \phi$.

Then K is conditionally weakly compact in $L^1_{F,G}(\mathcal{B}, X)$. If, in addition, $c_0 \not\subset G$, then K is relatively weakly compact in $L^1_{F,G}(\mathcal{B}, X)$.

In the last case, for every sequence (H^n) from K, there exists a subsequence (H^{n_r}) such that $(\int H^{n_r} dX)_t$ converges weakly in L^p_G as $r \to \infty$, for every t.

The next theorem follows from Theorem AI.21.

3.35 THEOREM. *Let X be E-valued and p-summable realtive to $(\mathbb{R}, E)$. Let $K \subset L^1_{\mathbb{R},E}(\mathcal{B}, X)$ be a set satisfying the following conditions:*

(1) K is bounded in $L^1_{\mathbb{R},E}(\mathcal{B}, X)$;

(2) $\int_{A_n} H dI_X \to 0$ in L^p_E, uniformly for $H \in K$, whenever $A_n \in \mathcal{P}$ and $A_n \searrow \phi$.

Then K is conditionally weakly compact in $L^1_{\mathbb{R},E}(\mathcal{B},X)$. If, in addition, $c_0 \not\subset E$, then K is relatively weakly compact in $L^1_{\mathbb{R},E}(\mathcal{B},X)$.

In this last case, for any sequence (H^n) from K, there exists a subsequence (H^{n_r}) such that $(H^{n_r} \cdot X)_t$ converges weakly in L^p_E as $r \to \infty$, for each t.

Finally, we state a result about sequential weak convergence in $L^1_{\mathbb{R},E}(\mathcal{B},X)$. This theorem follows from Theorem AI.22.

3.36 THEOREM. *Let X be an E-valued process, p-summable relative to $(\mathbb{R},E)$. Let $(H^n)_{n\geq 0}$ be a sequence of scalar processes from $L^1_{\mathbb{R},E}(\mathcal{B},X)$. Suppose that $c_0 \not\subset E$. If*

$$\int 1_A H^n dX \to \int 1_A H^0 dX \text{ for every } A \in \mathcal{P},$$

then

$$H^n \to H^0, \text{ weakly in } L^1_{\mathbb{R},E}(\mathcal{B},X),$$

hence

$$(\int H^n dX)_t \to (\int H^0 dX)_t, \text{ weakly in } L^p_E, \text{ for each } t \geq 0.$$

4. Local summability and local integrability

Throughout this section, $X : \mathbb{R} \to E \subset L(F,G)$ is a cadlag, adapted process with $X_t \in L^p_E$ for each $t \in \mathbb{R}_+$. We shall study the properties of the stochastic integral $H \cdot X$ in the case X is locally p-summable and H is locally integrable with respect to X.

4.1 DEFINITIONS.

(a) We say that X is locally p-summable relative to (F,G) if there exists an increasing sequence (T_n) of stopping times with $T_n \nearrow \infty$, such that for each n, X^{T_n} is p-summable relative to (F,G).

If the set of measures $(I_{X^{T_n}})_{F,G}$ is uniformly σ-additive for each n, we say that the set of measures $(I_X)_{F,G}$ is lcoally uniformly σ-additive.

The sequence (T_n) is called a *determining sequence* for the local summability of X relative to (F,G).

Examples of locally summable processes are: locally square integrable processes, and processes with locally integrable variation.

(b) A predictable process $H : I\!R \to F$ is said to be locally integrable with respect to a process $X : I\!R \to E \subset L(E, F)$, which is locally p–summable relative to (F, G), if there exists an increasing sequence (T_n) of stopping times with $T_n \nearrow \infty$, such that for each n, X^{T_n} is p–summable relative to (F, G) and $1_{[0,T_n]}H$ is integrable with respect to X^{T_n}.

The sequence (T_n) is called a determining sequence for the local integrability of H with respect to X.

The set of all F–valued, predictable processes which are locally integrable with respect to X will be denoted by $L^1_{F,G}(X)_{\text{loc}}$.

(c) Let X be a locally summable process relative to (F, G) and let D be a Banach space. We denote by $\mathcal{F}_D(I_{F,G})_{\text{loc}}$ the space of all predictable D–valued processes H for which there exists a sequence of stopping times (T_n) with $T_n \nearrow \infty$, such that for each n, X^{T_n} is p–summable relative to (F, G), and $1_{[0,T_n]}H \in \mathcal{F}_D((I_{X^{T_n}})_{F,G})$, that is

$$(\widetilde{I}_{X^{T_n}})_{F,G}(1_{[0,T_n]}H) < \infty.$$

If $\mathcal{C}$ is any set of D–valued, *bounded*, predictable processes, we denote by $\mathcal{F}_D(\mathcal{C}, I_{F,G})_{\text{loc}}$ the set of all processes $H \in \mathcal{F}_D(I_{F,G})_{\text{loc}}$, such that for each stopping time T_n as above, we have $1_{[0,T_n]}H \in \mathcal{F}_D(\mathcal{C}, (I_{X^{T_n}})_{F,G})$.

Instead of writing $H \in \mathcal{F}_D(\mathcal{C}, I_{F,G})_{\text{loc}}$, we shall say that H *is locally in* $\mathcal{F}_D(\mathcal{C}, I_{F,G})$.

(d) If H^n and H are processes, we say $H^n \to H$ locally uniformly if there exists a sequence (T_k) of stopping times with $T_k \nearrow \infty$, such that for each k, $H^n \to H$ uniformly on $[0, T_k]$.

Basic properties

1. If X is p–summable relative to (F, G), then X is locally p-summable relative to (F, G).

2. If X is locally p–summable relative to (F, G), then X is locally p–summable relative to $(I\!R, E)$.

3. If (T_n) is a sequence of stopping times, determining for the local p–summability of X relative to (F, G), and if $S_n \nearrow \infty$ is another sequence of stopping times, then $(T_n \wedge S_n)$ is determining for the local p–summability of X relative to (F, G). A similar result holds for determining sequences for the local integrability of H with respect to X.

4. If X is locally p–summable relative to (F, G) and if T is a stopping time, then X^T is locally p–summable relative to (F, G).

5. If X is p–summable relative to (F, G) and if $H \in L^1_{F,G}(X)$, then H is locally integrable with respect to X.

Let (T_n) be a sequence, determining for the local integrability of H with respect to X. Then for each n, we have

$$((1_{[0,T_{n+1}]}H) \cdot X^{T_{n+1}})^{T_n} = (1_{[0,T_n]}H) \cdot X^{T_n}$$

outside an evanescent set. It follows that the limit

$$\lim_n (1_{[0,T_n]}H) \cdot X^{T_n}$$

exists pointwise outside an evanescent set. The limit is independent of the determining sequence. Moreover, this limit is cadlag and adapted.

This leads to the following definition:

4.2 DEFINITION. *If X is locally p–summable relative to (F, G) and if the F–valued process H is locally integrable with respect to X, then the stochastic integral of H with respect to X is a process denoted by $H \cdot X$ or $\int H dX$, and is defined up to an evanescent set by the equality*

$$H \cdot X = \int H dX = \lim_n (1_{[0,T_n]}H) \cdot X^{T_n},$$

for any sequence (T_n) of stopping times which is determining for the local integrability of H with respect to X.

It follows that for each n, we have

$$(H \cdot X)^{T_n} = (1_{[0,T_n]}H) \cdot X^{T_n},$$

The following theorem states that integrability and local integrability are equivalent for processes of $\mathcal{F}_{F,G}(X)$, in case X is p–summable.

4.3 THEOREM. *Let* X *be a* p*–summable process relative to* (F, G) *and* $H \in \mathcal{F}_{F,G}(X)$*. Then* H *is integrable with respect to* X *if and only if* H *is locally integrable with respect to* X*. In this case, the stochastic integral* $H \cdot X$ *is the same, whether* H *is considered integrable or locally integrable with respect to* X*.*

Proof. If H is integrable with respect to X, it easily follows that H is locally integrable with respect to X and the two integrals agree. The converse is proved by taking a determining sequence for the local integrability of H with respect to X and applying Theorem 3.16.

4.4 THEOREM. *If* X *is locally* p*–summable relative to* (F, G) *and has finite variation, then*

$$(H \cdot X)_t = \int_{[0,t]} H_s(\omega) dX_s(\omega),$$

as long as both sides are defined.

This follows from Theorem 3.3.

An important class of processes that are locally integrable with respect to any locally p–summable process is the class of σ–elementary processes, where the H_i used in defining the σ–elementary process are not assumed to be bounded. In Theorem 4.9 *infra*, we shall prove that all the caglad, adapted processes are locally integrable with respect to any locally p–summable process. We note that a σ–elementary process is not necessarily integrable with respect to a p–summable process.

4.5 THEOREM. *Let* H *be an* F*–valued* σ*–elementary process of the form*

$$H = H_0 1_{\{0\}} + \Sigma_{1 \le i < \infty} H_i 1_{(T_i, T_{i+1}]},$$

where the H_i *are not necessarily bounded. Then* H *is locally integrable with respect to any locally* p*–summable process* X *relative to* (F, G)*, and the stochastic integral can be computed pathwise by*

$$(H \cdot X)_t = H_0 X_0 + \Sigma_{1 \le i < \infty} H_i (X_{T_{i+1} \wedge t} - X_{T_i \wedge t}).$$

Proof. We note that for each t and ω, the above series reduces to a finite sum. For each n, consider the stopping time $S_n = \inf\{t : |H_{t+}| > n\}$. Since H_+ is cadlag, we have $S_n \nearrow \infty$. Also $1_{[0,S_n]}|H| \leq n$, since H is caglad. Note that $1_{[0,S_n]}|H_i| \leq n$, for each i. Now we observe that $1_{[0,S_n \wedge T_n]}H$ is an elementary process, hence it is integrable with respect to X. As a result, H is locally integrable with respect to X.

Let $U_n \nearrow \infty$ be a determining sequence of stopping times for the local p-summability of X. Set $R_n = U_n \wedge S_n \wedge T_n$. Then $1_{[0,R_n]}H$ is an elementary process and the stochastic integral can be computed pathwise by

$$((1_{[0,R_n]}H) \cdot X^{R_n})_t =$$
$$= 1_{[0,R_n]}H_0X_0 + \Sigma_{1\leq i\leq n}1_{[0,R_n]}H_i(X_{U_n\wedge S_n\wedge T_{i+1}\wedge t} - X_{U_n\wedge S_n\wedge T_i\wedge t}).$$

For fixed ω and t, we take n such that $t < R_n$. Then

$$(H \cdot X)_t(\omega) = \lim_n((1_{[0,R_n]}H) \cdot X^{R_n})_t(\omega)$$
$$= H_0(\omega)X_0(\omega) + \Sigma_{1\leq i<n}H_i(\omega)(X_{T_{i+1}\wedge t}(\omega) - X_{T_i\wedge t}(\omega)),$$

and the conclusion follows.

Convergence theorems

We shall need the following theorem:

4.6 THEOREM. *Assume X is locally p-summable relative to (F,G) and let H^n, $H \in L^1_{F,G}(X)_{\text{loc}}$ for $n \in I\!N$. Let $T_k \nearrow \infty$ be stopping times such that for each k, X^{T_k} is p-summable relative to (F,G), the processes $1_{[0,T_k]}H^n$ and $1_{[0,T_k]}H$ belong to $L^1_{F,G}(X^{T_k})$, and $1_{[0,T_k]}H^n \to 1_{[0,T_k]}H$ in $L^1_{F,G}(X^{T_k})$ as $n \to \infty$.*

Then

(a) For each t, $(H^n \cdot X)_t \to (H \cdot X)_t$ in probability;

(b) There exists a subsequence (n_r) such that $(H^{n_r} \cdot X)_t \to (H \cdot X)_t$, as $r \to \infty$, uniformly on compact time intervals.

Proof. (a) Let $t \geq 0$ and choose $\epsilon > 0$. Note that $P(\{T_k \leq t\}) \searrow 0$. Fix k_0 so that $P(\{T_{k_0} \leq t\}) < \epsilon$. If $\eta > 0$, we have

$$P(\{|(H^n \cdot X)_t - (H \cdot X)_t| > \eta\}) \leq \epsilon + P(\{t < T_{k_0}\} \cap \{|(H^n \cdot X)_t - (H \cdot X)_t| > \eta\}).$$

From the hypothesis we deduce that $1_{[0,t]}1_{[0,T_{k_0}]}H^n \to 1_{[0,t]}1_{[0,T_{k_0}]}H$, in $L^1_{F,G}(X^{T_{k_0}})$, which implies that $(H^n \cdot X)_t^{T_{k_0}} \to (H \cdot X)_t^{T_{k_0}}$ in L^p_G, hence in probability. There exists an N such that for $n \geq N$, we have $P(\{|(H^n \cdot X)_t^{T_{k_0}} - (H \cdot X)_t^{T_{k_0}}| > \eta\}) < \epsilon$, thus $P(\{|(H^n \cdot X)_t - (H \cdot X)_t| > \eta\}) < 2\epsilon$, for $n \geq N$, and this proves assertion (a).

(b) Since by hypothesis, $1_{[0,T_k]}H^n \to 1_{[0,T_k]}H^n$ in $L^1_{F,G}(X^{T_k})$, as $n \to \infty$, by Theorem 3.10, for each k there exists a subsequence $(n(r,k))_r$ such that

$$(1_{[0,T_k]}H^{n(r,k)} \cdot X^{T_k})_t \to (1_{[0,T_k]}H \cdot X^{T_k}),$$

as $r \to \infty$, uniformly on compact time intervals. One may assume that $(n(r,k+1))_r$ is a subsequence of $(n(r,k))_r$. By a diagonalization argument, and the fact that $T_k \nearrow \infty$, the conclusion follows.

Next we consider uniformly convergent sequences of locally integrable processes.

4.7 THEOREM. *Assume that X is locally summable relative to (F,G). Let (H^n) be a sequence from $L^1_{F,G}(X)_{\rm loc}$ and let H be an F–valued process such that $H^n \to H$ uniformly on $\mathbb{R}$.*

Then

(a) H is locally integrable with respect to X;

(b) For each t, $(H^n \cdot X)_t \to (H \cdot X)_t$ in probabilty;

(c) There is a subsequence (n_r) such that $(H^{n_r} \cdot X)_t \to (H \cdot X)_t$, as $r \to \infty$, uniformly on compact time intervals.

Proof. We choose N so that $|H^n - H^N| \leq 1$ for $n \geq N$. Let (T_k) be a determining sequence for the local integrability of H^N with respect to X. Since $1_{[0,T_k]}H^N \in L^1_{F,G}(X^{T_k})$, for each k, we deduce that $1_{[0,T_k]}H^n \in \mathcal{F}_{F,G}(X^{T_k})$, for $n \geq N$. Note that $1_{[0,T_k]}H^n$ is locally integrable with respect to X^{T_k}, hence by Theorem 4.3, $1_{[0,T_k]}H^n$ is integrable with respect to X^{T_k}. Since $1_{[0,T_k]}H^n \to 1_{[0,T_k]}H$ uniformly, as $n \to \infty$, by Theorem 3.13, it follows that, for each k, we have $1_{[0,T_k]}H \in L^1_{F,G}(X^{T_k})$ and $1_{[0,T_k]}H^n \to 1_{[0,T_k]}H$ in $L^1_{F,G}(X^{T_k})$, as $n \to \infty$. The conclusion follows by applying Theorem 4.6.

Another application of Theorem 4.6 is the Lebesgue theorem for locally integrable processes. A Vitali convergence theorem can also be proved along the same lines.

4.8 THEOREM. *(Lebesgue) Assume that X is locally p–summable relative to (F,G). Let (H^n) be a sequence of F–valued processes, which are locally integrable with respect to X, let H be a predictable, F–valued process and let $\phi \in \mathcal{F}_{\mathbb{R}}(\mathcal{B},(I_X)_{F,G})_{\text{loc}}$.*

Assume that

(1) $|H^n| \leq \phi$, for each n;

and either

(2) $H^n \to H$ locally uniformly;

or

(2′) $H^n \to H$ pointwise and the family of measures $(I_X)_{F,G}$ is locally uniformly σ–additive.

Then

(a) H is locally integrable with respect to X;

(b) For each t, we have $(H^n \cdot X)_t \to (H \cdot X)_t$ in probability;

(c) There is a subsequence (n_r) such that $(H^{n_r} \cdot X)_t \to (H \cdot X)_t$, as $r \to \infty$, a.s., uniformly on any compact time interval.

Proof. The proof uses a sequence (T_k) of stopping times which is determining for the local p–summability of X, and at the same time, for each k we have $H^n \to H$ uniformly on $[0,T_k]$ in the case of (2), and such that $(I_{X^{T_k}})_{F,G}$ is uniformly σ–additive, in the case of (2′). We may also assume that $\phi \in \mathcal{F}_{\mathbb{R}}(\mathcal{B},(I_{X^{T_k}})_{F,G})$ for each k. With this setting in place, the conclusions follow by applying Theorems 3.15 and 4.6.

As an application of Theorem 4.7, we shall deduce the local integrability of any caglad, adapted process, with respect to any locally p–summable process.

4.9 THEOREM. *Any F–valued, caglad, adapted process is locally integrable with respect to any process X which is locally p–summable relative to (F,G).*

More precisely, if X is locally p–summable relative to (F,G) and if $H : \mathbb{R} \to F$ is cadlag and adapted, then there exists a sequence (H^n) of

F–valued σ–elementary processes converging uniformly to H_-. For every t, we have $(H^n \cdot X)_t \to (H_- \cdot X)_t$ in probability. Moreover, there is a subsequence (n_r) such that $(H^{n_r} \cdot X)_t \to (H_- \cdot X)_t$ a.s. as $r \to \infty$, uniformly on every compact time interval.

Proof. Let $K : \mathbb{R} \to F$ be caglad and adapted. Then $H = K_+$ is cadlag, adapted and $K = H_-$. Let $b_n \searrow 0$ and define the stopping times $v(n,0) = 0$, and for $k = 1$,

$$v(n,k+1) = \inf\{t > v(n,k) : |H_t - H_{v(n,k)}| > b_n\} \wedge (b_n + v(n,k)).$$

These stopping times have the following properties:

(i) for each n we have $v(n,k) \nearrow \infty$, as $k \to \infty$;

(ii) $\lim_n \sup_k (v(n,k+1) - v(n,k)) = 0$;

(iii) $|H_t - H_{v(n,k)}| \le a_n$, for $t \in [v(n,k), v(n,k+1))$.

For each n, define the σ–elementary process

$$H^n = \Sigma_{k \ge 0} H_{v(n,k)} 1_{(v(n,k),v(n,k+1)]}$$

From properties (i), (ii), and (iii), it follows that $H^n \to H_-$ uniformly. The conclusion then follows from Theorem 4.7.

Additional properties

We shall state some properties that are extensions of corresponding properties proved in section 3 for integrable processes.

The following theorem follows from Theorem 3.7.

4.10 THEOREM. *Assume that X is locally p–summable relative to (F,G) and let $S \le T$ be stopping times. Then:*

(1) $(h1_{(S,T]}H) \cdot X = h[(1_{(S,T]}H) \cdot X]$ in each of the following two cases:

(a) h is a real valued, $\mathcal{F}_S$–measurable, random variable, and $H \in L^1_{F,G}(X)_{\text{loc}}$;

(b) h is an F–valued, $\mathcal{F}_S$–measurable, random variable and $H \in L^1_{\mathbb{R},E}(X)_{\text{loc}} \cap \mathcal{F}_{\mathbb{R}}(I_{F,G})_{\text{loc}}$.

(2) If, in addition, S is predictable and h is $\mathcal{F}_{S-}$–measureable in (a) and (b) above, then

$$(h1_{[S,T]}H) \cdot X = h[(1_{[S,T]}H) \cdot X].$$

For the proof of the next theorem, which states some properties of the stopped process, we use Theorem 3.9.

4.11 THEOREM. *Assume that X is locally p–summable relative to (F,G), and let T be a stopping time. Then:*

(a) X^T is locally p–summable relative to (F,G) and also relative to $(I\!R,E)$, and we have

$$X^T = 1_{[0,T]} \cdot X.$$

(a′) If T is predictable, then X^{T-} is locally p–summable relative to (F,G) and relative to $(I\!R,E)$ and

$$X^{T-} = 1_{[0,T)} \cdot X.$$

(b) An F–valued predictable process H belongs to $L^1_{F,G}(X^T)_{\text{loc}}$ if and only if $1_{[0,T]}H \in L^1_{F,G}(X)_{\text{loc}}$.

(b′) Assume T is predictable. An F–valued predictable process H belongs to $L^1_{F,G}(X^{T-})_{\text{loc}}$ if and only if $1_{[0,T)}H \in L^1_{F,G}(X)_{\text{loc}}$.

(c) If $H \in L^1_{F,G}(X)_{\text{loc}}$, then $H \in L^1_{F,G}(X^T)_{\text{loc}}$, and $1_{[0,T]}H \in L^1_{F,G}(X)_{\text{loc}}$, and we have

$$(H \cdot X)^T = H \cdot X^T = (1_{[0,T]}H) \cdot X.$$

(c′) If T is predictable and if $H \in L^1_{F,G}(X)_{\text{loc}}$, then $H \in L^1_{F,G}(X^{T-})_{\text{loc}}$ and $1_{[0,T)}H \in L^1_{F,G}(X)_{\text{loc}}$ and we have

$$(H \cdot X)^{T-} = H \cdot X^{T-} = (1_{[0,T)}H) \cdot X.$$

Next we state the associativity property of the stochastic integral.

4.12 THEOREM. *Let $X : I\!R \to E \subset L(F,G)$ be a cadlag, adapted process.*

I) Assume that X is locally p–summable relative to (F,G) (hence relative to $(I\!R,E)$) and let $H \in L^1_{I\!R,E}(X)_{\text{loc}} \cap \mathcal{F}_{I\!R}(I_{F,G})_{\text{loc}}$. Assume there is a sequence (T_n) of stopping times, determining for the local integrability of H with respect to X, such that, for each n and each $A \in \mathcal{P}$ we have $\int_A 1_{[0,T_n]} H \, dI_{X^{T_n}} \in L^p_E$. Then:

(a) $H \cdot X$ is locally p–summable relative to (F,G);

(b) An F–valued, predictable process K belongs to $L^1_{F,G}(H\cdot X)_{\rm loc}$ if and only if $KH \in L^1_{F,G}(X)_{\rm loc}$. In this case we have

$$K\cdot(H\cdot X) = (KH)\cdot X.$$

II) Assume that X is locally p–summable relative to (F,G) and let $H \in L^1_{F,G}(X)_{\rm loc}$. Assume there is a sequence (T_n) of stopping times, determining for the local integrability of H with respect to X, such that, for each n and each $A \in \mathcal{P}$ we have $\int_A 1_{[0,T_n]} H dI_{X^{T_n}} \in L^p_G$. Then:

(a) $H\cdot X$ is locally p–summable relative to $(\mathbb{R}, G)$.

(b) If K is a real valued, predictable process, and $KH \in L^1_{F,G}(X)_{\rm loc}$, then $K \in L^1_{\mathbb{R},G}(H\cdot X)_{\rm loc}$. In this case we have

$$K\cdot(H\cdot X) = (KH)\cdot X.$$

We use Theorem 3.19 to deduce that $1_{[0,T_n]}H\cdot X^{T_n}$ is p–summable, and that the associativity holds locally.

The formula for the jumps of the stochastic integral can be established using Theorem 3.20.

4.13 THEOREM. *Assume that X is locally p–summable relative to (F,G) and let $H \in L^1_{F,G}(X)_{\rm loc}$. Then*

$$\Delta(H\cdot X) = H\Delta X.$$

The property of being a local martingale is inherited by the stochastic integral, if X is a local martingale.

4.14 THEOREM. *(a) Assume that X is locally p–summable relative to (F,G) and let $H \in L^1_{F,G}(X)_{\rm loc}$. If X is a local martingale, then $H\cdot X$ is a local martingale.*

(b) If X is a martingale and if for each $t \in \mathbb{R}_+$, X^t is p–summable relative to (F,G) and $1_{[0,t]}H \in L^1_{F,G}(X^t)$, then $H\cdot X$ is a martingale.

The proof of the above theorem uses an appropriate sequence of stopping times and Theorem 3.21.

Semi–summable process

As we have seen in this section and in section 2, the stochastic integral $H \cdot X$ can be defined when X belongs to one of the following two classes of processes: (1) locally p–summable processes; (2) processes with finite variation.

Putting these two classes together we obtain the following definition.

4.15 DEFINITION. *We say that a process $Z : \mathbb{R} \to E \subset L(F, G)$ is semi–p–summable relative to (F, G), if it is of the form $Z = X + Y$, where X is locally p–summable relative to (F, G) and Y is a cadlag, adapted process with finite variation. If $p = 1$ above, we say that Z is semi–summable relative to (F, G).*

An F–valued process H is said to be locally integrable with respect to a semi–p–summable process Z, if there exists a decomposition $Z = X + Y$, as above, such that both integrals $H \cdot X$ and $H \cdot Y$ are defined. In this case, we define the stochastic integral $H \cdot Z$ by

$$H \cdot Z = H \cdot X + H \cdot Y.$$

The definition of the stochastic integral is independent of the decomposition $Z = X + Y$.

4.16 THEOREM. *If E is a Hilbert space then any semimartingale is semi–summable relative to any embedding $E \subset L(F, G)$ with G a Hilbert space.*

A real valued process is a semi–summable if and only if it is a semimartingale.

Proof. If Z is an E–valued semimartingale, then $Z = M + V$, where M is a locally square integrable martingale and V is a process of finite variation. Then M is locally summable; hence Z is semi–summable.

Conversely, suppose Z is a real valued, semi–summable process. Let $Z = X + Y$, where X is locally summable relative to $(\mathbb{R}, \mathbb{R})$ and Y has finite variation. We can assume X is summable, by stopping it at a convenient sequence $T_n \nearrow \infty$ of stopping times. Taking $G \equiv 1$ in Theorem 2.5(5), we deduce that X is a quasimartingale on $(0, \infty]$, hence Z is a semimartingale.

Remark. For a Banach space — or even a Hilbert space — the concept of semi-summability is more general than that of semimartingale, as it can be seen from the following example:

Example. Let $\Omega = \{\omega\}$ consist of one element and $\mathcal{F}_t = \mathcal{F} = \{\Omega, \phi\}$ for each $t \geq 0$. Then any local martingale is constant (hence of finite variation). Let E be any infinite dimensional Banach space; then $L^1_E(P) = E$. Let (x_n) be a sequence in E such that the series Σx_n is unconditionally convergent, but $\Sigma |x_n| = \infty$. Such a sequence exists by the Dvoretzky–Rogers theorem. For each n set $e_n = \Sigma_{i \geq n} x_i$; then $\lim e_n = 0$ and $x_n = e_{n+1} - e_n$.

Let $s_n \nearrow 1$ with $s_1 = 0$ and define the process

$$X = \Sigma_n e_n 1_{[s_n, s_{n+1})}.$$

This process is cadlag and has infinite variation, equal to the sum of the norms of the jumps:

$$|X| = |e_1| + \Sigma_n |e_{n+1} - e_n| = |e_1| + \Sigma_n |x_n| = \infty.$$

It follows that X is not a semimartingale.

Now we show that X is summable relative to $(\mathbb{R}, E)$. For each interval $(a, b] \subset [0, 1]$, let n and m be such that $n \leq a < n + 1 \leq m \leq b < m + 1$ and set $\Delta(a, b] = \{n, n + 1, \ldots, m - 1\}$. Then $I_X((a, b]) = e_m - e_n = \Sigma_{i \in \Delta(a,b]} x_i$. If $A \in \mathcal{R}$ and $A = \cup_{1 \leq i \leq k}(a_i, b_i]$ with $(a_i, b_i]$ mutually disjoint, set $\Delta(A) = \cup_{1 \leq i \leq k} \Delta(a_i, b_i]$. Then $I_X(A) = \Sigma_{i \in \Delta(A)} x_i$. If now (A_n) is a sequence of mutually disjoint sets from $\mathcal{R}$, then $\Sigma_n I_X(A_n) = \Sigma_n \Sigma_{i \in \Delta(A_n)} x_i = \Sigma_{i \in \cup_n \Delta(A_n)} x_i$ and this series is convergent in E, since the series $\Sigma_n x_n$ is unconditionally convergent. It follows that I_X is strongly additive on $\mathcal{R}$. If (A_n) is a sequence of disjoint sets from $\mathcal{R}$ with union $A \in \mathcal{R}$, then $\Delta(A) = \cup_n \Delta(A_n)$ therefore $\Sigma_n I_X(A_n) = \Sigma_{i \in \Delta(A)} x_i = I_X(A)$, hence I_X is σ–additive on $\mathcal{R}$. By Theorem AI.1, I_X can be extended to a σ–additive measure on $\mathcal{P}$; hence I_X is bounded on $\mathcal{P}$, therefore I_X has finite semivariation relative to $(\mathbb{R}, L^1_E)$. It follows that X is summable relative to $(\mathbb{R}, E)$.

A stochastic integral $H \cdot X$ can be defined by using our approach, while the classical approach cannot be applied in this case.

4.17 THEOREM. *If $Z : \mathbb{R} \to E \subset L(F, G)$ is semi–p–summable relative to (F, G), then any F–valued, caglad, adapted process is locally integrable with respect to Z.*

This follows from the fact that the caglad adapted processes are locally bounded.

All properties stated in sections 1 and 3, that are common to processes of finite variation and to locally p–summable processes, are obviously valid for semi–p–summable processes. Among these properties we mention the associativity property $K \cdot (H \cdot X) = (KH) \cdot X$ and the jumps property $\Delta(H \cdot X) = H\Delta X$.

Appendix I: General integration theory in Banach spaces.

In this section we shall present a theory of integration in which both the integrand and the measure are Banach–valued. The measure will be countably additive with finite semivariation. The basis for this theory is essentially found in [BD.2]; however, in order to apply the general theory to stochastic integration, a further development and new results were required. In this section, the necessary extension of the general theory is presented.

The framework for this section consists of a nonempty set S, a ring $\mathcal{R}$ of subsets of S and the σ–algebra Σ generated by $\mathcal{R}$. We assume that $S = \cup_{1 \leq n} S_n$, with $S_n \in \mathcal{R}$. We shall use the notation established in section 1.

Strong additivity

Let $m : \mathcal{R} \to E$ be a finitely additive measure. We say that m is *strongly additive* if for any sequence (A_n) of disjoint sets from $\mathcal{R}$, the series $\Sigma m(A_n)$ is convergent in E (or equivalently, if $m(A_n) \to 0$, for any sequence (A_n) of disjoint sets from $\mathcal{R}$).

The reader is referred to [BD.1] for a more complete study of strong additivity. We list below some properties that will be used in the sequel:

1) m is strongly additive iff for any increasing (respectively decreasing) sequence (A_n) from $\mathcal{R}$, $\lim_n m(A_n)$ exists in E.

2) A σ–additive measure defined on a σ–algebra is strongly additive; but if its domain is simply a ring, this need not be true.

3) A strongly additive measure on a ring is bounded; if E does not contain a copy of c_0, then the converse is true (cf. Theorem AI.2).

4) Any finitely additive measures with bounded variation is strongly additive

Strong additivity plays an important role in the problem of the extension of a measure from $\mathcal{R}$ to Σ (see Theorem AI.1).

Uniform σ–additivity

A family $(m_\alpha)_{\alpha\in I}$ of E–valued measures on the ring $\mathcal{R}$ is said to be *uniformly σ–additive* if for any sequence (A_n) of mutually disjoint sets from $\mathcal{R}$ with union in $\mathcal{R}$ we have

$$m_\alpha(\cup_n A_n) = \Sigma_n m_\alpha(A_n)$$

where the series is uniformly convergent with respect to α; or, equivalently if for every decreasing sequence $A_n \searrow \phi$ of sets from $\mathcal{R}$ we have

$$\lim_n m_\alpha(A_n) = 0$$

uniformly with respect to α.

A finitely additive measure $m : \mathcal{R} \to E$ is σ–additive iff the family $\{x^*m; x^* \in E_1^*\}$ of scalar measures is uniformly σ–additive. The measure $x^*m : \mathcal{R} \to \mathbb{R}$ is defined by

$$(x^*m)(A) = \langle m(A), x^*\rangle, \text{ for } A \in \mathcal{R}.$$

A family $(m_\alpha)_{\alpha\in I}$ of E–valued measures on a σ–algebra Σ is uniformly σ–additive iff there is a control measure λ, that is a positive, σ–additive measure λ on Σ such that $m_\alpha \ll \lambda$ uniformly with respect to α and $\lambda(A) \leq \sup_\alpha \tilde{m}_\alpha(A)$, for $A \in \Sigma$, where $\tilde{m}_\alpha$ is the semivariation of m_α (see [B–D.1]).

In particular, any σ–additive measure $m : \Sigma \to E$ has a control measure λ, such that $m \ll \lambda \leq \tilde{m}$.

Measures with finite variation

Let $m : \mathcal{R} \to E$ be a finitely additive measure. The ***variation*** of m is a set function $|m| : \mathcal{R} \to \overline{I\!R}_+$ defined for every set $A \in \mathcal{R}$ by

$$|m|(A) = \sup \Sigma |m(A_i)|,$$

where the supremum is taken over all finite families (A_i) of disjoint subsets from $\mathcal{R}$ with union A.

The variation $|m|$ is additive; $|m|$ is σ–additive iff m is σ–additive. The measure m has ***finite variation*** (resp. ***bounded variation***) on $\mathcal{R}$ if $|m|(A) < \infty$ for every $A \in \mathcal{R}$ (respectively $\sup\{|m|(A) : A \in \mathcal{R}\} < \infty$). Note that if m is ***real valued*** and ***bounded***, then m has bounded variation.

Now let $m : \Sigma \to E$ be σ–additive with finite variation $|m|$. We say that a set or a function is m–negligible, m–measurable, or m–integrable if it has the same property with respect to $|m|$. For any Banach space F, we denote $L^1_F(m) = L^1_F(|m|)$, and endow this space with the seminorm $\|f\|_1 = \int |f| d|m|$. If G is another Banach space such that $E \subset L(F, G)$, then for $f \in L^1_F(m)$, we can define the integral $\int f dm \in G$ and we have

$$|\int f dm| \leq \int |f| d|m| = \|f\|_1.$$

This is done by defining the integral in the obvious way for simple functions which are dense in $L^1_F(m)$, and then extending the integral by continuity to the whole space $L^1_F(m)$.

Stieltjes measures

An important particular case of measures with finite variation are the Stieltjes measures on a subinterval of $I\!R$.

Let $I \subset I\!R$ be an interval containing its left endpoint, of the form $[a, b)$ or $[a, b]$ with $a < b \leq \infty$ and let $f : I \to E$ be a function.

We say that f has ***finite variation*** on I if the variation $V_{[s,t]}(f)$ of f on any compact interval $[s, t] \subset I$ is finite. We say f has ***bounded variation*** on I if

$$V_I(f) := \sup\{V_{[s,t]}(f) : [s, t] \subset I\} < \infty.$$

If f has finite variation on I we define the *variation function* of f to be the function $|f| : I \to \mathbb{R}_+$ defined by

$$|f|(t) = |f(a)| + V_{[a,t]}(f), \text{ for } t \in I.$$

The variation $|f|$ of f is increasing and satisfies

$$||f|(t) - f(s)| \leq |f|(t) - |f|(s), \text{ for } s < t.$$

Moreover, f is right (or left) continuous iff $|f|$ has the same property (see [D.3]).

Let $\mathcal{R}$ be the ring generated by the intervals of the form $[a, t] \subset I$. We define a measure $\mu_f : \mathcal{R} \to E$ by

$$\mu_f([a,t]) = f(t) - f(a).$$

Then $\mu_f((s,t]) = f(t) - f(s)$, if $(s,t] \subset I$. The measure μ_f has finite (resp. bounded) variation $|\mu_f|$ on $\mathcal{R}$ iff f has finite (resp. bounded) variation $|f|$ on I. In this case we have

$$|\mu_f|(A) = \mu_{|f|}(A), \text{ for } A \in \mathcal{R}$$

(see [D.1], p. 363).

If f is *right continuous* and has bounded variation $|f|$, then μ_f and $\mu_{|f|}$ have σ–additive extensions on the σ–algebra $\mathcal{B}(I)$, denoted by the same letters, and we still have $|\mu_f| = \mu_{|f|}$ on $\mathcal{B}(I)$.

Assume that f is right continuous and has *bounded variation* and assume $E \subset L(F,G)$. A function $g : I \to F$ is said to be *Stieltjes integrable* with respect to f if it is integrable with respect to μ_f, that is, with respect to $|\mu_f| = \mu_{|f|}$. In this case the integral $\int g d\mu_f$ is called the Stieltjes integral of g with respect to f and is denoted $\int g df$ or $\int_I g df$. To say that g is Stieltjes integrable with respect to f means that g is μ_f–measurable and $\int |g| d|f| < \infty$. In this case

$$|\int g df| \leq \int |g| d|f|.$$

Extensions of measures

If $m : \mathcal{R} \to E$ is σ–additive with bounded variation $|m|$, then it has a unique σ–additive extension $m' : \Sigma \to E$, with bounded variation $|m'|$, and $|m'|$ is the unique σ–additive extension of $|m|$ from $\mathcal{R}$ to Σ (see [D.1], p. 62).

If m is σ–additive but does not have finite variation on $\mathcal{R}$, then a σ–additive extension to Σ does not necessarily exist.

We now present some extension theorems for Banach–valued measures, which will be applied to stochastic measures. These theorems are an improvement over the existing extension theorems (which were stated for the particular case when $Z = E^*$).

AI.1 THEOREM. *Let $m : \mathcal{R} \to E$ be a finitely additive measure. Suppose that $Z \subset E^*$ is norming for E. The following assertions are equivalent:*

(a) m is strongly additive on $\mathcal{R}$ and for every $x^* \in Z$, the scalar measure x^*m is σ–additive on $\mathcal{R}$;

(b) m is strongly additive and σ–additive on $\mathcal{R}$;

(c) m can be extended uniquely to a σ–additive measure $m : \Sigma \to E$.

Proof. The proof is done in the following way: $a \Longrightarrow b \Longrightarrow c \Longrightarrow a$. Assume (a) and prove (b), that is prove that m is σ–additive on $\mathcal{R}$. Let $A_n \in \mathcal{R}$ such that $A_n \searrow \phi$. Since m is strongly additive, $\lim_n m(A_n) = x$ exists in E. Let $x^* \in Z$; since x^*m is σ–additive, it follows that $x^*x = 0$, hence $x = 0$. Thus m is σ–additive on E, that is (b).

Assume now (b) and prove (c). We deduce first that the family of scalar measures $\{x^*m : x^* \in Z_1\}$ is uniformly σ–additive on $\mathcal{R}$. Since m is strongly additive, it is bounded on $\mathcal{R}$. Then each scalar measure x^*m is bounded on $\mathcal{R}$, hence it has bounded variation on $\mathcal{R}$; being also σ–additive on $\mathcal{R}$, it can be extended uniquely to a σ–additive measure m_{x^*} on Σ with bounded variation $|m_{x^*}|$, which is equal to the extension of $|x^*m|$ to Σ.

Now we assert that the family of measures $\{|m_{x^*}| : x^* \in Z_1\}$ is uniformly σ–additive on Σ. If not, there exists an $\epsilon > 0$, a sequence of sets $A_n \in \Sigma$ with $A_n \searrow \phi$, and a sequence (x_n^*) from Z_1 such that if we denote $\mu_n = |m_{x_n^*}|$, then $\mu_n(A_n) > \epsilon$ for each n. Let $\mathcal{R}_0$ be a countable subring of $\mathcal{R}$ such that all the A_n belong to $\sigma(\mathcal{R}_0)$, the σ-algebra generated by $\mathcal{R}_0$. Let $\lambda = \Sigma_n 2^{-n}\mu_n$. Then each μ_n is absolutely continuous with respect to the σ-additive measure λ, and the sequence (μ_n) is uniformly absolutely continuous with respect to λ on $\mathcal{R}_0$, since the μ_n are uniformly σ-additive on $\mathcal{R}_0$. Then, for $\epsilon > 0$ above, there is a $\delta > 0$ such that if $B \in \mathcal{R}_0$ and $\lambda(B) < \delta$, then $\mu_n(B) < \eta$ for each n. Let now $A \in \sigma(\mathcal{R}_0)$ with $\lambda(A) < \delta$. There is a sequence of disjoint sets $B_n \in \mathcal{R}_0$ such that $A \subset \cup_n B_n$ and $\Sigma_n \lambda(B_n) < \delta$. Let $C_k = \cup_{1\leq i\leq k} B_i$. Then $\lambda(C_k) < \delta$, hence $\mu_n(C_k) < \epsilon$ for each n. Thus

$$\mu_n(A) \leq \mu_n(\lim_k C_k) = \lim_k \mu_n(C_k) \leq \epsilon,$$

for each n. In particular, taking $A = A_n$ we obtain $\mu_n(A_n) \leq \epsilon$ for each n. But by our choice of A_n and μ_n, we have $\mu_n(A_n) > \epsilon$ for each n, and we reached a contradiction. Hence the family of measures $\{|m_{x^*} : x^* \in Z_1\}$ is uniformly σ-additive on Σ.

For each $A \in \Sigma$, define $m_1(A) : Z \to I\!R$ by $\langle z, m_1(A)\rangle = m_z(A)$, for $z \in Z$. Then $m_1(A)$ is a linear functional on Z and

$$|\langle z, m_1(A)\rangle| = |m_z|(A)| \leq |m_z|(S)$$
$$\leq 2\sup\{|zm(B)| : B \in R\} \leq 2|z|c,$$

where $c = \sup\{|m(B)| : B \in \mathcal{R}\} < \infty$. Thus $m_1(A) \in Z^*$. Note that $m_1 = m$ on $\mathcal{R}$. Since $\{m_z : z \in Z_1\}$ is uniformly σ-additive on Σ, it follows that m_1 is σ-additive on Σ. Finally, we observe that m_1 takes its values in $E \subset Z^*$. To see this, let $\mathcal{C}$ denote the class of subsets A from Σ such that $m_1(A) \in E$. Since $\mathcal{C}$ is a monotone class which contains $\mathcal{R}$, we deduce that $\mathcal{C} = \Sigma$. Thus m_1 is a σ-additive extension of m to Σ. The uniqueness of the extension follows by using a monotone class argument; therefore (c) is proved.

The implication $c \Longrightarrow a$ is evident and this proves the theorem.

As we mentioned earlier, any strongly additive measure on a ring is bounded. We next prove a partial converse.

AI.2 THEOREM. *If $m : \mathcal{R} \to E$ is a bounded finitely additive measure, and if E does not contain a copy of c_0, then m is strongly additive.*

Proof. Let (A_n) be a sequence of disjoint sets from $\mathcal{R}$. It suffices to show that the series $\Sigma_n m(A_n)$ is convergent in E. For each $x^* \in E^*$, the scalar measure

x^*m is bounded on $\mathcal{R}$, hence it has bounded variation $|x^*m|$. Thus

$$\Sigma_{1\leq i\leq n}|x^*m(A_i)| \leq |x^*m|(\cup_{1\leq i\leq n}A_i)$$
$$\leq \sup\{|x^*m|(B) : B \in \mathcal{R}\} < \infty.$$

Hence the series $\Sigma_{1\leq i<\infty}x^*m(A_i)$ is unconditionally convergent. Since $E \not\supset c_0$, by the Bessaga–Pelczinski theorem [B–P], the series $\Sigma_{1\leq i<\infty}m(A_i)$ converges. Thus m is strongly additive.

Combining the preceding two theorems, we obtain the following extension theorem.

AI.3 THEOREM. *Assume that E does not contain a copy of c_0 and let $Z \subset E^*$ be a norming space for E. If $m : \mathcal{R} \to E$ is a bounded finitely additive measure, and if x^*m is σ–additive on $\mathcal{R}$ for each $x^* \in Z$, then m can be extended uniquely to a σ–additive E–valued measure on Σ.*

The following particular case of the preceding theorem is used in the construction of the stochastic integral.

AI.4 THEOREM. *Assume that E does not contain a copy of c_0 and let $Z \subset E^*$ be a norming space for E. Let $(\Omega, \mathcal{F}, \mu)$ be a measure space, $a \leq p < \infty$, and let $m : \mathcal{R} \to L^p_E(\mu)$ be a finitely additive measure. For each $z \in Z$ define the measure $zm : \mathcal{R} \to L^p(\mu)$ by $(zm)(A) = \langle m(A), z\rangle$, for $A \in \mathcal{R}$. If m is bounded on $\mathcal{R}$ and if for each $z \in Z$, the measure zm is σ–additive on $\mathcal{R}$, then m can be uniquely extended to a σ–additive measure $m_1 : \Sigma \to L^p_E(\mu)$.*

Proof. By a theorem of Kwapien [Kw], L^p_E does not contain a copy of c_0 if E does not contain a copy of c_0. Let M be the space of Z–valued $\mathcal{R}$–measurable simple functions. Then $M \subset L^q_{E^*}(\mu) \subset (L^p_E(\mu))^*$, and M is norming for $L^p_E(\mu)$. Let $f \in M$. Consider the scalar measure fm defined on $\mathcal{R}$ by

$$(fm)(A) = \int \langle m(A), f\rangle d\mu.$$

Note that if $A_n \in \mathcal{R}$ and $A_n \searrow \phi$, then for each $z \in Z$, we have $\langle m(A_n), z\rangle \to 0$ in $L^p(\mu)$, as $n \to \infty$. Hence $(fm)(A_n) \to 0$ as $n \to \infty$; that is, fm is σ–additive on $\mathcal{R}$. We can then apply the preceding theorem, replacing E and Z by $L^p_E(\mu)$ and M respectively.

The semivariation

Let $m : \mathcal{R} \to E \subset L(F,G)$ be finitely additive. For every set $A \in \mathcal{R}$, we define the *semivariation* $\tilde{m}_{F,G}(A)$ of m on A, relative to the pair (F,G), by

$$\tilde{m}_{F,G}(A) = \sup |\Sigma_{i\in I} m(A_i)x_i|,$$

where the supremum is taken over all finite families $(A_i)_{i\in I}$ of disjoint sets from $\mathcal{R}$, with union A, and all finite families $(x_i)_{i\in I}$ of elements from F_1. We thus obtain a set function $\tilde{m}_{F,G} : \mathcal{R} \to [0,+\infty]$. Sometimes the semivariation $\tilde{m}_{F,G}$ is denoted by $\text{svar}_{F,G} m$. Note that

$$\tilde{m}_{F,G}(A) = \sup |\int s dm|,$$

where the supremum is taken over all F–valued simple $\mathcal{R}$–measurable functions s, such that $|s| \leq 1_A$, where the integral $\int s dm$ is defined in the usual manner.

We say that m has *finite* (respectively *bounded*) *semivariation* relative to (F,G) if $\tilde{m}_{F,G}(A) < \infty$ for every $A \in \mathcal{R}$ (respectively $\sup\{\tilde{m}_{F,G}(A) : A \in \mathcal{R}\} < \infty$).

If $E = L(\mathbb{R}, E)$, we sometimes write $\text{svar}\, m$, or $\tilde{m}$, instead of $\tilde{m}_{\mathbb{R},E}$, and we call it simply the semivariation of m. In this case m has bounded semivariation on $\mathcal{R}$ if and only if m is bounded on $\mathcal{R}$; more precisely, for every $A \in \mathcal{R}$, we have

$$\tilde{m}(A) \leq 2 \sup\{|m(B)| : B \in \mathcal{R}, B \subset A\}.$$

If $E \subset L(F,G)$, then we have ([D], pages 51, 54):

$$\tilde{m}_{\mathbb{R},E} \leq \tilde{m}_{F,G} \leq \tilde{m}_{E^*,\mathbb{R}} = |m|.$$

In particular, for a real measure $m : \mathcal{R} \to \mathbb{R} = L(\mathbb{R},\mathbb{R})$ we have $\tilde{m} = |m|$.

Computation of the semivariation

Let $m : \mathcal{R} \to E \subset L(F,G)$ be a finitely additive measure, and let $Z \subset G^*$ be a norming space for G. For each $A \in \mathcal{R}$ we have $m(A) : F \to G$. Consider the adjoint $m(A)^* : G^* \to F^*$. For each $x \in F$ and $y^* \in G^*$, we have

$$\langle m(A)x, y^* \rangle = \langle x, m(A)^* y^* \rangle.$$

We denote by $m_{y^*} : \mathcal{R} \to F^*$ the finitely additive measure defined by $m_{y^*}(A) = m(A)^* y^*$ for $A \in \mathcal{R}$. In particular, for $z \in Z$, we have, for $m_z : \mathcal{R} \to F^*$,

$$\langle m(A)x, z \rangle = \langle x, m_z(A) \rangle, \text{ for } x \in F \text{ and } A \in \mathcal{R}.$$

One can show ([D.1], page 55) that

$$\tilde{m}_{F,G} = \sup\{|m_z| : z \in Z_1\},$$

where $|m_z|$ is the variation set function of m_z. Note that the above equality is independent of the norming space $Z \subset G^*$. In particular, we have

$$\tilde{m}_{\mathbb{R},E} = \sup\{|x^* m| : x^* \in Z_1\},$$

where $Z \subset E^*$ is a norming space for E^*.

If $\tilde{m}_{F,G}$ is bounded, then each $|m_z|$ is bounded, for $z \in Z$. In this case we define the set $m_{F,G}$ of positive measures by

$$m_{F,G} = \{|m_z| : z \in Z_1\}.$$

Note that $m_{F,G}$ depends upon $Z \subset G^*$.

We have the following property of the semivariation of the extension of m to Σ.

AI.5 PROPOSITION. *Let $m : \mathcal{R} \to E \subset L(F,G)$ be a finitely additive measure with bounded semivariation $\tilde{m}_{F,G}$. If m has a σ-additive extension m' to Σ, then m' has bounded semivariation $\tilde{m}'_{F,G}$ on Σ and $\tilde{m}'_{F,G}$ is the extension of $\tilde{m}_{F,G}$.*

For the proof we use the fact that for every $z \in Z$, the measure m'_z is the extension of m_z and $|m'_z|$ is the extension of $|m_z|$.

Measures with bounded semivariation

From now on we shall assume that $m : \Sigma \to E \subset L(F,G)$ is σ-additive and has bounded semivariation $\tilde{m}_{F,G}$, and that $Z \subset G^*$ is a norming space for G. To develop the stochastic integral, we shall use an integration theory with respect to m, for functions $f : S \to F$. Observe that since m is σ-additive, it is bounded on Σ and hence $\tilde{m}_{\mathbb{R},E}$ is bounded on Σ.

We say that a set $A \in \Sigma$ is *m–negligible* if $m(B) = 0$ for every $B \subset A$, $B \in \Sigma$. Thus A is m–negligible if and only if $\tilde{m}_{F,G}(A) = 0$.

If D is any Banach space, we say that a function $f : S \to D$ is *m–negligible* (or that $f = 0$, m–a.e.) if it vanishes outside an m–negligible set. This notion is independent of the embedding $E \subset L(F,G)$. A subset $Q \subset S$ is said to be $m_{F,G}$–negligible if for each $z \in Z$, Q is contained in an $|m_z|$–negligible set. Note that Q need not belong to Σ.

A function $f : S \to D$ is said to be *$m_{F,G}$–measurable* if it is m_z–measurable for every $z \in Z$. We say $f : S \to D$ is *m–measurable* if it is the m–a.e. limit of a sequence of D–valued, Σ–measurable simple functions.

If f is m–measurable, then it is $m_{F,G}$–measurable. The converse is true if $m_{F,G}$ is uniformly σ–additive, as the next proposition shows.

AI.6 PROPOSITION. *Suppose that $m_{F,G}$ is uniformly σ–additive. Then a function $f : S \to D$ is m–measurable if and only if f is $m_{F,G}$–measurable.*

Proof. Suppose f is $m_{F,G}$–measurable. Since $m_{F,G}$ is uniformly σ–additive, there exists control measure λ on Σ, of the form $\lambda = \Sigma c_n \mu_n$, for some $c_n \geq 0$ with $\Sigma c_n = 1$, and some $\mu_n \in m_{F,G}$ (see [BD.1] Lemma 3.1). Let (f_{1n}) be a sequence of Σ–measurable simple functions converging to f on $S - S_1$, where $S_1 \in \Sigma$ and $\mu_1(S_1) = 0$; we can assume all the $f_{1n} = 0$ on S_1. Let (f_{2n}) be a sequence of Σ–measurable simple functions converging to f on $S_1 - S_2$, where $S_2 \in \Sigma$ and $\mu_2(S_2) = 0$; we can assume that all the $f_{2n} = 0$ on S_2. Continue in this fashion and obtain for each i, a sequence $(f_{in})_{1 \leq n < \infty}$ of Σ–measurable simple functions converging to f on $S_{i-1} - S_i$, where $\mu_i(S_i) = 0$; we can assume that all the $f_{in} = 0$ on S_i. If $S_0 = \cap_{1 \leq i < \infty} S_i$, then $\lambda(S_0) = 0$, hence S_0 is m–negligible. The sequence $(\Sigma_{1 \leq i \leq n} f_{in})_{1 \leq n < \infty}$ of Σ–measurable simple function converges to f m–a.e., hence f is m–measurable.

Remark. Although the set of measures $m_{F,G}$ depends upon $Z \subset G^*$, the uniform σ–additivity of $m_{F,G}$ is equivalent to $\tilde{m}_{F,G}(A_n) \to 0$ whenever $A_n \searrow \phi$ (using a control measure as in the above proof), and as a result, the uniform σ–additivity of $m_{F,G}$ is independent of Z.

We shall now extend the definition of $\tilde{m}_{F,G}$ to functions. Recall that

$m : \Sigma \to E \subset L(F, G)$ is σ-additive with bounded semivariation $\tilde{m}_{F,G}$.

For each $f : S \to D$ (or $\overline{I\!R}$) which is $m_{F,G}$-measurable, define

$$\tilde{m}_{F,G}(f) = \tilde{m}_{F,G}(|f|) = \sup\{|\int s dm|\},$$

where the supremum is extended over all F-valued, Σ-measurable simple functions s such that $|s| \leq |f|$ on S. Note that if $A \in \Sigma$, then $\tilde{m}_{F,G}(A) = \tilde{m}_{F,G}(1_A)$. We shall use this equality to extend the definition of $\tilde{m}_{F,G}(A)$ to any $m_{F,G}$-measurable set $A \subset S$.

We can also define $\tilde{m}_{F,G}(f)$ in terms of an arbitrary norming subspace $Z \subset G^*$:

AI.7 Proposition. *Let $f : S \to D$ be any $m_{F,G}$-measurable function and let $Z \subset G^*$ be a norming subspace for G. Then*

$$\tilde{m}_{F,G}(f) = \sup\{\int |f| d|m_z| : z \in Z_1\}.$$

Proof. If $s : S \to F$ is a Σ-measurable simple function such that $|s| \leq |f|$, and if $z \in Z_1$, then

$$|\langle \int s dm, z\rangle| \leq \int |s| d|m_z| \leq \sup\{\int |f| d|m_z| : z \in Z_1\}.$$

Since Z is norming for G, we conclude that

$$\tilde{m}_{F,G}(F) \leq \sup\{\int |f| d|m_z| : z \in Z_1\}.$$

Conversely, let $\epsilon > 0$ and choose $a \in I\!R$ such that $a < \sup\{\int |f| d|m_z| : z \in Z_1\}$. There is a scalar Σ-measurable simple function $\varphi \leq |f|$ such that $a < \int \varphi d|m_z|$, for some $z \in Z_1$. Let $\varphi = \Sigma_{1 \leq i \leq n} 1_{A_i} \alpha_i$, where the A_i are disjoint sets from Σ and $\alpha_i > 0$. There exists a finite family $(B_{ij})_{i,j}$ of disjoint sets from Σ, such that $A_i = \cup_j B_{ij}$ and

$$|m_z|(A_i) < \Sigma_j |m_z(B_{ij})| + \frac{\epsilon}{n\alpha_i}.$$

We choose elements $x_{ij} \in F_1$ such that

$$\Sigma_j |m_z(B_{ij})| < \Sigma_j \langle x_{ij}, m_z(B_{ij})\rangle + \frac{\epsilon}{n\alpha_i}.$$

If $s = \Sigma_{ij}\alpha_i x_{ij} 1_{B_{ij}}$, then

$$a < \int \varphi d|m_z| = \langle \int s dm, z\rangle + \epsilon$$
$$\leq |\int s dm| + \epsilon \leq \tilde{m}_{F,G}(f) + \epsilon,$$

since $|s| \leq |f|$. Since $\epsilon > 0$ and a were arbitrary, the result follows.

We now list some properties, whose proofs we omit. For simplicity, write $N = \tilde{m}_{F,G}$.

(1) N is subadditive and positively homogeneous on the space of $m_{F,G}$–measurable functions.

(2) $N(f) = N(|f|)$

(3) $N(f) \leq N(g)$ if $|f| \leq |g|$.

(4) $N(f) = \sup\{N(f1_A) : A \in \Sigma\} = \sup_n\{N(f1_{\{|f|\leq n\}})\}$.

(5) $N(\sup f_n) = \sup N(f_n)$, for every increasing sequence (f_n) of positive $m_{F,G}$–measurable functions.

(6) $N(\Sigma f_n) \leq \Sigma N(f_n)$, for every sequence of positive $m_{F,G}$–measurable functions.

(7) $N(\liminf f_n) \leq \liminf N(f_n)$, for every sequence of positive $m_{F,G}$–measurable functions.

(8) If $N(f) < \infty$, then f is finite $m_{F,G}$–a.e.

(9) If $f : S \to D$ is $m_{F,G}$–measurable and $c > 0$, then

$$N(\{|f| > c\}) \leq \frac{1}{c} N(f).$$

If $f_n, f : S \to D$ are $m_{F,G}$–measurable, we say $f_n \to f$ in $m_{F,G}$–measure if for every $\epsilon > 0$, we have

$$\tilde{m}_{F,G}(\{|f_n - f| > \epsilon) \to 0, \text{ as } n \to \infty.$$

(10) If $N(f_n - f) \to 0$, then $f_n \to f$ in $m_{F,G}$–measure and there exists a subsequence (f_{n_k}) converging $m_{F,G}$–a.e. to f (use property (6)).

The Egorov thoerem is not valid in general. However, using a control measure, it is valid whenever $m_{F,G}$ is uniformly σ–additive.

AI.7 THEOREM. *(Egorov) Assume that $m_{F,G}$ is uniformly σ–additive and let f_n, f be D–valued, $m_{F,G}$–measurable functions such that $f_n \to f$ $m_{F,G}$–a.e.*

Then

(a) for every $m_{F,G}$–measurable set A, and $\epsilon > 0$, there exists a set $B \in \Sigma$ with $B \subset A$, such that $\tilde{m}_{F,G}(A - B) < \epsilon$ and $f_n \to f$ uniformly on B;

(b) $f_n \to f$ in $m_{F,G}$–measure.

The space of integrable functions

We maintain the framework of a σ–additive measure $m : \Sigma \to E \subset L(F, G)$ with finite semivariation $\tilde{m}_{F,G}$, and $Z \subset G^*$ a norming space for G. Let D be a Banach space. We denote by $\mathcal{F}_D(m_{F,G})$ the set of all $m_{F,G}$–measurable functions $f : S \to D$ such that $\tilde{m}_{F,G}(f) < \infty$. The mapping $f \to \tilde{m}_{F,G}(f)$ is a seminorm on the vector space $\mathcal{F}_D(m_{F,G})$ which is complete (use property (6)). Note that $\mathcal{F}_D(m_{F,G}) \subset L^1_D(|m_z|)$ continuously, for each $z \in Z$.

The set $\mathcal{B}_D$ of bounded, D–valued, $m_{F,G}$–measurable functions is contained in $\mathcal{F}_D(m_{F,G})$. In particular, the sets $\mathcal{S}_D(\mathcal{R})$ and $\mathcal{S}_D(\Sigma)$, the D–valued, $\mathcal{R}$–measurable, respectively Σ–measurable, simple functions are contained in $\mathcal{F}_D(m_{F,G})$. However, unlike the classical case, these sets are not necessarily dense in $\mathcal{B}_D$ for the seminorm $\tilde{m}_{F,G}$. This is due to the fact that the Lebesgue dominated convergence theorem, valid for convergence in $\tilde{m}_{F,G}$–measure, is not valid, in general, for pointwise convergence, unless $m_{F,G}$ is uniformly σ–additive.

For any subspace $\mathcal{C} \subset \mathcal{F}_D(m_{F,G})$, we denote by $\mathcal{F}_D(\mathcal{C}, m_{F,G})$ the closure of $\mathcal{C}$ in $\mathcal{F}_D(m_{F,G})$, which is also complete. We write $\mathcal{F}_D(\mathcal{B}, m_{F,G})$, $\mathcal{F}_D(\mathcal{S}(\mathcal{R}), m_{F,G})$, and $\mathcal{F}_D(\mathcal{S}(\Sigma), m_{F,G})$ when $\mathcal{C}$ is $\mathcal{B}_D$, $\mathcal{S}_D(\mathcal{R})$, or $\mathcal{S}_D(\Sigma)$ respectively. Since $\mathcal{R}$ generates Σ, we shall see later (AI.11 *infra*) that

$$\mathcal{F}_D(\mathcal{S}(\mathcal{R}), m_{\mathbb{R},E}) = \mathcal{F}_D(\mathcal{B}, m_{\mathbb{R},E}),$$

and if $m_{F,G}$ is uniformly σ–additive, then

$$\mathcal{F}_D(\mathcal{S}(\mathcal{R}), m_{F,G}) = \mathcal{F}_D(\mathcal{B}, m_{F,G}).$$

We shall now list some properties of functions in $\mathcal{F}_D(m_{F,G})$ without proofs.

AI.8 THEOREM. *(a) If $f \in \mathcal{F}_D(\mathcal{B}, m_{F,G})$, then $\tilde{m}_{F,G}(f1_A) \to 0$ as $\tilde{m}_{F,G}(A) \to 0$. The converse is also true if $m_{F,G}$ is uniformly σ–additive.*

(b) If $f \in \mathcal{F}_D(m_{F,G})$ and if $\tilde{m}_{F,G}(f1_{A_n}) \to 0$ for any sequence of $m_{F,G}$–measurable sets $A_n \searrow \phi$, then $f \in \mathcal{F}_D(\mathcal{B}, m_{F,G})$.

(c) If $f : S \to D$ is $m_{F,G}$–measurable and if $|f| \le g \in \mathcal{F}_{\mathbb{R}}(\mathcal{B}, m_{F,G})$, then $f \in \mathcal{F}_D(\mathcal{B}, m_{F,G})$.

(d) A function $f : S \to D$ belongs to $\mathcal{F}_D(\mathcal{B}, m_{F,G})$ if and only if f is $m_{F,G}$–measurable and $|f| \in \mathcal{F}_{\mathbb{R}}(\mathcal{B}, m_{F,G})$.

(e) Suppose (f_n) is a sequence of functions from $\mathcal{F}_D(m_{F,G})$ such that $f_n \to f$ uniformly on S. Then $f \in \mathcal{F}_D(m_{F,G})$ and $f_n \to f$ in $\mathcal{F}_D(m_{F,G})$.

(f) If m has finite variation $|m|$ on Σ, then $m_{F,G}$ is uniformly σ–additive and $\mathcal{F}_D(\mathcal{B}, m_{F,G}) = \mathcal{F}_D(m_{F,G})$.

AI.9 THEOREM. *(Vitali) Let (f_n) be a sequence from $\mathcal{F}_D(m_{F,G})$ and let $f : S \to D$ be $m_{F,G}$–measurable. If condition (1) below and either of conditions (2a) or (2b) are satisfied, then $f \in \mathcal{F}_D(m_{F,G})$ and $f_n \to f$ in $\mathcal{F}_D(m_{F,G})$.*

(1) $\tilde{m}_{F,G}(f_n 1_{A_n}) \to 0$ as $\tilde{m}_{F,G}(1_A) \to 0$, uniformly in n;

(2a) $f_n \to f$ in $m_{F,G}$–measure;

(2b) $f_n \to f$ pointwise, and $m_{F,G}$ is uniformly σ–additive.

Conversely, if $f_n \to f$ in $\mathcal{F}_D(\mathcal{B}, m_{F,G})$, then conditions (1) and (2a) are satisfied.

For the proof, see [B–D.2], Theorem 2.5.

The next theorem follows from Vitali's theorem and Theorem AI.8(a).

AI.10 THEOREM. *(Lebesgue) Let (f_n) be a sequence from $\mathcal{F}_D(\mathcal{B}, m_{F,G})$, let $f : S \to D$ be an $m_{F,G}$–measurable function and $g \in \mathcal{F}_{\mathbb{R}}(\mathcal{B}, m_{F,G})$. If*

(1) $|f_n| \le g$, $m_{F,G}$–a.e. for each n, and any one of the conditions (2a) or (2b) below is satisfied:

(2a) $f_n \to f$ in $m_{F,G}$–measure;

(2b) $f_n \to f$ pointwise and $m_{F,G}$ is uniformly σ–additive,

then $f \in \mathcal{F}_D(\mathcal{B}, m_{F,G})$ and $f_n \to f$ in $\mathcal{F}_D(m_{F,G})$.

We now state without proof some closure properties:

AI.11 PROPOSITION. *(a) $\mathcal{F}_{\mathbb{R}}(\mathcal{S}(\Sigma), m_{F,G}) = \mathcal{F}_{\mathbb{R}}(\mathcal{B}, m_{F,G})$.*

(b) If $m_{F,G}$ is uniformly σ–additive, and if $\Sigma = \sigma(\mathcal{R})$, then

$$\mathcal{F}_D(\mathcal{S}(\mathcal{R}), m_{F,G}) = \mathcal{F}_D(\mathcal{S}(\Sigma), m_{F,G}) = \mathcal{F}_D(\mathcal{B}, m_{F,G}).$$

In particular,

$$\mathcal{F}_D(\mathcal{S}(\mathcal{R}), m_{\mathbb{R},E}) = \mathcal{F}_D(\mathcal{S}(\Sigma), m_{\mathbb{R},E}) = \mathcal{F}_D(\mathcal{B}, m_{\mathbb{R},E}).$$

The integral

Let $m : \Sigma \to E \subset L(F,G)$ be σ–additive with finite semivariation $\tilde{m}_{F,G}$ and take $Z = G^*$. In the special case $D = F$, we can define an integral $\int f dm$ for functions f belonging to $\mathcal{F}_F(m_{F,G})$. To simplify the notation, we shall denote $\mathcal{F}_{F,G}(m) = \mathcal{F}_F(m_{F,G})$.

The construction is as follows. If $f \in \mathcal{F}_{F,G}(m)$, then $f \in L^1_F(|m_z|)$ for each $z \in G^*$, hence the real number $\int f dm_z$ is defined. The mapping $z \to \int f dm_z$ is a linear continuous mapping from G^* into $\mathbb{R}$:

$$|\int f dm_z| \le \int |f| d|m_z| \le |z| \tilde{m}_{F,G}(f),$$

hence, this mapping belongs to G^{**}; we denote this mapping by $\int f dm$. Thus

$$\langle z, \int f dm \rangle = \int f dm_z, \text{ for } z \in G^*$$

and

$$|\int f dm| \le \tilde{m}_{F,G}(f).$$

If $Z \subset G^*$, we can regard $\int f dm \in Z^*$, by considering the restriction of $\int f dm$ to Z.

Note that since $\tilde{m}_{\mathbb{R},E}$ is finite, we can define $\int \varphi dm \in E^{**}$, for $\varphi \in \mathcal{F}_{\mathbb{R},E}(m)$, and we have

$$\langle \int \varphi dm, x^* \rangle = \int \varphi d(x^* m), \text{ for } x^* \in E^*,$$

and

$$|\int \varphi dm| \le \tilde{m}_{\mathbb{R},E}(\varphi).$$

We are particularly interested in the case when $\int f dm \in G$. Of course this holds when G is reflexive. In general, if $\mathcal{C}$ is a subset of $\mathcal{F}_{F,G}(m)$ such that $\int f dm \in G$, whenever $f \in \mathcal{C}$, then by continuity of the integral, it follows that $\int f dm \in G$, whenever $f \in \mathcal{F}_{F,G}(\mathcal{C}, m)$. For example, if $\mathcal{C} = \mathcal{S}(\Sigma)$, then we have $\int f dm \in G$ for any $f \in \mathcal{S}(\Sigma)$, hence also for $f \in \mathcal{F}_{F,G}(\mathcal{S}(\Sigma), m)$. Since the Σ–measurable functions are not necessarily dense in $\mathcal{F}_{F,G}(\mathcal{B}, m)$, the integral of bounded measurable functions need not belong to G; however, if $m_{F,G}$ is uniformly σ–additive, this property holds. In particular, since $m_{\mathbb{R},E}$ is uniformly σ–additive, it follows that $\int f dm \in E$, whenever $f \in \mathcal{F}_{\mathbb{R},E}(\mathcal{B}, m)$.

Since the integral is continuous on $\mathcal{F}_{F,G}(m)$, any theorem insuring convergence $f_n \to f$ in $\mathcal{F}_{F,G}(m)$ can be completed by stating convergence of the integrals $\int f_n dm \to \int f dm$ in G^{**}. In particular, whenever the f_n and f satisfy the hypotheses of the Vitali or Lebegue theorems, we have $\int f_n dm \to \int f dm$ in G^{**}.

Remark. If $m : \Sigma \to E \subset L(F, G)$ has finite total variation $|m|$, then $\tilde{m}_{F,G}$ is finite and $m_{F,G}$ is uniformly σ–additive. Moreover, $L^1_F(m) \subset \mathcal{F}_{F,G}(m)$. Using simple functions, we see that for every $f \in L^1_F(m)$, the integral $\int f dm$ is the same relative to either $L^1_F(m)$ or $\mathcal{F}_{F,G}(m)$.

The indefinite integral

We still assume the same conditions on m hold, namely, $m : \Sigma \to E \subset L(F, G)$ is σ–additive and $\tilde{m}_{F,G}$ is finite. For $f \in \mathcal{F}_{F,G}(m)$ and $A \in \Sigma$, we define $\int_A f dm = \int 1_A f dm$. Set $n(A) = \int_A f dm$. Then $n : \Sigma \to G^{**}$. We call n the indefinite integral of f with respect to m, or the measure with density f and base m; we also denote this finitely additive measure by fm. In general, fm is not countably additive.

AI.12 PROPOSITION. *Let $f \in \mathcal{F}_{F,G}(m)$. Then fm is σ–additive on Σ in each of the following cases:*

(a) $\int_A f dm \in G$, for every $A \in \Sigma$; in particular if f is a Σ–step funtion.

(b) $f \in \mathcal{F}_{F,G}(\mathcal{B}, m)$ and $m_{F,G}$ is uniformly σ–additive (this is the case if $F = \mathbb{R}$); in this case we have $\int g dm \in G$ for every $g \in \mathcal{F}_{F,G}(\mathcal{B}, m)$.

(c) G does not contain a copy of c_0; in this case we have $\int g dm \in G$ for every $g \in \mathcal{F}_{F,G}(m)$.

Proof. (a) follows from the Pettis theorem (any weakly σ–additive measure is strongly σ–additive), since the set function $\langle \int_{(\cdot)} f dm, z \rangle = \int_{(\cdot)} f dm_z$ is σ–additive, for each $z \in G^*$. (b) follows from Theorem AI.11(b). To prove (c), assume first that g is a σ-step function in $\mathcal{F}_{F,G}(m)$, of the form $g = \Sigma_{1\leq n<\infty} 1_{A_n} x_n$ with $x_n \in F$ and A_n mutually disjoint and $m_{F,G}$-measurable. Let $z \in G^*$. Since $g \in L^1_F(|m_z|)$, we have

$$\Sigma_n |\langle m(A_n)x_n, z\rangle| \leq \Sigma_n |m_z|(A_n)|x_n| = \int |g| d|m_z| < \infty,$$

hence $\Sigma_n m(A_n)x_n$ is weakly unconditionally convergent. Since $c_0 \notin G$, by the Bessaga–Pelczynski theorem [B–P], the series $\Sigma_n m(A_n)x_n$ converges to an element $y \in G$. Thus

$$\langle \int g dm, z\rangle = \langle \Sigma_n m(A_n)x_n, z\rangle = \langle y, z\rangle$$

for $z \in G^*$, consequently $\int g dm = y \in G$.

If g is arbitrary in $\mathcal{F}_{F,G}(m)$, there is a sequence (g_n) of $m_{F,G}$-measurable, σ-step functions such that $g_n \to g$ uniformly and $|g_n| \leq |g|$ for each n. Then $g_n \in \mathcal{F}_{F,G}(m)$, hence, by the above, $\int g_n dm \in G$ for each n. By Theorem AI.8(e), we have $g_n \to g$ in $\mathcal{F}_{F,G}(m)$, hence $\int g_n dm \to \int g dm$ in G^{**}, consequently $\int g dm \in G$. We take then $g = f 1_A$ with $A \in \Sigma$ and obtain (a).

Relationship between the spaces $\mathcal{F}_D(m_{F,G})$

Next we show that the inequality $\tilde{m}_{\mathbb{R},E}(A) \leq \tilde{m}_{F,G}(A)$, valid for $A \in \Sigma$, can be extended to real functions.

AI.13 THEOREM. *Let $m : \Sigma \to E \subset L(F,G)$ be a σ-additive measure with finite semivariation $\tilde{m}_{F,G}$. Then*

$$\mathcal{F}_{\mathbb{R}}(m_{F,G}) \subset \mathcal{F}_{\mathbb{R}}(m_{\mathbb{R},E})$$

and

$$\tilde{m}_{\mathbb{R},E}(\varphi) \leq \tilde{m}_{F,G}(\varphi), \text{ for } \varphi \in \mathcal{F}_{\mathbb{R}}(m_{F,G}).$$

Proof. Suppose φ is a Σ-measurable, scalar valued, simple function. Let $n = \varphi m$. Then n is σ-additive on Σ into E, and for $A \in \Sigma$, $z \in G^*$ and $y \in F$, we have

$$\langle n_z(A), y\rangle = \langle n(A)y, z\rangle = \langle \int_A \varphi dm_z, y\rangle;$$

hence

$$n_z(A) = \int_A \varphi dm_z,$$

and by [D.1], Theorem 7, p. 278, $|n_z|(A) = \int_A |\varphi| d|m_z|$.

In particular, regarding E as $L(I\!R, E)$, we have, for $x^* \in E^*$,

$$|n_{x^*}|(A) = |x^* n|(A) = \int_A |\varphi| d|x^* m|.$$

Thus

$$\begin{aligned}
\tilde{m}_{I\!R,E}(\varphi) &= \sup\{\int |\varphi| d|x^* m| : x^* \in E_1^*\} \\
&= \sup\{|n_{x^*}|(S) : x^* \in E_1^*\} = \tilde{n}_{I\!R,E}(S) \\
&\leq \tilde{n}_{F,G}(S) = \sup\{|n_z|(S) : z \in G_1^*\} \\
&= \sup\{\int |\varphi| d|m_z| : z \in G_1^*\} = \tilde{m}_{F,G}(\varphi).
\end{aligned}$$

In general, if $\varphi \in \mathcal{F}_{I\!R}(m_{F,G})$, we choose a sequence (φ_n) of Σ–measurable simple functions such that $\varphi_n \to \varphi$ pointwise and $|\varphi_n| \leq |\varphi|$. Using the dominated convergence theorem relative to the measures $|x^* m|$ and $|m_z|$, for $x^* \in E^*$ and $z \in G^*$, we obtain

$$\begin{aligned}
\int |\varphi| d|x^* m| &= \lim_n \int |\varphi_n| d|x^* m| \leq \limsup_n \tilde{m}_{I\!R,E}(\varphi_n) \\
&\leq \limsup_n \tilde{m}_{F,G}(\varphi_n) \leq \tilde{m}_{F,G}(\varphi).
\end{aligned}$$

Thus $\tilde{m}_{I\!R,E}(\varphi) \leq \tilde{m}_{F,G}(\varphi)$, and the theorem is proved.

If $m : \Sigma \to L(F, G)$ is a σ–additive measure and $y \in F$, then we denote by $my : \Sigma \to G$ the σ–additive measure defined by

$$(my)(A) = m(A)y, \text{ for } A \in \Sigma.$$

AI.14 THEOREM. *Let $m : \Sigma \to E \subset L(F, G)$ be a σ–additive measure with finite semivariation $\tilde{m}_{F,G}$, and let $y \in F$. Then $\mathcal{F}_{I\!R}(m_{F,G}) \subset \mathcal{F}_{I\!R}((my)_{I\!R,G})$, $y\mathcal{F}_{I\!R}(m_{F,G}) \subset \mathcal{F}_F(m_{F,G})$, and for $\varphi \in \mathcal{F}_{I\!R}(m_{F,G})$, we have*

$$\begin{aligned}
(my)\tilde{}_{I\!R,G}(\varphi) &\leq |y| \tilde{m}_{F,G}(\varphi), \\
\tilde{m}_{F,G}(\varphi y) &= |y| \tilde{m}_{F,G}(\varphi),
\end{aligned}$$

and

$$\int \varphi y dm = \int \varphi d(my).$$

If in addition, $\int \varphi dm \in E$, *then*

$$(\int \varphi dm)y = \int \varphi y dm = \int \varphi d(my).$$

Proof. Let $z \in G_1^*$ and let $A \in \Sigma$. Then

$$\begin{aligned}|(my)_z(A)| &= |\langle m(A)y, z\rangle| = |\langle y, m_z(A)|\\ &\leq |y||m_z(A)| \leq |y||m_z|(A),\end{aligned}$$

hence $|(my)_z| \leq |y||m_z|$. As a result, for any $\varphi \in \mathcal{F}_{\mathbb{R}}(m_{F,G})$, we have

$$\int |\varphi| d|(my)_z| \leq |y| \int |\varphi| d|m_z| \leq |y| \widetilde{m}_{F,G}(\varphi) < \infty,$$

and the first inequality in the conclusion follows. The second equality is immediate.

Suppose now that $\varphi = 1_A$, with $A \in \Sigma$. Then $(\int \varphi dm)y = \int \varphi y dm$ and $m(A)y = \int \varphi d(my)$, hence the equalities in the conclusion hold when φ is a measurable simple function. For the general case, let $\varphi \in \mathcal{F}_{\mathbb{R}}(m_{F,G})$ and let (φ_n) be a sequence of Σ–measurable simple functions such that $\varphi_n \to \varphi$ pointwise and $|\varphi_n| \leq |\varphi|$. Since $\varphi \in L^1_{\mathbb{R}}((my)_z)$, we can use the dominated convergence theorem to conclude that $\langle \int \varphi_n d(my), z\rangle \to \langle \int \varphi d(my), z\rangle$; similarly, since $\varphi y \in L^1_F(m_z)$, we have $\langle \int \varphi_n y dm, z\rangle \to \langle \int \varphi y dm, z\rangle$. Since $\int \varphi_n y dm = \int \varphi_n d(my)$, for each n, we conclude that $\int \varphi y dm = \int \varphi d(my)$.

Assume now that $\int \varphi dm \in E$. From Theorem AI.13, we have $\mathcal{F}_{\mathbb{R}}(m_{F,G}) \subset \mathcal{F}_{\mathbb{R}}(m_{\mathbb{R},E})$, hence $\varphi \in \mathcal{F}_{\mathbb{R}}(m_{\mathbb{R},E})$. For $x^* \in E^*$, we apply the dominated convergence theorem to deduce that $\int \varphi_n d(x^*m) \to \int \varphi d(x^*m)$. Hence $\langle \int \varphi_n dm, x^*\rangle \to \langle \int \varphi dm, x^*\rangle$. Now we choose $x^* \in E^*$ to be defined by $x^*(x) = \langle x(y), z\rangle$, for $x \in E$. Since $\int \varphi dm \in E$, the convergence $\int \varphi_n d(x^*m) \to \int \varphi d(x^*m)$ can be written $\langle (\int \varphi_n dm)y, z\rangle \to \langle (\int \varphi dm)y, z\rangle$. Since for each n we have $(\int \varphi_n dm)y = \int \varphi_n y dm$, we deduce $(\int \varphi dm)y = \int \varphi y dm$, and the theorem is proved.

Associativity properties

The following theorem concerns the "associativity" of the integral: $f(gm) = (fg)m$. We shall consider the case when one of the functions f and g is real valued.

AI.15 THEOREM. *Let $m : \Sigma \to E \subset L(F,G)$ be a σ-additive measure with finite semivariation $\tilde{m}_{F,G}$.*

I) Let $\varphi \in \mathcal{F}_{\mathbb{R}}(m_{F,G}) \subset \mathcal{F}_{\mathbb{R}}(m_{\mathbb{R},E})$ and assume that $\int_A \varphi dm \in E$, for every $A \in \Sigma$. Consider the measure $\varphi m : \Sigma \to E$ defined by $(\varphi m)(A) = \int_A \varphi dm$, for $A \in \Sigma$.

(a) The measure φm is σ-additive with finite semivariation $(\varphi m)\tilde{}_{F,G}$;

(b) If $f \geq 0$ is Σ-measurable, then

$$(\varphi m)\tilde{}_{\mathbb{R},E}(f) = \tilde{m}_{\mathbb{R},E}(\varphi f),$$

and

$$(\varphi m)\tilde{}_{F,G}(f) = \tilde{m}_{F,G}(\varphi f);$$

(c) We have $f \in \mathcal{F}_{F,G}(\varphi m)$ if and only if $f\varphi \in \mathcal{F}_{F,G}(m)$, and in this case

$$f(\varphi m) = (f\varphi)m;$$

(d) Suppose $m_{F,G}$ is uniformly σ-additive. Then $(\varphi m)_{F,G}$ is uniformly σ-additive if and only if $\varphi \in \mathcal{F}_{\mathbb{R}}(\mathcal{B}, m_{F,G})$.

II) Let $f \in \mathcal{F}_{F,G}(m)$ and assume that $\int_A f dm \in G$ for every $A \in \Sigma$. Consider the measure $fm : \Sigma \to G$ defined by $(fm)(A) = \int_A f dm$, for $A \in \Sigma$.

(a) fm is σ-additive and has finite semivariation $(fm)\tilde{}_{\mathbb{R},G}$;

(b) If $\varphi \geq 0$ is Σ-measurable, we have

$$(fm)\tilde{}_{\mathbb{R},G}(\varphi) \leq \tilde{m}_{F,G}(f\varphi) = (|f|m)\tilde{}_{F,G}(\varphi)$$

(c) If φ is real valued and Σ-measurable and if $\varphi f \in \mathcal{F}_{F,G}(m)$, then $\varphi \in \mathcal{F}_{\mathbb{R},G}(fm)$, and in this case we have

$$\varphi(fm) = (\varphi f)m;$$

(d) Suppose $m_{F,G}$ is uniformly σ-additive. Then $(fm)_{\mathbb{R},G}$ is uniformly σ-additive if $f \in \mathcal{F}_F(\mathcal{B}, m_{F,G})$.

Proof. I. Let $n = \varphi m$.

(a) n is weakly σ-additive, therefore it is strongly σ-additive (see Proposition AI.12). The finiteness of $\tilde{n}_{F,G}$ will follow from (b).

(b) Let $z \in G^*$, $y \in F$, and $A \in \Sigma$. Then

$$\langle y, n_z(A)\rangle = \langle n(A)y, z\rangle = \langle (\int 1_A \varphi dm)y, z\rangle$$
$$= \langle \int 1_A \varphi y dm, z\rangle = \int 1_A \varphi y dm_z = \langle y, \int_A \varphi dm_z\rangle.$$

Thus $n_z(A) = \int_A \varphi dm_z$ for $A \in \Sigma$, therefore

$$|n_z|(A) = \int_A |\varphi| d|m_z|, \text{ for } A \in \Sigma.$$

If $f \geq 0$ is Σ–measurable, then $\int f d|n_z| = \int |f\varphi| d|m_z|$; taking the supremum over G_1^*, we have $\tilde{n}_{F,G}(f) = \tilde{m}_{F,G}(f\varphi)$. The other equality follows by taking $F = I\!R$ and $G = E$.

(c) From (b) we deduce that $f \in \mathcal{F}_{F,G}(n)$ if and only if $f\varphi \in \mathcal{F}_{F,G}(m)$, and from the proof of (b), for each $z \in G^*$ we have $n_z = \varphi m_z$; hence if $f \in \mathcal{F}_{F,G}(n)$ and $A \in \Sigma$, then $f1_A \in L_F^1(n_z)$ and $\int_A f dn_z = \int_A f\varphi dm_z$. This implies that $\langle \int_A f dn, z\rangle = \langle \int_A f\varphi dm, z\rangle$, which yields the conclusion in (c).

Assertion (d) follows from AI.8a and b. The proof of II is similar.

AI.16 Corollary. *Let $m : \Sigma \to E \subset L(F,G)$ be a σ–additive measure with finite semivariation $\tilde{m}_{F,G}$ and let $f \in \mathcal{F}_{F,G}(m)$ be Σ–measurable. If $\int f d(1_A m) \in G$ for every $A \in \Sigma$, then $\int_A f dm \in G$ for every $A \in \Sigma$.*

Proof. Let $\varphi = 1_A$. We note that since $\varphi f \in \mathcal{F}_{F,G}(m)$, we have $f \in \mathcal{F}_{F,G}(\varphi m)$ and

$$(f(\varphi m))(S) = ((f\varphi)m)(S) = \int_A f dm.$$

Since $f(\varphi m)(S) \in G$, by hypothesis, the conclusion follows.

Weak completeness and weak compactness in $\mathcal{F}_{F,G}(\mathcal{B}, m)$

One of the main goals in [B–D.2] was to obtain sufficient conditions for weak completeness and weak compactness in $\mathcal{F}_{F,G}(\mathcal{B}, m)$. To establish these results, a characterization of elements in $(\mathcal{F}_{F,G}(\mathcal{B}, m))^*$ was given, using techniques of Köthe spaces. This theory can be applied to stochastic integration theory to yield new convergence theorems. In this section we shall present the necessary tools for this application.

A crucial property in establishing weak compactness criteria is the following "Beppo Levi property."

Let $m : \Sigma \to E \subset L(F, G)$ be a σ-additive measure. We say that $m_{F,G}$ has the *Beppo Levi property* if every increasing sequence (f_n) of positive Σ-measurable simple functions, with $\sup_n \tilde{m}_{F,G}(f_n) < \infty$, is a Cauchy sequence in $\mathcal{F}_{\mathbb{R}}(\mathcal{B}, m_{F,G})$ (hence $\sup_n f_n \in \mathcal{F}_{\mathbb{R}}(\mathcal{B}, m_{F,G})$).

We remark that if $m_{F,G}$ has the Beppo Levi property, then $m_{F,G}$ is uniformly σ-additive.

One of the main theorems in [B–D$_1$] (Theorem 8.8), gives sufficient conditions that $m_{F,G}$ has the Beppo Levi property.

AI.17 THEOREM. *Let $m : \Sigma \to E \subset L(F, G)$ be σ-additive. Suppose that m has finite semivariation $\tilde{m}_{F,G}$ and that $m_{F,G}$ is uniformly countably additive. If $\int f dm \in G$ for every Σ-measurable function $f \in \mathcal{F}_{F,G}(m)$, then $m_{F,G}$ has the Beppo Levi property.*

For applications to the theory of stochastic integration, we shall strengthen Corollary 8.10 in [B–D$_1$]. The following result will be used repeatedly in the sequel.

AI.18 COROLLARY. *Suppose $m : \Sigma \to E \subset L(F, G)$ is σ-additive with finite semivariation $\tilde{m}_{F,G}$. If $m_{F,G}$ is uniformly σ-additive and if G does not contain a copy of c_0, then $m_{F,G}$ has the Beppo Levi property.*

Proof. Since $c_0 \not\subset G$, we have $\int f dm \in G$ for every $f \in \mathcal{F}_{F,G}(m)$, by Proposition AI.12(c). We can then apply Theorem AI.17.

We shall now present, without proofs, the main results in [B–D.2] concerning the weak completeness of $\mathcal{F}_{F,G}(\mathcal{B}, m)$ and criteria for weak compactness of subsets of $\mathcal{F}_{F,G}(\mathcal{B}, m)$. We shall state these results in a slightly different form from the results given in [B–D.2], by using Corollary AI.18 above.

Recall that a set K in a Banach space is *conditionally weakly compact* if every sequence of elements from K contains a subsequence which is weakly Cauchy; and that K is *relatively weakly compact* if its weak closure is weakly compact.

To avoid repetition, we shall assume in the sequel that $m : \Sigma \to E \subset L(F,G)$ is σ-additive, and has finite semivariation $\tilde{m}_{F,G}$.

AI.19 THEOREM. *Assume that $m_{F,G}$ is uniformly σ-additive, F is reflexive and G does not contain a copy of c_0. Then $\mathcal{F}_{F,G}(\mathcal{B},m)$ is weakly sequentially complete.*

AI.20 THEOREM. *Assume that $m_{F,G}$ is uniformly σ-additive and F is reflexive. Let $K \subset \mathcal{F}_{F,G}(\mathcal{B},m)$ be a set satisfying the following conditions:*

(1) K is bounded;

(2) $\lim_n \tilde{m}_{F,G}(f1_{A_n}) = 0$ uniformly for $f \in K$ whenever $A_n \in \Sigma$ and $A_n \searrow \phi$.

Then K is conditionally weakly compact in $\mathcal{F}_{F,G}(\mathcal{B},m)$. If, in addition, G does not contain a copy of c_0, then K is relatively weakly compact.

AI.21 THEOREM. *Let $K \subset \mathcal{F}_{\mathbb{R},E}(\mathcal{B},m)$ be a set satisfying the following conditions:*

(1) K is bounded;

(2) $\int_{A_n} f dm \to 0$ uniformly for $f \in K$ whenever $A_n \in \Sigma$ and $A_n \searrow \phi$.

Then K is conditionally weakly compact. If, in addition, E does not contain a copy of c_0, then K is relatively weakly compact.

AI.22 THEOREM. *Assume that E does not contain a copy of c_0. Let $(f_n)_{n\geq 0}$ be a sequence of elements from $\mathcal{F}_{\mathbb{R},E}(\mathcal{B},m)$. If $\int_A f_n dm \to \int_A f_0 dm$, for every $A \in \Sigma$, then $f_n \to f_0$ weakly in $\mathcal{F}_{\mathbb{R},E}(\mathcal{B},m)$.*

Appendix II: Quasimartingales

In this section we shall present some basic properties of Banach-valued quasimartingales, which are used in section 2 concerning summability. This material is taken from [B-D.5] and [Ku.1].

In this section, we assume that $X : \mathbb{R} \to E$ is a cadlag, adapted process, such that $X_t \in L^p_E$, for every $t \geq 0$. If X has a limit at ∞, we denote it by $X_{\infty-}$. We extend X at ∞ with $X_\infty = 0$.

Rings of subsets of $\overline{\mathbb{R}}_+ \times \Omega$

We shall consider five rings of subsets of $\overline{\mathbb{R}}_+ \times \Omega$:

(1) $\mathcal{A}[0] = \{0\} \times \mathcal{F}_0 = \{[0_A] : A \in \mathcal{F}_0\}$, where $[0_A] = \{0\} \times A$ is the graph of the stopping times 0_A which is zero on A and $+\infty$ on A^c.

(2) $\mathcal{A}(0, \infty)$ is the ring of all finite unions of predictable rectangles $(s,t] \times A$, with $0 \le s < t < \infty$, and $A \in \mathcal{F}_s$.

(3) $\mathcal{A}[0, \infty) = \mathcal{A}[0] \cup \mathcal{A}(0, \infty)$.

(4) $\mathcal{A}(0, \infty]$ is the ring of all finite unions of predictable rectangles $(s,t] \times A$, with $0 \le s \le t \le \infty$, and $A \in \mathcal{F}_s$.

(5) $\mathcal{A}[0, \infty] = \mathcal{A}[0] \cup \mathcal{A}(0, \infty]$; $\mathcal{A}[0, \infty]$ is an algebra of subsets of $\overline{\mathbb{R}}_+ \times \Omega$, and contains, along with $\mathcal{A}[0, \infty)$, predictable rectangles of the form $(t, \infty] \times A$, where $A \in \mathcal{F}_t$.

The Doléans function

Since $L^p_E \subset L^1_E$, we have $X_t \in L^1_E$ for every $t \ge 0$. We define the additive measure $\mu_X : \mathcal{A}[0, \infty] \to E$, called the *Doléans function* of the process X, first for predictable rectangles, and then extend it in an additive fashion to $\mathcal{A}[0, \infty]$. For $[0_A] \in \mathcal{A}[0, \infty]$ and $(s, t] \times B \in \mathcal{A}[0, \infty]$, we set

$$\mu_X([0_A]) = E(1_A X_0),$$

and

$$\mu_X((s,t] \times A) = E(1_A(X_t - X_s)).$$

Note that

$$\mu_X([0,t] \times A) = E(1_A X_t)$$

and

$$\mu_X((s,\infty] \times A) = -E(1_A X_s).$$

We also have $\mu_X([0,\infty] \times A) = 0$ and $\mu_X(A) = E(I_X(A))$, where I_X is the stochastic measure defined in section 2. The restriction of μ_X to $\mathcal{A}[0]$ is bounded and σ–additive. Hence μ_X is bounded (respectively σ–additive) on $\mathcal{A}[0, \infty)$ or on $\mathcal{A}[0, \infty]$ if and only if μ_X has the same property on $\mathcal{A}(0, \infty)$ or $\mathcal{A}(0, \infty]$ respectively.

Quasimartingales

We say X is a quasimartingale on $(0,\infty)$ (respectively on $(0,\infty]$, or $[0,\infty)$, or $[0,\infty]$) if the measure μ_X has bounded variation on $\mathcal{A}(0,\infty]$ (respectively on $\mathcal{A}(0,\infty]$, or $\mathcal{A}[0,\infty)$, or $\mathcal{A}[0,\infty]$). Since μ_X has bounded variation on $\mathcal{A}[0]$, X is a quasimartingale on $(0,\infty)$ or $(0,\infty]$ if and only if it is a quasimartingale on $[0,\infty)$ or $[0,\infty]$ respectively.

We now list some properties of quasimartingales.

1. X is a quasimartingale on $(0,\infty]$ if and only if X is a quasimartingale on $(0,\infty)$ and $\sup_t \|X_t\|_1 < \infty$.

2. If X is a quasimartingale on $(0,\infty)$ or on $(0,\infty]$, then so is the process $|X| = (|X_t|)_{t\geq 0}$.

3. Any process with integrable variation is a quasimartingale on $(0,\infty]$.

4. X is a martingale if and only if $\mu_X = 0$ on $\mathcal{A}(0,\infty)$; a martingale X is a quasimartingale on $(0,\infty)$; it is a quasimartingale on $(0,\infty]$ if and only if $\sup_t \|X_t\|_1 < \infty$.

5. X is a submartingale if and only if $\mu_X \geq 0$ on $\mathcal{A}(0,\infty)$. Any negative submartingale and any positive supermartingale is a quasimartingale on $(0,\infty]$.

6. If X is a quasimartingale on $(0,\infty]$, then for every stopping time T, we have $X_T \in L^1_E$.

7. If X is a quasimartingale on $(0,\infty]$ and if (T_n) is a decreasing sequence of stopping times such that $T_n \searrow T$, then $X_{T_n} \to X_T$ in L^1_E.

8. X is a quasimartingale of class (D) on $(0,\infty]$ if and only if μ_X is σ-additive and has bounded variation on $\mathcal{A}(0,\infty]$.

9. If X is a real valued quasimartingale on $(0,\infty]$, then $X = M + V$, where M is a local martingale and V is a predictable process with integrable variation (cf. [Ku, Theorem 9.15]). If, in addition, X is of class (D), then M is a martingale class (D). In this case we have

$$\mu_X = \mu_V \text{ on } \mathcal{A}(0,\infty).$$

10. If X is a real valued quasimartingale, then X is summable if and only if $X^* = \sup_t |X_t|$ is integrable.

References

[B–P] C. Bessaga and A. Pelczynski, *On bases and unconditional convergence of series in Banach spaces*, Studia Math. **5** (1974), 151–164.

[B–D.1] J.K. Brooks and N. Dinculeanu, *Strong additivity, absolute continuity and compactness in spaces of measures*, J. Math. Anal. and Appl. **45** (1974), 156–175.

[B–D.2] ——, *Lebesgue–type spaces for vector integration, linear operators, weak completeness and weak compactness*, J. Math. Anal. and Appl. **54** (1976), 348–389.

[B–D.3] ——, *Weak compactness in spaces of Bochner integrable functions and applications*, Advances in Math. **24** (1977), 172–188.

[B–D.4] ——, *Projections and regularity of abstract process*, Stochastic Analysis and Appl. **5** (1987), 17–25.

[B–D.5] ——, *Regularity and the Doob–Meyer decomposition of abstract quasimartingales*, Seminar on Stochastic Processes, Birkhaüser, Boston (1988), 21–63.

[B–D.6] ——, *Stochastic integration in Banach spaces*, Advances in Math. **81** (1990), 99–104.

[B–D.7] ——, *Itô's Formula for stochastic integration in Banach spaces*, Conference on diffusion processes, Birkhaüser (to appear).

[D–M] C. Dellacherie and P.A. Meyer, *Probabilities and Potential*, North–Holland, (1978), (1980).

[D.1] N. Dinculeanu, *Vector Measures*, Pergamon Press, 1967.

[D.2] ——, *Vector valued stochastic processes I. Vector measures and vector valued stochastic processes with finite variation*, J. Theoretical Probability **1** (1988), 149–169.

[D.3] ——, *Vector valued stochastic processes V. Optional and predictable variation of stochastic measures and stochastic processes*, Proc. A.M.S. **104** (1988), 625–631.

[D–S] N. Dunford and J. Schwartz, *Linear Operators, Part I*, Interscience, New York, 1958.

[G–P] B. Gravereaux and J. Pellaumail, *Formule de Itô pour des processus à valeurs dans des espaces de Banach*, Ann. Inst. H. Poincaré **10** (1974), 399–422.

[K] H. Kunita, *Stochastic integrals based on martingales taking their values in Hilbert spaces*, Nagoya Math J. **38** (1970), 41–52.

[Ku.1] A.U. Kussmaul, *Stochastic integration and generalized martingales*, Pitman, London, 1977.

[Ku.2] ——, *Regularität und stochastische Integration von Semimartingalen mit Werten in einem Banachraum*, Dissertation, Stuttgart (1978).

[Kw] S. Kwapien, *On Banach spaces containing* c_0, Studia Math. **5** (1974), 187–188.

[M.1] M. Métivier, *The stochastic integral with respect to processes with values in a reflexive Banach space*, Theory Prob. Appl. **14** (1974), 758–787.

[M.2] ——, *Semimartingales*, de Gruyter, Berlin, 1982.

[M–P] M. Métivier and J. Pellaumail, *Stochastic Integration*, Academic Press, New York, 1980.

[P] J. Pellaumail, *Sur l'intégrale stochastique et la décomposition de Doob–Meyer*, S.M.F., Asterisque **9** (1973).

[Pr] M. Pratelli, *Intégration stochastique et Géometrie des espaces de Banach*, Séminaire de Probabilitiés, Springer Lecture Notes, New York (1988).

[Pro] P. Protter, *Stochastic integration and differential equations*, Springer-Verlag, New York, 1990.

[Y.1] M. Yor, *Sur les intégrales stochastiques à valeurs dans un espace de Banach*, C.R. Acad. Sci. Paris Ser. A **277** (1973), 467–469.

[Y.2] ———, *Sur les intégrales stochastiques à valeurs dans un espace de Banach*, Ann. Inst. H. Poincaré **X** (1974), 31–36.

J.K. BROOKS
Department of Mathematics
University of Florida
Gainesville, FL 32611–2082
USA

N. DINCULEANU
Department of Mathematics
University of Florida
Gainesville, FL 32611–2082
USA

Absolute Continuity of the Measure States in a Branching Model with Catalysts

DONALD A. DAWSON[1], KLAUS FLEISCHMANN

and SYLVIE ROELLY

1. INTRODUCTION

Spatially homogeneous measure-valued branching Markov processes X on the real line $\mathbb{R}$ with certain motion processes and branching mechanisms with finite variances have *absolutely continuous states* with respect to Lebesgue measure, that is, roughly speaking,

$$X(t,dy) = \eta(t,y)dy$$

for some random density function $\eta(t)=\eta(t,\cdot)$. Results of this type are established in Dawson and Hochberg (1979), Roelly-Coppoletta (1986), Wulfsohn (1986), Konno and Shiga (1988), and Tribe (1989).

More generally, if the branching mechanism does not necessarily has finite second moments, a similar absolute continuity result is valid in $\mathbb{R}^d$ for all dimensions d smaller than a critical value which depends on the underlying motion process and the branching mechanism. This critical value can take on any positive value. We refer to Fleischmann (1988, Appendix).

The simplest case, namely, a *continuous critical super-Brownian motion* $X = [X,\mathbb{P}^{\rho}_{s,\mu};s\in\mathbb{R},\mu\in\mathcal{M}_f]$ in $\mathbb{R}$ is re-

[1]Supported by an NSERC grant.

lated to the parabolic partial differential equation

$$(1.1)\qquad \frac{\partial}{\partial s}v(s,t,x) = -\kappa\frac{\partial^2}{\partial x^2}v(s,t,x) + \rho v^2(s,t,x),$$

$s\le t$, $x\in\mathbb{R}$, where $\kappa>0$ is the diffusion constant and $\rho\ge 0$ the *constant branching rate*. In fact, the Laplace transition functional of X is given by

$$(1.2)\qquad \mathbb{E}^{\rho}_{s,\mu}\exp(X(t),-\varphi) = \exp(\mu,-v(s,t)),$$
$$s\le t,\ \mu\in M_f,\ \varphi\in F_+,$$

where v solves (1.1) with *final* condition $v(t,t) = \varphi$. Here M_f is the set of all finite measures μ on $\mathbb{R}$, and F_+ is some set of continuous non-negative test functions on $\mathbb{R}$, defined in Section 2 below. Moreover, $(m,h) := \int m(dx)h(x)$, and $\mathbb{E}^{\rho}_{s,\mu}$ denotes expectation with respect to $\mathbb{P}^{\rho}_{s,\mu}$, the law of the process X with branching rate ρ and starting at time $s\in\mathbb{R}$ with the measure μ.

(We mention that we adopt time-inhomogeneous notation and a backward formulation of the equation, in order to facilitate the generalization later to time-inhomogeneous Markov processes.)

Intuitively it is clear that the absolute continuity result for the states of the process X will remain true if the constant branching rate ρ is replaced by a bounded non-negative function, smoothly varying in time and space (*varying medium* ρ).

However it is not immediately clear what will happen if ρ degenerates to a generalized function, for instance, to the weighted δ-function $a\delta_0$, $a>0$. In this case one can interpret $\rho=a\delta_0$ as a *point catalyst with action weight* a and located at 0. In other words, branching does not occur except at the origin. From the viewpoint of an approximating particle system, a particle will split only if it approaches 0 within a distance $\varepsilon<<1$, and then the branching rate is given by the scaled action weight $a/2\varepsilon$.

Actually, it is possible to give (1.1) a precise meaning in the degenerate case $\rho=a\delta_0$, namely in terms of the *integral equation*

$$(1.3)\qquad v(s,t,x) = \int dy\ p(s,t,x,y)\varphi(y) - a\int_s^t dr\ p(s,r,x,0)v^2(r,t,0), \qquad s\le t,\ x\in\mathbb{R}$$

where $p(s,t,x,y)=p(t-s,y-x)$, $s<t$, $x,y\in\mathbb{R}$, is the continuous transition density function of the heat flow corresponding to $\kappa\Delta$, and formally we set $p(0,y)=\delta_0(y)$.

In Dawson and Fleischmann (1990a), it is shown, that there exists a continuous F_+-valued curve $v(\cdot,t)\ge 0$ which solves equation (1.3), for each given $t\in\mathbb{R}$ and $\varphi\in F_+$ (so-called *mild solution* of (1.1)). It is constructed by approximating $\rho=a\delta_0$ by the smooth functions $\rho_\varepsilon=ap(\varepsilon,\cdot)$ as $\varepsilon\to 0$. Using this type of approximation and continuity properties of the Laplace transition functional in (1.2), a *superprocess* X *with singular branching rate* $\rho=a\delta_0$ can be defined which is related to (1.3) by (1.2).

To give a feeling for this process X, we provide some *moment calculations*. To this end, fix $s<t$ and $\mu=\delta_x$ (unit mass at x). In (1.2) replace φ by $\theta\varphi$ with $\theta>0$, (formally) differentiate with respect to θ at $\theta=0+$, and proceed in the same way with equation (1.3). Then it turns out that the first moment measure of X(t) with respect to $\mathbb{P}^\rho_{s,x}:=\mathbb{P}^\rho_{s,\delta_x}$ is given by

$$\mathbb{E}^\rho_{s,x}X(t,dy) = p(s,t,x,y)dy.$$

Consequently, since the branching term, i.e. the nonlinear term in (1.1), does not effect the expectation of the process, we get the same first moment density as in the classical model of constant branching rate, namely $p(s,t,x,\cdot)$.

Following an analogous procedure, for the covariance measure of X(t) with respect to $\mathbb{P}^\rho_{s,x}$ we obtain

$$\mathrm{Cov}^\rho_{s,x}[X(t,dy),X(t,dz)] = \left[2a\int_s^t dr\ p(s,r,x,0)p(t-r,y)p(t-r,z)\right]dydz.$$

Hence, this process has a finite smooth covariance density function, except at 0, the position of the catalyst. Indeed, letting $y=z$, the latter integral behaves like

(1.4) $$\text{const} \int_s^t dr\, p^2(r,y) \sim \text{const}\, \big|\log|y|\big| \quad \text{as } y\to 0$$

(recall that $s<t$ and x are fixed). Such behavior is in a sharp contrast to the "classical" models in constant media ρ.

On the other hand, despite this singularity, as in the classical models above *this superprocess* X *has absolutely continuous states*, since the singularity (1.4) at the catalyst's position $y=0$ is (locally) integrable with respect to Lebesgue measure (see Meidan (1980)). More precisely, there is a second order random function $\eta(t,\cdot)= \eta(t)$ such that

$$\mathbb{E}^{\rho}_{s,x}\left|\int X(t,dy)f(y) - \int dy\ \eta(t,y)f(y)\right|^2 = 0, \qquad f\in\mathbf{F}_+.$$

However, by (1.4) this L^2*-random density function* $\eta(t)$ *is singular* at $y=0$ since, by (1.4),

$$\mathbb{E}^{\rho}_{s,x}\eta^2(t,y) \longrightarrow \infty \quad \text{as } y\to 0, \quad s<t,\ x\in\mathbb{R}.$$

In this case of a single non-moving catalyst we now consider an *alternative approach* to the problem. Rewrite equation (1.3) in the following way:

$$v(s,t,x) = E_{s,x}\Big[\varphi(w(t)) - a\int_s^t L_0(dr)v^2(r,t,0)\Big],$$

where $[[w,L_0],P_{s,x};s,x\in\mathbb{R}]$ is a *Wiener process* w in $\mathbb{R}$ with transition density function p, and its *local time* L_0 at 0 (and $E_{s,x}$ denotes expectation with respect to $P_{s,x}$, the law of w starting at time s at x). Since the latter equation can further be reformulated as

$$v(s,t,x) = E_{s,x}\Big[\varphi(w(t)) - \int_s^t aL_0(dr)v^2(r,t,w(s))\Big],$$

$s\le t$, $x\in\mathbb{R}$, we obtain a special case of equation (1.23) in Dynkin (1990). Thus, this superprocess X corresponding to a single non-moving catalyst is a member of the general family of superprocesses constructed by Dynkin (1990).

By the way, this also illuminates the reason why the point catalyst model discussed above has to be restricted to the *space dimension one* since it involves the local time L_0 of the Brownian motion w at the catalyst's position 0, whereas for the Brownian motion in dimensions d>1 a single point set is polar and does not carry a positive local time.

In the general model we investigate, the branching rate ρ is given by a *dense set of point catalysts*, which are also *allowed to move* in space, and whose action weights are *not locally bounded above*.

To worsen the situation, we can think of a more general branching mechanism which does not necessarily have finite second moments. Consequently, in this case as a rule the *covariance measure does not exist*, i.e. it is not a locally finite measure. This in fact raises the question as to whether a process with densities exists at all in such a general situation.

It is the *main purpose of the present paper* to demonstrate that even in such a general situation of a superprocess X without second moments and in a highly singular varying medium ρ, the absolute continuity results remain true.

To be precise, we will consider a superprocess X related to the following *integral equation*

$$(1.5)\quad v(s,t,x) = \int dy\; p(s,t,x,y)\varphi(y) - \int_s^t dr \int \rho(r,dy)\, p(s,r,x,y)\, |v|^{1+\beta}(r,t,y),$$

$s\le t$, $x\in\mathbb{R}$, $\varphi\in F_+$. Here $p(s,t,x,y)$, $s<t$, $x,y\in\mathbb{R}$, is now the continuous transition density function of a symmetric stable flow with index $\alpha\in(1,2]$ corresponding to the fractional Laplacian $\kappa\Delta_\alpha:=-\kappa(-\Delta)^{\alpha/2}$, the critical continuous state splittings have index $1+\beta\in(1,2]$, and ρ is some *branching rate kernel*.

The latter is a measurable kernel ρ of $\mathbb{R}$ into the set of all locally finite measures on $\mathbb{R}$ with the following property:

$$(1.6)\qquad \sup_{r\in[s,t]} (\rho(r),f^\beta) < \infty, \qquad s\le t, \quad f\in F_+.$$

To mention an example, set $\rho(r)\equiv\mu$ where μ is any finite measure on $\mathbb{R}$. Here $\mu(dx)$ is the time-independent branching rate at x.

Note that by a formal differentiation, (1.5) can be

written as

$$\frac{\partial}{\partial s}v(s,t,x) = -\kappa\Delta_\alpha v(s,t,x) + \rho(s,dx)|v|^{1+\beta}(s,t,x), \tag{1.7}$$

$s\le t$, $x\in\mathbb{R}$, with *final state* $v(t,t,\cdot)=\varphi\in F_+$.

A rigorous setting of equation (1.5) is given in Dawson and Fleischmann (1990a). Based on this, actually a *superprocess* $X = [X,\mathbb{P}^\rho_{s,\mu};s\in\mathbb{R},\mu\in\mathcal{M}_f]$ related to (1.7) can be constructed:

PROPOSITION 1.8. *To each branching rate kernel* ρ *there exists an* $\mathcal{M}_f$*-valued time-inhomogeneous superprocess* $X = [X,\mathbb{P}^\rho_{s,\mu};s\in\mathbb{R},\mu\in\mathcal{M}_f]$ *with Laplace transition functional*

$$\mathbb{E}^\rho_{s,\mu}\exp\langle X(t),-\varphi\rangle = \exp\langle\mu,-v(s,t)\rangle, \tag{1.9}$$

$s\le t$, $\mu\in\mathcal{M}_f$, $\varphi\in F_+$, *where* v *solves* (1.5).

To formulate the results of the present paper, we introduce the following definition.

Definition 1.10. Fix $J:=(s',t)$, $s'<t$. The *restricted branching rate kernel* $\rho_J:=\{\rho(r);r\in J\}$ is called *admissible*, if there exists a Borel subset $N(\rho_J)$ of $\mathbb{R}$ of Lebesgue measure 0 such that the following holds. For each $z\in\mathbb{R}\setminus N(\rho_J)$

$$\sup_{r\in J}\langle\rho(r),p^\beta(r,t,z,\cdot)\rangle < \infty, \tag{1.11}$$

$$\sup_{r\in J}\langle\rho(r),p(s',r,z,\cdot)\rangle < \infty, \tag{1.12}$$

as well as

$$\limsup_{n\to\infty}\ \sup_{r\in J}\langle\rho(r),p^\beta(r,t+\varepsilon_n(z),z,\cdot)\rangle < \infty \tag{1.13}$$

for some sequence $\underline{\varepsilon}(z):=\{\varepsilon_n(z)\in(0,1);n\ge1\}$ satisfying $\varepsilon_n(z)\longrightarrow0$ as $n\to\infty$. The zero set $N(\rho_J)$ is called an *exceptional set for* ρ_J.

A trivial example is given by the branching rate kernel $\rho(r,dy) \equiv a\delta_0(dy)$ as discussed above. In fact, in this case the restricted branching rate kernels ρ_J are admissible for any J since we can set $N(\rho_J) \equiv \{0\}$, and the conditions hold whatever the $\underline{\varepsilon}$-sequence is. Note that here the exceptional set just represents the position of the non-moving catalyst.

A more interesting class of admissible ρ_J will be provided in Proposition 1.18 below.

Our first result can be formulated as follows. Recall that $X=[X,\mathbb{P}^{\rho}_{s,\mu};s\in\mathbb{R},\mu\in\mathcal{M}_f]$ is the superprocess with branching rate kernel ρ.

THEOREM 1.14. *Fix* $J=(s',t)$, $s'<t$, *and let the restricted branching rate kernel* ρ_J *be admissible. Then with respect to* $\mathbb{P}^{\rho}_{s,\mu}$, $s<s'$, $\mu\in\mathcal{M}_f$, *the random measure* $X(t)$ *is absolutely continuous a.s., that is, there exists a random density function* $\eta(t)=\eta(t,\cdot)$ *such that*

$$\mathbb{P}^{\rho}_{s,\mu}\left\{X(t,dy) = \eta(t,y)dy\right\} = 1.$$

Consequently, if the branching rate kernel ρ is such that its restriction ρ_J to $J=(s',t)$ is admissible, then the superprocess X corresponding to ρ and starting before s' has with probability one an absolutely continuous state at time t.

The key of proof of that result is the following *Basic Lemma*.

LEMMA 1.15. *Let* ν *be a random element in* $\mathcal{M}_f$ *and assume that*

(i) *there is a Borel subset* N *of* $\mathbb{R}$ *of Lebesgue measure* 0 *such that for each* $z\in\mathbb{R}\setminus N$ *there is a sequence* $\underline{\varepsilon}(z):=\{\varepsilon_n(z)\in(0,1);n\geq1\}$ *with* $\varepsilon_n(z)\longrightarrow0$ *as* $n\to\infty$, *and* $\nu([z-\varepsilon_n(z),z+\varepsilon_n(z)])/2\varepsilon_n(z)$ *converges in distribution to a random variable* $\eta(z)$ *as* $n\to\infty$,

(ii) *the expectation* $\mathbf{E}(\nu,f)$ *coincides with* $\mathbf{E}\int_{\mathbb{R}\setminus N}dz\;\eta(z)f(z)$ *for all* $f\in F_+$.

Then with probability one, ν *is an absolutely continuous measure.*

Roughly speaking, if (ν,δ_z) exists and has full expectation, then ν is absolutely continuous and has density (ν,δ_z).

In order to apply this lemma to the random measure $\nu = X(t)$ and having in mind the relation (1.9) with equation (1.5), it is necessary to develop a formulation of the nonlinear equation (1.5) which in particular applies for *generalized* final functions $v(t-,t)=\varphi=\delta_z$, (*mild basic solutions to* (1.7)). This is one of the technical developments which has to be carried out in this paper. The essence is the following result.

THEOREM 1.16. *Let* J *and* ρ_J *as well as* $N(\rho_J)$ *as described in Definition* 1.10. *Then to each* $z \in \mathbb{R} \setminus N(\rho_J)$ *there exists a continuous* $\mathbf{F}_+$*-valued curve* $v(\cdot,t)$ *on* J *which solves equation* (1.5) *on* $J \times \mathbb{R}$ *with* $\int dy\, p(s,t,\cdot,y)\varphi(y)$ *replaced by* $p(s,t,\cdot,z)$.

We note that this approach to proving the absolute continuity via basic solutions of the nonlinear equation was already used in Fleischmann (1988), namely for superprocesses with constant branching rate (and this approach differs from that used in the other papers quoted above).

An interesting class of branching rate kernels satisfying Definition 1.10 is obtained by sampling ρ from an *α-stable moving system* Γ *of* *γ-stable point catalysts* described as follows. At time 0 the *random catalytic medium* Γ is given by the *stable random measure* $\Gamma(0) = \sum_i a_i \delta_{x(i)}$ on $\mathbb{R}$ with index $\gamma \in (0,1)$. It is determined by its Laplace functional

$$(1.17) \qquad E\exp(\Gamma(0),-f) = \exp[-\int dx\, f^{\gamma}(x)], \qquad f \in \mathbf{F}_+.$$

(Note that $\Gamma(0)$ has a *dense* set of atoms.) Then, as time t goes forward or backward, the point catalysts $a_i \delta_{x(i)}$ perform, independently of each other, symmetric stable motions with index $\alpha \in (1,2]$ and "diffusion" constant κ carrying their action weights a_i with them. This results in a measure-valued Markov process $\Gamma = \{\Gamma(t); t \in \mathbb{R}\}$, the *catalyst process*. Note that the law of Γ is shift invariant in time and space. Recall that $1<\alpha\le 2$, $0<\beta\le 1$, and

$0<\gamma<1$.

PROPOSITION 1.18. *Let* α, β, *and* γ *as introduced above. If* $\alpha<2$ *holds, we additionally require that* $(\beta\gamma)^{-1}<1+\alpha$ *is fulfilled. Then with probability one,* Γ *is a branching rate kernel. Moreover, for each given* $J=(s',t)$, $s'<t$, *with probability one the restricted process* $\Gamma_J := \{\Gamma(r); r\in J\}$ *is an admissible restricted branching rate kernel.*

Combining both Proposition 1.18 and Theorem 1.14 we recognize that for almost all realizations Γ of the catalyst process the superprocess $X = [X, \mathbb{P}^{\Gamma}_{s,\mu}; s\in\mathbb{R}, \mu\in\mathcal{M}_f]$ with branching rate kernel Γ exists. By mixing over Γ we then get the probability laws $\mathcal{P}_{s,\mu} := \mathbb{E}\mathbb{P}^{\Gamma}_{s,\mu}$, $s\in\mathbb{R}$, $\mu\in\mathcal{M}_f$ corresponding to a *superprocess* X *in the random medium* Γ (which of course is no longer a Markov process). Our *main result* then reads as follows.

THEOREM 1.19. *Let* α,β,γ *be given as in Proposition* 1.18 *and* X *be the superprocess in the random medium* Γ. *If* $t\in\mathbb{R}$ *is a fixed time point, then the random measure* $X(t)$ *is absolutely continuous with* $\mathcal{P}_{s,\mu}$*-probability one, for all* $s<t$ *and* $\mu\in\mathcal{M}_f$.

We note that for the continuous critical super-Brownian motion with *constant* branching rate considered above, Konno and Shiga (1988) obtained a stronger result, namely that with probability one the absolute continuity property holds simultaneously for all times $t>0$.

It can be noted that if the motion of catalysts is allowed to have *oscillitory discontinuities*, then the admissibility conditions in Definition 1.10 may fail. In fact, consider the following simple counter example.

Example 1.20. Set $\rho(t) := \delta_{\sin(1/(1-t))}$ for $t\in J:=(0,1)$ and $\rho(t):\equiv 0$ otherwise. This is obviously a branching rate kernel. But for this J, condition (1.11) is violated on the set $(-1,+1)$ of positive Lebesgue measure. Indeed, for each $z\in(-1,+1)$, in J we find a sequence $r_n \longrightarrow 1$

such that $\sin(1/(1-r_n)) \equiv z$ holds. Then

$$(\rho(r_n), p^\beta(r_n,1,z,\cdot)) = p^\beta(1-r_n,0) = \text{const } (1-r_n)^{-\beta/\alpha}$$

(see Lemma A.14 in the Appendix) which goes to infinity as $n\to\infty$.

To prove Proposition 1.18 we will heavily exploit *scaling properties* of stable distributions. This is the main reason why from the beginning we restricted ourselves to an α-stable mass flow, to α-stable motions of the catalysts, and to $(1+\beta)$-continuous state branching.

However, the results should not depend on these special properties, because they are of a local nature. It is clear that certain perturbations can be allowed, for instance, the Laplacian could be replaced by a uniformly elliptic differential operator. The symmetric stable processes could also be replaced by more general infinitely divisible processes whose Lévy measures have a similar behavior near the origin.

The *plan of the paper* is as follows. First we mention that all theorems and propositions of the Introduction will be reformulated in the sequel. In Section 2 we start by proving the Basic Lemma 1.15 and introducing the function space $\mathbf{F}_+$ and measure space $\mathcal{M}^{\mathbf{F},\beta}$. Then in Theorem 3.5 a precise setting for equation (1.7) is given including basic solutions and some continuity properties. In Section 4 first an existence proof of the superprocess X is sketched. Then the absolute continuity result for a fixed admissible branching rate kernel ρ_J follows (Theorem 4.4). After providing some estimates involving stable flows and densities related to the interplay of both stable motion laws, the catalyst process Γ (including a simplified Poisson version) is introduced in Section 6. Its properties are derived in Section 7, ending up with our main absolute continuity result which is formulated in Theorem 7.14 for the superprocess in the random medium Γ. Comprehensive facts on the stable semi-group used in the present paper are compiled in an Appendix.

2. PRELIMINARIES

Before giving a more precise description of the model, we will prove the Basic Lemma.

Proof of the Basic Lemma 1.15. Assume that $[\Omega,\mathcal{F},P]$ is a probability space and that ν is a measurable map of $[\Omega,\mathcal{F}]$ into $[\mathcal{M}_f,\mathfrak{M}_f]$. Here $\mathfrak{M}_f$ is the smallest σ-algebra of subsets of $\mathcal{M}_f$, the set of all finite measures on $\mathbb{R}$, such that for each interval I the mapping $m \mapsto m(I)$ of $\mathcal{M}_f$ into $\mathbb{R}$ is measurable.

For each $\omega\in\Omega$, we can decompose the measure $\nu(\omega,dx)$ into its absolutely continuous and singular parts, $\nu_{ac}(\omega,dx)$ and $\nu_s(\omega,dx)$, respectively. Then again ν_{ac} and ν_s are measurable maps of $[\Omega,\mathcal{F}]$ into $[\mathcal{M}_f,\mathfrak{M}_f]$; see, for instance, Cutler (1984), Theorem 2.1.6.

Furthermore, for each $\omega\in\Omega$, the limit

$$\lim_{\varepsilon\to 0+} (1/2\varepsilon)\ \nu(\omega,[z-\varepsilon,z+\varepsilon]) =: \eta_{ac}(\omega,z) \tag{2.1}$$

exists for all $z\in\mathbb{R}\setminus N(\omega)$, where $N(\omega)$ is a Borel subset of $\mathbb{R}$ of Lebesgue measure 0, and $\eta_{ac}(\omega,\cdot)$ is a version of the Radon-Nikodym derivative of $\nu_{ac}(\omega,dx)$ with respect to Lebesgue measure; see, for instance, [8], Theorem III.12.6. Moreover, from the proof there, it can be seen that $\eta_{ac}=\{\eta_{ac}(\omega,z);\omega\in\Omega,z\in\mathbb{R}\setminus N(\omega)\}$ is measurable with respect to the σ-algebra $\mathcal{F}\otimes\mathcal{R}$ corresponding to $\Omega\times\mathbb{R}$.

Hence (2.1) holds almost everywhere with respect to the product measure $P(d\omega)dz$ on $\mathcal{F}\otimes\mathcal{R}$. In particular, for almost all z, the limit relation (2.1) is true with respect to convergence in distribution. Then by assumption (i) in the lemma, we conclude that $\eta_{ac}(\cdot,z)$ coincides in distribution with $\eta(z)$, for almost all $z\in\mathbb{R}$. Therefore, by the statement (ii) of the lemma,

$$\int P(d\omega)\int dz\ \eta_{ac}(\omega,z)f(z) = \int P(d\omega)\int \nu(\omega,dx)\ f(x)$$

holds for all $f\in F_+$. Thus, we obtain $E\nu_{ac} = E\nu$. But then the natural inequality $\nu_{ac}(\omega) \le \nu(\omega)$, $\omega\in\Omega$, is with probability one even an equality. Consequently, ν is absolutely continuous a.s., and the proof of the Basic Lemma is finished. □

We continue with some terminology. For constants $\kappa>0$ and $\alpha\in(1,2]$, let $S:=\{S_t;t\ge0\}$ denote the *contraction semigroup* of a symmetric *stable* Markov process on the real line $\mathbb{R}$ with index α and generator $\kappa\Delta_\alpha=-\kappa(-\Delta)^{\alpha/2}$ where Δ is the *one*-dimensional Laplacian. That process possesses continuous transition probability density functions

$$p(s,t,x,y)=p(t-s,y-x)=p_\alpha(\kappa(t-s),y-x), \quad s<t,\ x,y\in\mathbb{R},$$

with p_α taken from the Appendix. Note that we include $\alpha=2$, the case of a *Wiener process* with generator $\kappa\Delta$.

Let $\mathbf{F}$ denote the set of all real-valued continuous functions f on $\mathbb{R}$ with the property that there exist positive constants c and τ (possibly depending on f) such that $|f(x)| \le c\ p(\tau,x)$ holds for all $x\in\mathbb{R}$. We equip the linear space $\mathbf{F}$ with the supremum norm $\|\cdot\|_\infty$ of uniform convergence.

In other words, $\mathbf{F}$ contains all those continuous functions $f(x)$, $x\in\mathbb{R}$, which, as $x\to\infty$, have at least an exponential decay $c_1\exp(-c_2x^2)$ (for positive constants c_1 and c_2 possibly depending on f) provided that $\alpha=2$; otherwise that exponential decay has to be replaced by a potential decay $c|x|^{-1-\alpha}$, $c>0$; see Lemma A.8 (in the Appendix).

LEMMA 2.2. *The space* $\mathbf{F}$ *is closed with respect to convolutions. For* $f\in\mathbf{F}$ *and* $T>0$, *all functions* S_tf, $0\le t\le T$, *belong to* $\mathbf{F}$ *and are dominated by some* h *in* $\mathbf{F}$, *i.e.* $|S_tf|\le h$, $0\le t\le T$.

Proof. See, for instance, Dawson and Fleischmann (1990a), Examples 3.1 and 3.3. □

Fix a number $\beta\in(0,1]$. Let $\mathcal{M}^{\mathbf{F},\beta}$ denote the set of those (non-negative) measures μ defined on the σ-field of all Borel subsets of $\mathbb{R}$ for which (μ,f^β) is finite for all $f\in\mathbf{F}_+$. Here the lower index $+$ at a set A refers to the collection of all non-negative members of A.

We endow $\mathcal{M}^{F,\beta}$ with the coarsest topology such that, for each $f\in F_+$, the mapping $\mu\mapsto(\mu,f^\beta)$ of $\mathcal{M}^{F,\beta}$ into $\mathbb{R}$ will be continuous. Of course, $\mathcal{M}_f$ is a subset of $\mathcal{M}^{F,\beta}$.

3. BASIC SOLUTIONS OF THE UNDERLYING SINGULAR EQUATION

In this section we will deal with equation (1.7) in the setting needed in the present paper. To this end we may fix a finite nonempty open time interval $J:=(L,T)\subset\mathbb{R}$, and write $\underline{J}$ and $\overline{J}$ for $[L,T)$ and $[L,T]$, respectively. Let $F^{\underline{J}}$ denote the set of all continuous mappings u of $\underline{J}$ into F such that

$$\|u\|_J := \int_J ds\ \|u(s)\|_\infty < \infty.$$

We will look for solutions to (1.7) in the normed space $[F^{\underline{J}},\|\cdot\|_J]$.

Next we introduce *possible final states* for solutions. Let Θ denote the set of all finite measures ϑ defined on $\mathbb{R}$ which are either degenerate (i.e. concentrated at a point) or absolutely continuous with a density function h such that $h\le f$ for some $f\in F_+$ possibly depending on ϑ. We equip Θ with the topology of weak convergence.

In particular, for each $\varepsilon\in(0,1)$ the *uniform distribution* on the closed interval $[-\varepsilon^{1/\alpha},\varepsilon^{1/\alpha}]$ belongs to Θ. Its density function is denoted by $\varphi(\varepsilon)$.

First of all we shall deal with the trivial case in which the nonlinear term of equation (1.7) disappears. We will use the notation $\vartheta*h(x) := \int\vartheta(dy)h(x-y)$, $x\in\mathbb{R}$, and set

$$S^{\underline{J}}\vartheta(s) := \vartheta*p(T-s), \quad s\in\underline{J}.$$

LEMMA 3.1. $S^{\underline{J}}\vartheta$ *belongs to* $F_+^{\underline{J}}$, *for each* $\vartheta\in\Theta$, *and it continuously depends on* ϑ. *Moreover, in* Θ *we have the weak convergence* $S^{\underline{J}}\vartheta(s)(x)dx \longrightarrow \vartheta(dx)$ *as* $s\to T$.

Proof. Fix $\vartheta\in\Theta$. For each $s\in\underline{J}$, obviously $S^{\underline{J}}\vartheta(s)$ is a continuous function on $\mathbb{R}$. Since the continuous density functions $p(T-s)$ belong to F_+ and F_+ is closed with respect to convolutions, $S^{J}\vartheta(s)$ belongs to F_{+}. Moreover it continuously depends on s since the stable density functions p are uniformly continuous on $[\varepsilon,\infty)\times\mathbb{R}$, for each $\varepsilon>0$. Finally, $S^{J}\vartheta$ belongs to F_{+}^{J} because of

$$\|S^{J}\vartheta\|_{J} \le \text{const } \|\vartheta\| \int_{J} ds\,(T-s)^{-1/\alpha} < \infty \tag{3.2}$$

where $\|\vartheta\|$ denotes the total mass of ϑ, and Lemma A.14 was used.

By the way, here we exploited the assumption that $\alpha>1$, which is essential since we intend to deal with point catalysts (recall that a symmetric stable process with index α has a positive local time if and only if $\alpha>1$).

We now assume the weak convergence $\vartheta_n \longrightarrow \vartheta$ in Θ and consider

$$\int_0^{T-L} ds \sup_{x\in\mathbb{R}} \left| \int \vartheta_n(dy)p(s,x-y) - \int \vartheta(dy)p(s,x-y) \right| .$$

By estimates as in (3.2) we see that we may assume that s is bounded away from zero, i.e. we suppose $s\in[\varepsilon,T-L]$ for some $\varepsilon>0$. There the stable density functions are uniformly bounded (cf. Lemma A.14), and by the weak convergence $\vartheta_n \longrightarrow \vartheta$ we may additionally assume that y varies only in a bounded set. Finally, $p(s,x-y)$ converges to 0 as $x\to\infty$, uniformly for such s and y (cf. Lemma A.8), thus it suffices to take the supremum over a bounded set of x.

Now it is enough to show that for fixed s and each bounded sequence $\{x_n;n\ge1\}$ in $\mathbb{R}$

$$\left| \int \vartheta_n(dy)p(s,x_n-y) - \int \vartheta(dy)p(s,x_n-y) \right| \xrightarrow[n\to\infty]{} 0 \tag{3.3}$$

holds. Consider a subsequence of $\{x_n;n\ge1\}$. Then it has a further subsequence converging to some x. But by continuous convergence along the latter subsequence, both terms in (3.3) tend to $\langle\vartheta,p(s,x-\cdot)\rangle$ (see, for instance, [1], Theorem 5.5). This then implies the full convergence statement (3.3). Thus $S^{J}\vartheta$ continuously depends on ϑ.

Finally, the weak convergence $S^{J}\vartheta(s)(x)dx \longrightarrow \vartheta(dx)$ as $s\to T$ is also easy to see by considering such integrals

$\int dx\, S^{\underline{J}}\vartheta(s)(x)h(x)$ where h is any uniformly continuous bounded function on $\mathbb{R}$. □

Now we will also take into consideration the nonlinear term in the equation. To this end, let $\mathcal{K}(\bar{\underline{J}})$ denote the set of all kernels ξ of $\bar{\underline{J}}:=[L,T]$ into $\mathcal{M}^{F,\beta}$ such that $\xi(t,\cdot)$ belongs to $\mathcal{M}^{F,\beta}$ for all $t\in\bar{\underline{J}}$, and $\xi(\cdot,I)$ is a measurable function defined on $\bar{\underline{J}}$, for all intervals I in $\mathbb{R}$, as well as

$$\|\xi\|_f := \sup_{r\in\bar{\underline{J}}} (\xi(r), f^\beta) < \infty, \qquad f\in F_+ \tag{3.4}$$

is true. Our results on the equation will be collected in the following theorem. Recall that $\varphi(\varepsilon)$ introduced before Lemma 3.1 is the density function of the uniform distribution on some ε-neighborhood of the origin.

THEOREM 3.5. *Let* ξ *belong to* $\mathcal{K}(\bar{\underline{J}})$ *and* ϑ *to* Θ. *If* ϑ *is absolutely continuous, then there exists a unique element* $v:=V^{\underline{J}}[\vartheta,\xi]$ *in* $F_+^{\underline{J}}$ *which satisfies the integral equation*

$$v(s,x) = \vartheta * p(T-s)(x) - \int_s^T dr \int \xi(r,dy)\, p(s,r,x,y)\, |v|^{1+\beta}(r,y), \tag{3.6}$$

$s\in\underline{J}$, $x\in\mathbb{R}$. *If* ϑ *has an atom at* z *and*

$$\sup_{r\in J} (\xi(r), [\delta_z * p(T-r)]^\beta) < \infty \tag{3.7}$$

holds, then there exists at most one element $v:=V^{\underline{J}}[\vartheta,\xi]$ *in* $F_+^{\underline{J}}$ *which satisfies* (3.6). *If* $\{\varepsilon(n);n\ge1\}$ *is a sequence with* $\varepsilon(n)\in(0,1)$, $n\ge1$, *and converging to* 0 *as* $n\to\infty$, *and in addition to* (3.7)

$$\lim_{n\to\infty} \sup_{r\in J} (\xi(r), [\delta_z * p(T-r+\varepsilon(n))]^\beta) < \infty \tag{3.8}$$

is true (where z *is the position of the atom of* ϑ*), then there exists a solution* $v=V^{\underline{J}}[\vartheta,\xi]\in F_+^{\underline{J}}$ *to* (3.6), *and the convergence*

$$v_n := V^{\underline{J}}[\vartheta * \varphi(\varepsilon(n))(x)dx, \xi] \xrightarrow[n\to\infty]{} V^{\underline{J}}[\vartheta,\xi] = v \tag{3.9}$$

takes place in F_+^J. *Finally, we have*

$$v_n(L,\mathbb{b}) \xrightarrow[n\to\infty]{} v(L,\mathbb{b}) \tag{3.10}$$

for all $\mathbb{b}\in\mathbb{R}$ *satisfying*

$$\sup_{r\in J} (\xi(r), \delta_{\mathbb{b}}*p(r-L)) < \infty. \tag{3.11}$$

Note that (3.6) can formally be written as in (1.7), with final condition expressed by the weak convergence

$$v(s,x)dx \longrightarrow \vartheta(dx) \quad \text{as} \quad s\to T.$$

Proof. Fix $\vartheta\in\Theta$ and $\xi\in\mathcal{K}(\bar{J})$. In order to prove uniqueness, assume that in F_+^J we have two solutions v_1 and v_2 of (3.6) which correspond to these data, i.e. we have

$$v_i(s) = \vartheta*p(T-s) - \int_s^T dr\int\xi(r,dy)p(s,r,\cdot,y)|v_i|^{1+\beta}(r,y),$$

$i=1,2$. Applying Lemma A.14, the following elementary inequality

$$|a^{1+\beta} - b^{1+\beta}| \le (a^\beta+b^\beta)|a-b|, \quad a,b \ge 0, \tag{3.12}$$

and

$$0 \le v_i(s) \le \vartheta*p(T-s), \quad s\in\underline{J}, \quad i=1,2, \tag{3.13}$$

for the following nonempty subinterval $\underline{J}'=[L',T)$ of $\underline{J}$ we get

$$0 < \|v_1-v_2\|_{J'} \le 2\int_{J'}ds \int_s^T dr \int\xi(r,dy)(r-s)^{-1/\alpha}$$
$$[\vartheta*p(T-r)]^\beta(y) \ \|v_1(r)-v_2(r)\|_\infty.$$

By a change of order of integration we may continue with

$$\le \text{const} \ \|v_1-v_2\|_{J'} \ (T-L')^{1-1/\alpha} \sup_{r\in J} (\xi(r),[\vartheta*p(T-r)]^\beta). \tag{3.14}$$

First we assume that ϑ is *degenerate* and has an atom at z. Then by (3.7) the latter supremum term is finite. Moreover, by the estimate (3.13) and Lemma 3.1 the norm expression is finite, too. Hence, since $\alpha>1$, for L' sufficiently close to T we get the contradiction

$\|v_1-v_2\|_{J'} < \|v_1-v_2\|_{J'}$ unless $\|v_1-v_2\|_{J'}=0$. In other words, for a degenerate ϑ and on a sufficiently small interval $\underline{J}'$ we get *uniqueness*.

Now we will prepare for the corresponding existence proof. For $\varepsilon\in(0,1)$ we consider the function $\vartheta * p(\varepsilon) =: \vartheta_\varepsilon$. By Lemma 3.1, it belongs to F_+, and it determines a measure in Θ which we denote by the same symbol ϑ_ε.

From the Existence Theorem 2.6 in Dawson and Fleischmann (1990a) we know that with probability one to each ϑ_ε there exists a continuous mapping v_ε of $\overline{\underline{J}}=[L,T]$ into F_+ which solves (3.6) on $\overline{\underline{J}}$. In fact, by time reversibility of the stable semigroup S and time reversibility of the condition (3.4), the forward formulation of the equation in [3] can easily be transferred to the backward formulation in the present paper.

Obviously, v_ε restricted to $\underline{J}$ (which we denote by the same symbol) belongs to $F_+^{\underline{J}}$. We will now apply these constructions to the sequence $\{\varepsilon(n):n\geq 1\}$ of the theorem. Our next task is to prove that

$$(3.15) \qquad \lim_{m,n\to\infty} \|v_{\varepsilon(m)} - v_{\varepsilon(n)}\|_{J'} = 0$$

for each sufficiently small subinterval $\underline{J}'=[L',T)$ of $\underline{J}$. From equation (3.6), for $s\in\underline{J}'$ and $x\in\mathbb{R}$ we get

$$|v_{\varepsilon(m)} - v_{\varepsilon(n)}|(s,x) \leq \|S^{\underline{J}}\vartheta_{\varepsilon(m)} - S^{\underline{J}}\vartheta_{\varepsilon(n)}\|_J$$
$$+ \int_s^T dr \int \xi(r,dy)p(s,r,x,y)\left|v_{\varepsilon(m)}^{1+\beta}(r,y) - v_{\varepsilon(n)}^{1+\beta}(r,y)\right|.$$

Applying again Lemma A.14, (3.12), and (3.13), as in the estimate (3.14) we obtain

$$(3.16) \qquad \|v_{\varepsilon(m)} - v_{\varepsilon(n)}\|_{J'} \leq \|S^{\underline{J}}\vartheta_{\varepsilon(m)} - S^{\underline{J}}\vartheta_{\varepsilon(n)}\|_J$$
$$+ \text{const}\ \|v_{\varepsilon(m)} - v_{\varepsilon(n)}\|_{J'}\ (T-L')^{1-1/\alpha}$$
$$\sup_{r\in J}\left(\xi(r),[\vartheta_{\varepsilon(m)} * p(T-r)]^\beta + [\vartheta_{\varepsilon(n)} * p(T-r)]^\beta\right).$$

If ϑ has an atom at z, then by Lemma 3.1 and assumption (3.8) we deduce (3.15) for sufficiently small $\underline{J}'$.

Now we will complete the *existence proof*. By the assertion (3.15), $\{v_{\varepsilon(n)};n\geq 1\}$ is a Cauchy sequence in $F_+^{\underline{J}'}$. However by construction, $F_+^{\underline{J}'}$ coincides with the Banach space $L^1[\underline{J}',F_+,ds]$ but restricted to continuous functions. Hence, $v_{\varepsilon(n)}$ converges in $L^1[\underline{J}',F_+,ds]$ to some limit $v\geq 0$ as $n\to\infty$. If again ϑ has an atom at z, using condition (3.8) and proceeding as in the derivation of (3.14) or (3.15), we conclude that

$$\int_{(\cdot)}^{T} dr\int\xi(r,dy)p(\cdot,r,\cdot,y)v_{\varepsilon(n)}^{1+\beta}(r,y)$$

$$\longrightarrow \int_{(\cdot)}^{T} dr\int\xi(r,dy)p(\cdot,r,\cdot,y)v^{1+\beta}(r,y)$$

also holds in $L^1[\underline{J}',F_+,ds]$ as $n\to\infty$. Noting that

$$s \longmapsto \int_s^T dr\int\xi(r,dy)p(s,r,\cdot,y)v^{1+\beta}(r,y)$$

is a continuous mapping of $\underline{J}'$ into F_+, and combining this with Lemma 3.1 we get that v is a continuous element in $L^1[\underline{J}',F_+,ds]$ which solves (3.6) on $\underline{J}'$. This gives the *existence claim* in the case of a degenerate ϑ and a sufficiently small interval $\underline{J}'$.

So far we proved uniqueness and existence on $\underline{J}'$ for degenerate ϑ. If now ϑ is *absolutely continuous* with density function $h\leq f'\in F_+$, then the supremum in the estimate (3.14) can be bounded above by

$$\leq \text{const} \sup_{r\in J} (\xi(r),[f'*p(T-r)]^{\beta}) \leq \text{const} \|\xi\|_f < \infty$$

where $f\in F_+$ is a dominating function for

$$f'*p(T-r)=S_{T-r}f', \quad r\in J$$

(and the norm $\|\cdot\|_f$ was defined in (3.4)). Such f actually exists by Lemma 2.2. Hence the uniqueness proof carries over to such ϑ.

For the same reasons, the supremum in (3.16) is finite, uniformly in m and n. Therefore the existence proof also remains valid for absolutely continuous ϑ.

Summarizing, for sufficiently small intervals $\underline{J}'$ uniqueness and existence hold, and in this case we turn to the *continuity assertion* (3.9).

Recall that $q(\varepsilon)$ is the density function of a uniform distribution. Let u_n denote the solution corresponding to $\vartheta * q(\varepsilon(n))$. Since

$$(3.17) \qquad q(\varepsilon) \le \text{const } p(\varepsilon), \quad 0<\varepsilon<1$$

is true (which follows from a simple scaling argument), we may proceed as in the proof of (3.15) to show that

$$\|u_n - v\|_{J'} \longrightarrow 0 \quad \text{as} \quad n\to\infty$$

holds where $v=V^{\underline{J}}[\vartheta,\xi]$, for both choices of ϑ (i.e. degenerate or absolutely continuous measure etc.).

In summary, if for the moment we exclude (3.10), then all assertions in the theorem hold, provided we replace $\underline{J}$ by a sufficiently small subinterval $\underline{J}'=[L',T)$.

Since the bounds used do not depend on $\underline{J}'$, an extension of the proved assertions from $\underline{J}'$ to the whole interval $\underline{J}$ can be established by the usual iteration scheme. Note, in particular, that $v(L')$ which will serve as the final state of the next iteration step, determines an absolutely continuous measure in Θ with density function in F_+. Therefore, the conditions (3.7) and (3.8) are only needed for the initial step of iteration.

For a *proof of* (3.10) we write $\vartheta=\lambda\delta_z$ and take a $\mathscr{b}$ satisfying (3.11). From (3.6), (3.12), (3.13), and (3.17) we get

$$|v_n(L,\mathscr{b})-v(L,\mathscr{b})| \le \lambda|\delta_z * q(\varepsilon(n)) * p(T-L)(\mathscr{b}) - \delta_z * p(T-L)|$$

$$+ \text{ const} \int_L^T dr \int \xi(r,dy) p(L,r,\mathscr{b},y)\,|v_n-v|(r,y)$$

$$[p^\beta(T-r+\varepsilon(n),y-z) + p^\beta(T-r+,y-z)].$$

Clearly, the first summand at the right hand side of this inequality approaches 0 by the weak convergence of $q(\varepsilon(n))(x)dx$ to $\delta_0(dx)$, as $n\to\infty$. The second summand may be estimated above by $\le$ const $\|v_n-v\|_J$ times the expression

$$\sup_{r\in J}\Big(\xi(r),\delta_{\mathscr{b}} * p(r-L)[p^\beta(T-r+\varepsilon(n),\cdot-z)+p^\beta(T-r+,\cdot-z)]\Big).$$

To show the boundedness of the latter term, we fix a time point $s\in J$. Then by Lemma A.14,

$$\delta_{\&}*p(r-L)(x) \le \text{const } (s-L)^{-1/\alpha} = \text{const}, \quad r\in[s,T),\ x\in\mathbb{R},$$

and we may apply (3.7) and (3.8). But analogously we can proceed on the remaining interval (L,s) by using (3.11).

Summarizing, the second summand may be estimated above by $\le$ const $\|v_n-v\|_J$ which by (3.9) converges to zero as $n\to\infty$.

This shows (3.10) and completes the proof of the theorem. □

4. SUPERPROCESS WITH ABSOLUTELY CONTINUOUS STATES

Let $\mathcal{K}(\mathbb{R})$ denote the set of all measurable kernels ρ of $\mathbb{R}$ into $\mathcal{M}^{\mathbf{F},\beta}$ (this set of measures was defined at the end of Section 2) such that (1.6) holds. In other words, $\mathcal{K}(\mathbb{R})$ is the set of all kernels of $\mathbb{R}$ into $\mathcal{M}^{\mathbf{F},\beta}$ such that their restrictions to any finite closed interval $\bar{J}=[s,t]$ belong to $\mathcal{K}(\bar{J})$.

For instance, if $\rho(r)\equiv\nu$ for a measure ν in $\mathcal{M}^{\mathbf{F},\beta}$, then ρ belongs to $\mathcal{K}(\mathbb{R})$.

Actually, each ρ in $\mathcal{K}(\mathbb{R})$ may serve as a *branching rate kernel* for a superprocess. (Recall that $\mathcal{M}_f$ is the set of all finite measures defined on $\mathbb{R}$.)

PROPOSITION 4.1. *To each* ρ *in* $\mathcal{K}(\mathbb{R})$, *there exists an* $\mathcal{M}_f$*-valued time-inhomogeneous superprocess* $X = [X,\mathbb{P}^{\rho}_{s,\mu};s\in\mathbb{R}_+,\mu\in\mathcal{M}_f]$ *with Laplace transition functional*

(4.2) $$\mathbb{E}^{\rho}_{s,\mu}\exp(X(t),-\varphi) = \exp(\mu,-v(s,t)),$$
$s<t$, $\mu\in\mathcal{M}_f$, $\varphi\in\mathbf{F}_+$, *where* $v(\cdot,t)=V^{\underline{J}}[\vartheta,\xi]$ *is the unique solution to equation* (3.6) *with* $\underline{J}=[s,t)$, $\vartheta(dx) = \varphi(x)dx$, *and* $\xi=\{\rho(r);r\in\bar{J}\}$.

Moreover, we have the following expectation formula:

(4.3) $$\mathbb{E}^{\rho}_{s,\mu}(X(t),\varphi) = (\mu,S_{t-s}\varphi), \quad s\le t,\ \varphi\in\mathbf{F}_+.$$

Sketch of Proof (for details we refer to Dawson and Fleischmann (1990b)). First of all we assume that ρ is absolutely continuous, i.e. $\rho(r,dx) = h(r,x)dx$, $x\in\mathbb{R}$, but where the measurable density function h on $\mathbb{R}\times\mathbb{R}$ is even bounded. Then there exists a time-inhomogeneous superprocess $X = [X, \mathbb{P}^{\rho}_{s,\mu}; s\in\mathbb{R}_+, \mu\in\mathcal{M}_f]$ with Laplace transition functional (4.2). See Dawson and Perkins (1990); compare also Fitzsimmons (1988) and (1989) for the time-homogeneous case.

To deal with a general $\rho\in\mathcal{K}(\mathbb{R})$, fix an interval $\underline{J}=[s,t)$, $s<t$. Then we will use continuity properties of solutions to equation (3.6) (with $\vartheta(dx) = \varphi(x)dx$ and $\xi=\{\rho(r); r\in\underline{J}\}$) as described in Dawson and Fleischmann (1990a, Theorems 2.11 and 2.13). There it was shown that under certain conditions the solutions $V^{\underline{J}}[\vartheta,\xi]$ to (3.6) depend continuously on ξ. Thus we can obtain them as the limit of a sequence $V^{\underline{J}}[\vartheta,\xi_n]$ where ξ_n, $n\geq1$, are approximations of ξ which are absolutely continuous with bounded density kernels as above. By dominated convergence then the corresponding right hand sides of (4.2) converge, and the limit will again be a Laplace functional. Since $\underline{J}$ is arbitrary, in this way we get Laplace transition functionals, which determine a time-inhomogeneous Markov process X with the desired properties.

The expectation formula (4.3) follows by a similar approximation procedure (or formally by differentiation as in the moment calculation in the Introduction). This finishes this sketch of the proof. □

Now we are in a position to formulate our absolute continuity result for a fixed admissible restricted branching rate kernel. Recall that $1<\alpha\leq2$ and $0<\beta\leq1$.

THEOREM 4.4. *Let* $\rho\in\mathcal{K}(\mathbb{R})$ *and* $[X, \mathbb{P}^{\rho}_{s,\mu}; s\in\mathbb{R}, \mu\in\mathcal{M}_f]$ *be a superprocess with branching rate kernel* ρ, *according to Proposition 4.1. Fix* $J=(s',t)$, $s'<t$, *and let the restricted kernel* ρ_J *be admissible as described in Definition 1.10. Then with respect to* $\mathbb{P}^{\rho}_{s,\mu}$, $s<s'$, $\mu\in\mathcal{M}_f$, *the random measure* $X(t)$ *is absolutely continuous a.s., that is, there exists a random density function* $\eta(t)$ *such that*

$$\mathbb{P}^{\rho}_{s,\mu}\Big\{X(t,dy) = \eta(t,y)dy\Big\} = 1.$$

Proof. Recall that $\varphi(\varepsilon)$ denotes the density function of a uniform distribution on some interval around the origin, as defined before Lemma 3.1.

Consider ρ, J, ρ_J and s, μ as in the theorem. Choose an exceptional set $N(\rho_J)$ for ρ_J and sequences

$$\underline{\varepsilon}(z) = \{\varepsilon_n(z) \in (0,1); n \geq 1\}, \qquad z \in \mathbb{R} \setminus N(\rho_J),$$

according to the Definition 1.10.

By the expectation formula (4.3), we get $\mathbb{E}^{\rho}_{s,\mu} X(s') = S_{s'-s}\mu$ which is an absolutely continuous measure. Hence, $X(s', N(\rho_J)) = 0$ with $\mathbb{P}^{\rho}_{s,\mu}$-probability one, because $N(\rho_J)$ is a Lebesgue zero set. By the Markov property it is therefore enough to show that $X(t)$ is absolutely continuous with $\mathbb{P}^{\rho}_{s',\mu}$-probability one, for all $\mu \in \mathcal{M}_f$ satisfying $\mu(N(\rho_J)) = 0$. We fix such a μ, and to simplify the notation we will write s instead of s'.

For $z \in \mathbb{R} \setminus N(\rho_J)$ and $\lambda \geq 0$, by (3.10) in Theorem 3.5, for

$$v^{\underline{J}}[\lambda \delta_z * \varphi(\varepsilon_n(z))(y)dy, \xi] = v_n \quad \text{and} \quad v^{\underline{J}}[\lambda \delta_z, \xi] = v_0$$

where ξ is the restriction of the branching rate kernel ρ to $\underline{J}$, we get

$$v_n(s, \mathscr{b}) \xrightarrow[n \to \infty]{} v_0(s, \mathscr{b}), \qquad \mathscr{b} \in \mathbb{R} \setminus N(\rho_J),$$

since (3.7), (3.8), and (3.11) are fulfilled (see the conditions (1.11), (1.13), and (1.12), respectively). By our assumption on μ and dominated convergence this implies

$$(\mu, v_n(s)) \xrightarrow[n \to \infty]{} (\mu, v_0(s)),$$

for all $\lambda \geq 0$. In fact, by (3.13), (3.17), and Lemma A.14, for all $n \geq 0$ (where we set $\varepsilon_0(z) \equiv 0$),

$$v_n(s) \leq \lambda \delta_z * p(\varepsilon_n(z) + t - s) \leq \text{const}\ \lambda (t-s)^{-1/\alpha}$$

which is a finite constant for the fixed λ, t, s. Using again this domination, we conclude that

$$0 \leq (\mu, v_0(s)) \leq \lambda (\mu, \delta_z * p(t-s)) \longrightarrow 0 \quad \text{as} \quad \lambda \to 0.$$

Therefore by Proposition 4.1 there exists a random variable $\eta(z) \geq 0$ such that

$$\mathbb{E}^{\rho}_{s,\mu} \exp(X(t), -\lambda \delta_z * \varphi(\varepsilon_n(z)))] = \exp(\mu, -v_n(s))$$

$$\xrightarrow[n\to\infty]{} \exp\langle\mu,-v_0(s)\rangle = \mathrm{E}\exp[-\lambda\eta(z)], \quad \lambda\ge 0,$$

holds. In other words, we have the convergence in distribution

$$(4.5) \qquad \langle X(t),\delta_z*\varphi(\varepsilon_n(z))\rangle \xrightarrow[n\to\infty]{\mathcal{D}} \eta(z),$$

for each $z\in\mathbb{R}\setminus N(\rho_J)$. According to the Basic Lemma 1.15, now it suffices to show that

$$\int dz\ \mathrm{E}\eta(z)f(z) = \mathbb{E}^{\rho}_{s,\mu}\langle X(t),f\rangle, \quad f\in F_+,$$

is true. But by (4.3), the right hand side coincides with $\langle S_{t-s}\mu,f\rangle$. Hence it is enough to prove that

$$\mathrm{E}\eta(z) = S_{t-s}\mu(z), \quad z\in\mathbb{R}\setminus N(\rho_J)$$

is valid. Now taking expectations in the convergence relation (4.5) and using the expectation formula (4.3) in Proposition 4.1 we get

$$(4.6) \qquad \mathbb{E}^{\rho}_{s,\mu}\langle X(t),\delta_z*\varphi(\varepsilon_n(z))\rangle = \langle\mu,\delta_z*\varphi(\varepsilon_n(z))*p(t-s)\rangle$$
$$\xrightarrow[n\to\infty]{} \langle\mu,\delta_z*p(t-s)\rangle \ge \mathrm{E}\eta(z).$$

On the other hand, by Jensen's inequality, for $\lambda>0$,

$$\exp\langle\mu,-v(s)\rangle \ge \exp[-\lambda\mathrm{E}\eta(z)].$$

Hence, by equation (3.6) and the estimate (3.13)

$$(4.7) \qquad \lambda\mathrm{E}\eta(z) \ \ge\ \langle\mu,\lambda\delta_z*p(t-s)\rangle$$
$$- \lambda^{\beta+1}\int\mu(dx)\int_s^t dr\int\xi(r,dy)p(s,r,x,y)p^{1+\beta}(t-r,z-y).$$

But the latter integral term may be estimated above by

$$\le \text{const} \int_s^t dr(r-s)^{-1/\alpha}(t-r)^{-1/\alpha} \sup_{s'\in J} \langle\xi(s'),p^{\beta}(t-s',z-\cdot)\rangle.$$

Since $\alpha>1$ and by (1.11), the latter expression is finite. In (4.7) we divide by λ and let λ tend to 0. Then together with the estimate (4.6) we are done. □

5. SOME ESTIMATES INVOLVING STABLE FLOWS AND DENSITIES

In this section we will collect some technical details later needed for catalyst processes.

Let S' be defined as S in Section 2, except replacing $\kappa>0$ by $\kappa'\geq 0$. We pay attention only to the cases $\kappa'=0$ and $\kappa'=\kappa$. (The former case will concern non-moving catalysts.)

Consider a constant $\gamma\in(0,1]$. If $\alpha<2$ holds, we additionally require that $(\beta\gamma)^{-1}<1+\alpha$. This condition guarantees that all functions f in F_+ are *$\beta\gamma$-fold integrable*, i.e. that $f^{\beta\gamma}$ is integrable with respect to Lebesgue measure.

LEMMA 5.1. *For* $K>0$, *the function*

$$x \mapsto \sup\left\{S'_s p^\beta(t,x)\,;\ 0\leq s\leq K,\ 0<t\leq K\right\}, \qquad x\neq 0,$$

is finite. Moreover, it is γ-fold integrable on the set $\{x;|x|>1\}$, *and if additionally* $\beta\gamma<1$ *holds, then it is also integrable on* $\{x;|x|<1\}$.

Proof. By Jensen's inequality

$$S'_s p^\beta(t,x) \leq [S'_s p(t,x)]^\beta. \tag{5.2}$$

But

$$S'_s p(t,x) = \begin{cases} p(t,x) & \text{if} \quad \kappa'=0 \\ p(s+t,x) & \text{if} \quad \kappa'=\kappa. \end{cases}$$

Hence, (5.2) can be continued with

$$\leq \sup_{0<r\leq 2T} p^\beta(r,x).$$

Then the statement directly follows from Lemma A.13 (with β' replaced by γ). □

LEMMA 5.3. *Under* $\beta<1$, *for* $K,T>0$ *the function*

$$x \mapsto \sup_{0\leq r\leq K} \int_0^T ds\ S_s\left|\Delta_\alpha p^\beta(r+T-s) + \frac{\partial}{\partial s}p^\beta(r+T-s)\right|(x), \qquad x\neq 0$$

is finite and γ-fold integrable.

Proof. For $x\neq 0$, we consider the integral

$$\int_0^T ds \int dy\ p(s,y-x)\,|\Delta_\alpha p^\beta + \frac{\partial}{\partial s}p^\beta|\,(r+T-s,y). \tag{5.4}$$

If we restrict the integration to $|y-x| \geq |x|/2$, then we get

$$\leq \int_0^T ds\ p(s,x/2)\ \|\Delta_\alpha p^\beta(r+T-s) + \frac{\partial}{\partial s}p^\beta(r+T-s)\|_1$$

where $\|\cdot\|_1$ denotes the L^1-norm. In view of (A.3),

$$p(s,x) \le \text{const } s^{-1/\alpha} p(T,x), \quad 0<s\le T,\ x\in\mathbb{R}.$$

On the other hand, by Lemma A.30, the norm expression can be estimated above by

$$\le \text{ const } (r+T-s)^{-1+(1-\beta)/\alpha} \le \text{ const } (T-s)^{-1+(1-\beta)/\alpha},$$

since $\alpha>1$ by assumption. Because we supposed $\beta<1$,

$$\int_0^T ds\ s^{-1/\alpha}(T-s)^{-1+(1-\beta)/\alpha} < \infty.$$

But $p(T,x)$ is finite, too, and γ-fold integrable, since $(\beta\gamma)^{-1}<1+\alpha$ implies that $\gamma^{-1}<1+\alpha$.

Now we restrict the integral (5.4) to $|y-x|<|x|/2$ which gives

$$(5.5) \qquad |x|/2 < |y| < 3|x|/2.$$

First of all, if additionally $|x|\ge 1$ is true, then by the Lemmas A.28 and A.22

$$|\Delta_\alpha p^\beta + \tfrac{\partial}{\partial s} p^\beta|(r+T-s,y) \le \text{ const } (r+T-s)^{-1} p^\beta(r+T-s,x/4),$$

and the restricted integral may be estimated to be

$$\le \text{ const } \int_0^{T+K} ds\ s^{-1} p^\beta(s,x/4)$$

$$= \text{ const } |x|^{-\beta} \int_0^{(T+K)|x|^{-\alpha}} ds\ s^{-1} p^\beta(s,1/4),$$

where we used (A.1). But if $\alpha<2$, by Lemma A.8 the latter inequality can be continued with

$$\le \text{ const } |x|^{-\beta(1+\alpha)},$$

which is finite and γ-fold integrable on $|x|\ge 1$. On the other hand, for $\alpha=2$ we also get a finite and γ-fold integrable bound.

Now assume $0<|x|<1$. By Lemma A.6 (with $K=|x|^{-1}$), for (5.4) restricted to (5.5) we can write

$$\int_0^T ds \int_{1/2<|y|<3/2} dy\ |x|\ p(s,|x|y-x)\ |x|^{-\alpha-\beta}$$

$$|\Delta_\alpha p^\beta + \tfrac{\partial}{\partial s} p^\beta|(|x|^{-\alpha}(r+T-s),y).$$

Using (A.1) we continue with

$$(5.6) \qquad \le |x|^{-\beta} \int_0^{T|x|^{-\alpha}} ds \int_{1/2<|y|<2} dy \; p(s, y-x|x|^{-1})$$

$$|\Delta_\alpha p^\beta + \tfrac{\partial}{\partial s} p^\beta|(|x|^{-\alpha}(r+T)-s, y).$$

$$\le |x|^{-\beta} \int_0^{(r+T)|x|^{-\alpha}} ds \int_{1/2<|y|<2} dy$$

$$p(|x|^{-\alpha}(r+T)-s, y-x|x|^{-1})\,|\Delta_\alpha p^\beta + \tfrac{\partial}{\partial s} p^\beta|(s,y).$$

Since y is bounded away from 0 and ∞, by all the Lemmas A.28, A.22, and A.30 we get

$$|\Delta_\alpha p^\beta + \tfrac{\partial}{\partial s} p^\beta|(s,y) \le h(s) := \begin{cases} \text{const } s^{\beta-1} & \text{if } 0<s<1 \\ \text{const } s^{-1-\alpha/\beta} & \text{if } s\ge 1. \end{cases}$$

Hence (5.6) may be estimated to

$$\le \text{const } |x|^{-\beta} \int_0^\infty ds\, h(s) = \text{const } |x|^{-\beta},$$

which is finite and γ-fold integrable around the origin. This ends the proof. □

6. CATALYST PROCESSES

Here we introduce some catalyst processes Γ, for details we refer to Dawson and Fleischmann (1990a), Sections 4 and 5.

The random quantities appearing in the following are all defined on some common probability space $[\Omega, \mathcal{F}, P]$. Recall that $\kappa'=0$ or $\kappa'=\kappa>0$.

Let $w^x := \{w^x(t); t\in\mathbb{R}\}$, $x\in\mathbb{R}$, be a *family of independent symmetric stable Markov processes* with generator $\kappa'\Delta_\alpha$, which at time $t=0$ go through the site $x\in\mathbb{R}$, i.e. $w^x(0)=x$, and having trajectories in $D[\mathbb{R},\mathbb{R}]$. Here $D[\mathbb{R},A]$ denotes the space of all functions of $\mathbb{R}$ into a topological space A which are right continuous and have left limits.

Recall that γ is a given parameter satisfying $0<\gamma\le 1$. If $\gamma=1$ holds, we consider a *Poisson random point measure* $\Gamma(0) = \sum_{i=1}^\infty \delta_{x(i)}$ on $\mathbb{R}$ with uniform density, determined by its Laplace functional

(6.1) $$\mathrm{E}\exp(\Gamma(0),-f) = \exp\left[\int dx\ [e^{-f(x)}-1]\right], \qquad f\in F_+ .$$

We assume that $\Gamma(0)$ is independent of the family $w := \{w^x; x\in\mathbb{R}\}$. Setting

$$\Gamma(t) := \sum_{i=1}^{\infty} \delta_{w^{x(i)}(t)}, \qquad t\in\mathbb{R},$$

we get a point measure-valued Markov process Γ.

Alternatively, if $\gamma<1$, again independently of w, consider a *stable random measure* $\Gamma(0)$ *with index* γ determined by the Laplace functional (1.17). As in the case of the Poisson point measure, this random measure $\Gamma(0)$ has *independent increments*. With probability one it can be represented as

$$\sum_{i=1}^{\infty} a_i\delta_{x(i)}, \qquad x(i)\neq x(j) \quad \text{for} \quad i\neq j.$$

We stress the fact that the supporting set $\{x(i); i\geq 1\}$ is now *dense* in $\mathbb{R}$. Finally, also in contrast to the Poisson point measure, $\Gamma(0)$ has *infinite asymptotic density*, i.e. $K^{-1}\Gamma(0,[-K,K]) \longrightarrow \infty$ as $K\to\infty$ with probability one. In this case

$$\Gamma(t) := \sum_{i=1}^{\infty} a_i\delta_{w^{x(i)}(t)}, \qquad t\in\mathbb{R}$$

yields a measure-valued Markov process Γ.

In both cases, $\gamma=1$ and $\gamma<1$, we call Γ a *catalyst process*. It describes a *random system of point catalysts moving independently according to* α*-stable processes*. Recall that the process Γ is defined on some basic probability space $[\Omega,\mathcal{F},P]$.

LEMMA 6.2. *The catalyst process* Γ *is (in distribution) stationary in time and space. With* P*-probability one the following expectation formula holds:*

$$\mathrm{E}\left\{(\Gamma(t),f)\,\middle|\,\Gamma(s)\right\} = (\Gamma(s),S'_{t-s}f), \qquad s\leq t,\ f\in F_+ .$$

Here stationary means that $\Gamma(r+\cdot,y+\cdot)$ has the same distribution as Γ, for all $r,y\in\mathbb{R}$.

Finally we quote the following result. Recall that we required $(\beta\gamma)^{-1}<1+\alpha$ if $\alpha<2$.

LEMMA 6.3. *The process* Γ *can be realized in* $D[\mathbb{R},\mathcal{M}^{F,\beta}]$, *and with probability one* Γ *is a branching rate kernel, i.e.* (1.6) *is satisfied.*

Remark 6.4. From the construction of solutions to (3.6) in the case of absolutely continuous final states ϑ provided in [3], Theorems 2.6, 2.11, and 2.13, it can be verified that the mapping $\rho \mapsto V^{\underline{J}}[\vartheta,\rho_{\underline{J}}]$ of $D[\mathbb{R},\mathcal{M}^{F,\beta}]$ into $F_{+}^{\underline{J}}$ is measurable in an appropriate sense, for each choice of J.

Then from the construction of our superprocess X (cf. Proposition 4.1) it can be shown that the map $\rho \mapsto \mathbb{P}^{\rho}_{s,\mu}$ is measurable in an appropriate sense, for each $s\in\mathbb{R}$ and $\mu\in\mathcal{M}_f$. This measurability property will be used below for defining the superprocess in a random medium.

7. FURTHER PROPERTIES OF THE CATALYST PROCESSES

First we recall that we assume $(\beta\gamma)^{-1}<1+\alpha$ in the case $\alpha<2$.

LEMMA 7.1. *With* P-*probability one,*

$$(7.2)\qquad \int\Gamma(0,dx)\ \sup\left\{S'_s p^{\beta}(t,x);\ 0\le s\le K,\ 0<t\le K\right\} < \infty$$

for all K>0. *Similarly, if* $\beta<1$, *for fixed* T>0 *with* P-*probability one,*

$$(7.3)\qquad \int\Gamma(0,dx)\ \sup_{0\le r\le K}\int_0^T ds\ S_s\left|\Delta_\alpha p^{\beta}(r+T-s) + \frac{\partial}{\partial s}p^{\beta}(r+T-s)\right|(x)$$

is finite, for all K>0.

Proof. First, by monotinicity in K, we may assume that K is fixed.

If in (7.2) we additionally introduce the indicator function $1\{|x|>1\}$, then by Lemma 5.1 the new integrand will be γ-fold integrable with respect to Lebesgue measure. Hence, from the formulas (6.1) and (1.17) (which can be extended to more general non-negative functions) we know that this restricted integral is finite a.s.

On the other hand, assume in addition that $|x|\le 1$. If $\gamma=1$, then with probability one the Poisson system $\Gamma(0)$ restricted to $\{|x|\le 1\}$ has finitely many points different from 0. Then by Lemma 5.1, the restricted integral is finite.

If $\gamma<1$, by Lemma 5.1 the integrand in (7.2) is γ-fold integrable on $\{|x|\le 1\}$ with respect to Lebesgue measure. Then we can employ (1.17) to get the a.s. finiteness of the integral in (7.2) restricted to $|x|\le 1$.

Thus, the assertion (7.2) is proved. Using Lemma 5.3, the proof to (7.3) is even simpler. □

Now we restrict our consideration to the fixed finite half-open interval $\underline{J}=[0,T)$. Recall that $\beta\le 1$.

LEMMA 7.4. *Fix* $r\ge 0$. *Given* $\Gamma(0)$,

$$M_t^r := (\Gamma(t),p^\beta(r+T-t)) - (\Gamma(0),p^\beta(r+T)) - \int_0^t ds\,(\Gamma(s),[\kappa\Delta_\alpha + \tfrac{\partial}{\partial s}]p^\beta(r+T-s)), \quad t\in\underline{J},$$

where the integral term must be deleted in the case $\beta=1$, *is a right continuous* $P\{\cdot|\Gamma(0)\}$*-martingale with respect to the filtration* $\mathcal{F}_t:=\sigma\{\Gamma(s);0\le s\le t\}$, $t\in\underline{J}$.

Proof. For $t\in\underline{J}$, by Lemma 6.2,

$$(7.5)\qquad E\Big\{(\Gamma(t),p^\beta(r+T-t))\Big|\Gamma(0)\Big\} = \int\Gamma(0,dx)S_t'p^\beta(r+T-t,x)$$
$$\le \int\Gamma(0,dx)\ \sup\Big\{S_s'p^\beta(s',x);\ 0\le s,s'\le r+T,\ s'\ne 0\Big\}.$$

But by Lemma 7.1, this expression is finite with probability one. Therefore, given $\Gamma(0)$, the first two terms in the definition of M_t^r are finite for all $t\in\underline{J}$.

Let $\beta=1$. Then (7.5) shows that these first two terms have the required martingale property. (Note also that $[\kappa\Delta_\alpha+\frac{\partial}{\partial s}]p(r+T-s)$ is identically zero in this case.)

Assume now that $\beta<1$. Let G_+ be the set of all functions $g\in F_+$ such that g^β belongs to the domain of $\kappa\Delta_\alpha$. Then we observe that by the expectation formula in Lemma 6.2, for g in G_+ and given $\Gamma(0)$,

$$(\Gamma(t),g^\beta) - (\Gamma(0),g^\beta) - \int_0^t ds\ (\Gamma(s),\kappa\Delta_\alpha g^\beta), \qquad t\in \underline{J},$$

is a right continuous martingale. Moreover, for sufficiently smooth mappings h of $\underline{J}$ into G_+,

$$(\Gamma(t),h^\beta(t)) - (\Gamma(0),h^\beta(0)) - \int_0^t ds\ (\Gamma(s),[\kappa\Delta_\alpha + \tfrac{\partial}{\partial s}]h^\beta(s)), \qquad t\in \underline{J},$$

is a right continuous martingale, too. From this the statement follows. □

Now let $\underline{J}$ denote the finite interval $[L,T)$.

LEMMA 7.6. *Fix* $z\in\mathbb{R}$ *and* $\tau\geq 0$. *Let* $r_n\to\tau$ *as* $n\to\infty$ *in* $[\tau,\tau+1)$ *be given. Then*

$$P\left\{\liminf_{n\to\infty} \sup_{t\in\underline{J}} (\Gamma(t),p^\beta(t,T+r_n,z,\cdot)) < \infty\right\} = 1.$$

Proof. Since the catalyst processes are stationary in time and space (see Lemma 6.2), without loss of generality we may assume that $z=0=L$.

If $\kappa'=0$, then by definition $\Gamma(t)\equiv\Gamma(0)$ a.s., and the expression under consideration can be estimated above by

$$\int\Gamma(0,dx) \sup_{0<s\leq T+\tau+1} p^\beta(s,x).$$

Then by Lemma 7.1 we directly get the statement.

From now on suppose that $\kappa'=\kappa$. Fix $r\in[\tau,\tau+1)$, let K be a natural number, and use the martingale M^r from Lemma 7.4 (with $L=0$). To this end we fix a $\Gamma(0)$ satisfying the assertions in the Lemmas 7.4 and 7.1.

If $\beta=1$,

$$(7.7) \qquad P\left\{\sup_{t\in\underline{J}} (\Gamma(t),p^\beta(r+T-t)) > K \,\middle|\, \Gamma(0)\right\}$$
$$\leq P\left\{\sup_{t\in\underline{J}}|M_t^r| > K/2 \,\middle|\, \Gamma(0)\right\} + 2K^{-1}(\Gamma(0),p^\beta(r+T)).$$

Applying Doob's inequality (which is also valid for the halfopen interval $\underline{J}$) yields

$$(7.8) \qquad \leq \text{const}\ K^{-1} \sup_{t\in\underline{J}} E\left\{(\Gamma(t),p^\beta(r+T-t)) \,\middle|\, \Gamma(0)\right\}.$$

If $\beta<1$, then (7.7) becomes true if at the right side we replace $K/2$ by $K/3$ and add the term

$$(7.9)\qquad + 3K^{-1}E\left\{\int_J ds\ \left(\Gamma(s), \left|\kappa\Delta_\alpha p^\beta + \frac{\partial}{\partial s}p^\beta\right|(r+T-s)\right)\Big|\Gamma(0)\right\}.$$

Changing here the order of expectation and integration, by the expectation formula in Lemma 6.2 the expressions (7.8) and (7.9) can be estimated above by

$$\le \text{const}\ K^{-1}\Big(\Gamma(0),\ \sup\left\{S_t p^\beta(r+T-t);\ t\in\underline{J},\ 0\le r<\tau+1\right\}$$
$$+ \sup_{0\le r<\tau+1}\int_J ds\ S_s\left|\kappa\Delta_\alpha p^\beta + \frac{\partial}{\partial s}p^\beta\right|(r+T-s)\Big)$$
$$=: K^{-1}H(\Gamma(0)).$$

Now $H(\Gamma(0))$ is finite with probability one by Lemma 7.1. Note that the exceptional set is independent of K and r.

Summarizing, we found that for each natural number K

$$P\left\{\sup_{t\in\underline{J}}(\Gamma(t), p^\beta(r+T-t)) > K\Big|\Gamma(0)\right\} \le K^{-1}H(\Gamma(0))$$

where $H(\Gamma(0))$ is finite a.s. with an exceptional set independent of K and r. We fix such a $\Gamma(0)$. Then for all natural numbers K, n, and k,

$$P\left\{\sup_{t\in\underline{J}}(\Gamma(t), p^\beta(r_n+T-t)) < K \text{ for some } n>k\Big|\Gamma(0)\right\}$$
$$\ge 1 - K^{-1}H(\Gamma(0)).$$

Hence, by monotinicity in K,

$$P\left\{\liminf_{n\to\infty}\sup_{t\in\underline{J}}(\Gamma(t), p^\beta(r_n+T-t)) \le K\Big|\Gamma(0)\right\}$$
$$\ge 1 - K^{-1}H(\Gamma(0)),$$

for all K. Finally,

$$P\left\{\liminf_{n\to\infty}\sup_{t\in\underline{J}}(\Gamma(t), p^\beta(r_n+T-t)) < \infty\Big|\Gamma(0)\right\} \ge 1, \quad \text{a.s.},$$

and we get

$$P\left\{\liminf_{n\to\infty}\sup_{t\in\underline{J}}(\Gamma(t), p^\beta(r_n+T-t)) < \infty\right\} = 1.$$

This completes the proof. □

LEMMA 7.10. *Fix* $J=(L,T)$, $L<T$. *Consider a sequence* $\underline{r}:=\{r_n; n\ge1\}$ *in* $[0,1)$ *with* $r_n\to0$ *as* $n\to\infty$. *Then with* P-

probability one the following holds. For all $\vartheta\in\Theta$ *except those which have an (weighted) atom at* z *for* z *in some set* $N(\Gamma_J,\underline{r})$ *of Lebesgue measure zero we have*

$$\liminf_{n\to\infty} \sup_{t\in J} (\Gamma(t),[\vartheta * p(r_n+T-t)]^{\beta}) < \infty.$$

Proof. Let $\vartheta\in\Theta$ and $r\in[0,1)$. By assumption on the space Θ,

$$\vartheta * p(r+T-t) \le \text{const } p(\tau+r+T-t,\cdot-z), \qquad t\in J,$$

for some $\tau\ge 0$ and $z\in\mathbb{R}$. In fact, either ϑ is concentrated at some point z (then take $\tau=0$) or it has a density function bounded by some function in $\mathbf{F}_+$ (then choose $z=0$). Hence, from Lemma 7.6 in connection with the spatial invariance of Γ, we see that for the given sequence $r_n\longrightarrow 0$ and each $\vartheta\in\Theta$

$$P\left\{\liminf_{n\to\infty} \sup_{t\in J} (\Gamma(t),[\vartheta * p(r_n+T-t)]^{\beta}) = \infty\right\} = 0.$$

If ϑ is absolutely continuous, we are done.

Assume now that ϑ is degenerate, and let z denote the atom of ϑ. Then by Lemma 7.6 we get

$$\int dz\; P\left\{\liminf_{n\to\infty} \sup_{t\in J} (\Gamma(t),[\delta_z * p(r_n+T-t)]^{\beta}) = \infty\right\} = 0.$$

Therefore, by Fubini's theorem, the limit inferior is infinite only on a zero set with respect to the product measure $P(d\omega)dz$, and once more by Fubini's theorem the claim follows. □

COROLLARY 7.11. *Fix again* $J=(L,T)$, $L<T$. *Then with* P-*probability one the following holds true. For all* $\vartheta\in\Theta$ *except those which have an (weighted) atom at* z *for* z *in some set* $N(\Gamma_J)$ *of Lebesgue measure zero we have*

$$\sup_{t\in J} (\Gamma(t),[\vartheta * p(t-L)]) < \infty.$$

Proof. First we observe that the right continuous version of the time reversed càdlàg process Γ has the same probability law as the original process. Moreover, the supremum expression in the corollary over the *open* interval J is insensitive to changes from left to right continuous versions. Hence, to get the claim we may use Lemma 7.10

with $r_n \equiv 0$, and the fact that $\gamma^{-1} < 1+\alpha$ from $(\beta\gamma)^{-1} < 1+\alpha$ follows. □

LEMMA 7.12. *For each given* $J=(L,T)$, $L<T$, *with* $\mathbb{P}$*-probability one the restricted branching rate kernel* $\Gamma_J :=$ $\{\Gamma(r); r\in J\}$ *is admissible.*

Proof. Fix $J=(L,T)$. Let $\underline{\varepsilon}$ be a sequence in $(0,1)$ with $\varepsilon_n \longrightarrow 0$ as $n\to\infty$. Applying Lemma 7.10 with $\underline{r}\equiv 0$ and also with $\underline{r}=\underline{\varepsilon}$, as well as Corollary 7.11 to obtain with probability one the existence of a Lebesgue zero set $N(\Gamma_J, \underline{\varepsilon})$ such that the following are satisfied:

$$\sup_{t\in J} (\Gamma(t), [\delta_z * p(T-t)]^\beta) < \infty,$$

$$\liminf_{n\to\infty} \sup_{t\in J} (\Gamma(t), [\delta_z * p(\varepsilon_n + T-t)]^\beta) < \infty,$$

$$\sup_{t\in J} (\Gamma(t), [\delta_z * p(t-L)]) < \infty,$$

for all $z\in\mathbb{R}\setminus N(\Gamma_J, \underline{\varepsilon})$. For each such z we may now choose a subsequence $\underline{\varepsilon}(z)$ of $\underline{\varepsilon}$ such that along this subsequence the latter limit inferior becomes a finite limit. Then all requirements in the Definition 1.10 are fulfilled, and the proof is complete. □

Combining the Lemmas 6.3 and 7.12, we immediately get the following result.

PROPOSITION 7.13. *With* $\mathbb{P}$*-probability one,* Γ *is a branching rate kernel. For each given* $J=(L,T)$, $L<T$, *with* $\mathbb{P}$*-probability one the restricted process* $\Gamma_J := \{\Gamma(r); r\in J\}$ *is an admissible restricted branching rate kernel.*

Since according to Remark 6.4 for all $s\in\mathbb{R}$ and $\mu\in\mathcal{M}_f$ the mapping $\rho \mapsto \mathbb{P}^\rho_{s,\mu}$ is measurable in an appropriate sense, and because of Proposition 7.13 with $\mathbb{P}$-probability one Γ is a branching rate kernel, by mixing we may form the probability measures $\mathcal{P}_{s,\mu} := \mathbb{E}\mathbb{P}^\Gamma_{s,\mu}$, $s\in\mathbb{R}$, $\mu\in\mathcal{M}_f$, describing a *superprocess* X *in the random medium* Γ.

THEOREM 7.14. *Let* X *be the superprocess in the random medium* Γ, *defined by the catalyst process* Γ. *If* $t\in\mathbb{R}$

is a fixed time point, then the random measure $X(t)$ *is absolutely continuous with* $\mathcal{P}_{s,\mu}$*-probability one, for all* $s<t$ *and* $\mu\in\mathcal{M}_f$.

Proof. We fix $s<t$ and $\mu\in\mathcal{M}_f$, choose an $s'\in(s,t)$, and set $J:=(s',t)$. By Proposition 7.13 with $\mathbb{P}$-probability one, Γ is a branching rate kernel and the restricted kernel $\Gamma_J:=\{\Gamma(r);r\in J\}$ is admissible. Therefore, given Γ_J, by Theorem 4.4 the random measure $X(t)$ is absolutely continuous with $\mathbb{P}^{\Gamma}_{s,\mu}$-probability one. But then it is also absolutely continuous with $\mathcal{P}_{s,\mu}$-probability one, and the proof is finished. □

APPENDIX: ON THE STABLE SEMI-GROUP

For convenience, here we compile some facts related to the stable semi-group and needed in the present paper. To this end, we *fix* the following constants:

$$\eta\in(0,1), \quad \alpha,\alpha'\in(0,2], \quad \text{and} \quad \beta,\beta'\in(0,1].$$

(Note that in the Appendix we do *not* impose restrictions as $\alpha>1$).

For $t>0$ let $q_\eta(t,.)$ denote the continuous density function of a *stable distribution on* $\mathbb{R}_+$ *with index* η determined by the Laplace transform

$$\int_0^\infty ds\ q_\eta(t,s)e^{-s\theta} = \exp[-t\theta^\eta\ ], \quad \theta\ge 0.$$

Similarly, let $p_\alpha(t,.)$ be the continuous density function of a *symmetric stable distribution with index* α given by the Fourier transform

$$\int dy\ p_\alpha(t,y)e^{iyx} = \exp[-t|x|^\alpha], \quad x\in\mathbb{R}.$$

In particular,

$$p_2(t,x) := (4\pi t)^{-1/2}\exp[-x^2/4t], \quad x\in\mathbb{R}.$$

We get the *self-similarity properties*

(A.0) $$q_\eta(t,s) = K\ q_\eta(K^\eta t,Ks),$$

(A.1) $$p_\alpha(t,x) = K\ p_\alpha(K^\alpha t,Kx),$$

$K>0$, and, in the case $\alpha<2$, the *subordination formula*

(A.2) $$p_\alpha(t,x) := \int_0^\infty ds\ q_{\alpha/2}(t,s)p_2(s,x).$$

Immediately from (A.1) we conclude

(A.3) $$p_\alpha(t,x) \le (t/c)^{-1/\alpha}\ p_\alpha(c,x), \quad 0<t\le c,\ x\in\mathbb{R},$$

for each $c>0$.

Let $S^\alpha:=\{S_t^\alpha;t\ge0\}$ denote the semi-group corresponding to the family $\{p_\alpha(t);t>0\}$:

$$S_t^\alpha h(x) := \int dy\ p_\alpha(t,y-x)h(y), \quad t>0,\ x\in\mathbb{R}$$

(provided that the integral exists). Its generator is given by the fractional power $-(-\Delta)^{\alpha/2}=\Delta_\alpha$ of the Laplacian Δ.

LEMMA A.4. *If $\alpha<2$, we have the representation*

$$\Delta_\alpha g = c_\alpha \int_0^\infty d\lambda\ \lambda^{-1-\alpha/2}[g - S_\lambda^2 g], \quad g\in\mathcal{D}(\Delta)$$

where c_α is some positive constant (determined by the gamma function) and $\mathcal{D}(\Delta)$ is the domain of definition of the one-dimensional Laplacian Δ.

Proof. See Yosida (1978), formula (9.11.5). □

Immediately from (A.1), for $g\in\mathcal{D}(\Delta)$, $t\ge0$, and $x\in\mathbb{R}$, we get

$$S_t^\alpha g(x) = S_{K^\alpha t}^\alpha\, g(K^{-1}\cdot)(Kx),$$

and therefore

(A.5) $$\Delta_\alpha g(x) = K^\alpha \Delta_\alpha g(K^{-1}\cdot)(Kx).$$

LEMMA A.6. *We have the following self-similarity formulas:*

$$\frac{\partial}{\partial t}p_\alpha{}^\beta(t,x) = K^{\alpha+\beta}[\frac{\partial}{\partial t}p_\alpha{}^\beta](K^\alpha t,Kx),$$

$$\frac{\partial}{\partial x}p_\alpha{}^\beta(t,x) = K^{1+\beta}[\frac{\partial}{\partial x}p_\alpha{}^\beta](K^\alpha t,Kx),$$

$$\Delta_{\alpha'}p_\alpha{}^\beta(t,x) = K^{\alpha'+\beta}[\Delta_{\alpha'}p_\alpha{}^\beta](K^\alpha t,Kx),$$

$t>0$, $x\in\mathbb{R}$, $K>0$.

Proof. The first two statements follow from (A.1) by differentiation, whereas the third one is a consequence of the identity (A.5) combined with (A.1). □

LEMMA A.7. $s^{1+\eta}q_\eta(1,s)$ *converges to some positive constant (depending on* η*) as* $s\to\infty$ *whereas* $\exp(1/s)q_\eta(1,s)$ *tends to* 0 *as* $s\to 0$.

Proof. See, e.g., Zolotariev (1983), formula (2.4.8) and Theorem 2.5.2. □

LEMMA A.8. *If* $\alpha<2$, *then* $t^{-1}|x|^{1+\alpha}p_\alpha(t,x)$ *is bounded in* $t>0$, $x\in\mathbb{R}$ *and, as* $x\to\infty$, *converges to some positive constant which is independent of* t. *On the other hand, for given* $k,K>0$, *it is bounded away from* 0 *on the set* $\{[t,x];\ 0<t\le K,\ |x|\ge k\}$.

Proof. By substitution in (A.2) and by the self-similarity properties (A.0) and (A.1) we get

$$t^{-1}|x|^{1+\alpha}p_\alpha(t,x)$$

$$=\int_0^\infty ds[t^{-2/\alpha}x^2s]^{1+\alpha/2}q_{\alpha/2}(1,t^{-2/\alpha}x^2s)s^{-1-\alpha/2}p_2(s,1).$$

The integral $\int_0^\infty ds\ s^{-1-\alpha/2}p_2(s,1)$ is finite. On the other hand, by Lemma A.7,

$$[t^{-2/\alpha}x^2s]^{1+\alpha/2}q_{\alpha/2}(1,t^{-2/\alpha}x^2s)$$

is bounded in s,t,x, which yields the first statement. Moreover, by the same lemma, for fixed s and t, as $x\to\infty$ it converges to a constant which is independent of s and t. Finally, by (A.1) we have

$$\text{(A.9)}\qquad p_\alpha(t,x) = t^{-1/\alpha}p_\alpha(1,t^{-1/\alpha}x),\quad t>0,\ x\in\mathbb{R},$$

hence

$$t^{-1}|x|^{1+\alpha}p_\alpha(t,x) = |t^{-1/\alpha}x|^{1+\alpha}p_\alpha(1,t^{-1/\alpha}x),$$

and the convergence implies the last statement. □

Recall that *const* always denotes a positive and finite constant.

LEMMA A.10. *Given* $k,K>0$, *we have*

$$p_\alpha(t,x) \le \text{const } p_\alpha(K,x), \quad 0<t\le K, \ |x|\ge k.$$

Proof. It is easy to see that the statement holds for $\alpha=2$, and in the case $\alpha<2$ we may apply Lemma A.8. □

LEMMA A.11. *Let be given* $\sigma\in[0,1+\alpha)$. *Then*

$t^{(1-\sigma)/\alpha}|x|^\sigma p_\alpha(t,x)$ *is bounded in* $t>0$, $x\in\mathbb{R}$.

Proof. First of all,

$$\sup_{r\ge 0} r^a e^{-r} < \infty, \quad a>0. \tag{A.12}$$

This already implies the statement in the case $\alpha=2$.

Now we assume that $\alpha<2$. From (A.2) and the proved statement in the case $\alpha=2$ we get

$$t^{(1-\sigma)/\alpha}|x|^\sigma p_\alpha(t,x)$$

$$\le \text{const } t^{(1-\sigma)/\alpha} \int_0^\infty ds\ q_{\alpha/2}(t,s)s^{-(1-\sigma)/2}.$$

By the self-similarity (A.0), the inequality can be continued with

$$= \text{const} \int_0^\infty ds\ q_{\alpha/2}(1,s)s^{-(1-\sigma)/2}.$$

But the latter integral is finite by Lemma A.7. □

LEMMA A.13. *Let* α,β,β' *be given as in the beginning of the Appendix and* $K>0$. *Then the function*

$$x \longmapsto [\sup_{0<t\le K} p_\alpha{}^\beta(t,x)]^{\beta'}, \quad x\ne 0$$

is finite. Moreover, it is integrable (with respect to Lebesgue measure) on the set $\{x;|x|>1\}$ *if in the case* $\alpha<2$ *additionally* $\beta\beta'(1+\alpha) > 1$ *is fulfilled, whereas on* $\{x;|x|<1\}$ *it is integrable if* $\beta\beta' < 1$ *holds.*

Proof. On $|x|>1$, we apply Lemma A.10, where in the case $\alpha<2$ we additionally employ Lemma A.8. On $|x|\le 1$, we may use Lemma A.11 with $\sigma=1$. □

LEMMA A.14. *For* $t>0$ *we have*

$$\|p_\alpha(t)\|_\infty = \text{const } t^{-1/\alpha},$$

$$\|\frac{\partial}{\partial t}p_\alpha(t)\|_\infty = \text{const } t^{-1-1/\alpha},$$

$$\|\frac{\partial}{\partial x}p_\alpha(t)\|_\infty = \text{const } t^{-2/\alpha},$$

and

$$\|\Delta p_\alpha(t)\|_\infty = \text{const } t^{-3/\alpha}.$$

Proof. The dependence in t results from Lemma A.6. By the Fourier inversion formula,

$$p_\alpha(t,x) = (2\pi)^{-1} \int dy \, \exp[-t|y|^\alpha]\cos(yx).$$

Hence

$$\frac{\partial}{\partial t}p_\alpha(t,x) = (2\pi)^{-1} \int dy \, [-|y|^\alpha]\exp[-t|y|^\alpha]\cos(yx),$$

and for all $x \in \mathbb{R}$,

$$|\frac{\partial}{\partial t}p_\alpha(t,x)| \le \text{const} \int dy \, |y|^\alpha \exp[-t|y|^\alpha] < \infty.$$

The remaining statements are quite analogous. □

LEMMA A.15. *For* $t>0$ *and* $x \in \mathbb{R}$,

$$|\frac{\partial}{\partial t}p_2^\beta(t,x)| + |\Delta p_2{}^\beta(t,x)| \le \text{const } t^{-1}[1+x^2/t]p_2{}^\beta(t,x)$$
$$\le \text{const } t^{-1}p_2{}^\beta(t,x/2).$$

Proof. First of all, for $0<\alpha\le 2$,

(A.16) $$\frac{\partial}{\partial t}p_\alpha{}^\beta(t,x) = \beta p_\alpha{}^{\beta-1}(t,x) \, \frac{\partial}{\partial t}p_\alpha(t,x),$$

and

(A.17) $$\Delta p_\alpha{}^\beta(t,x) = \beta(\beta-1)p_\alpha{}^{\beta-2}(t,x)[\frac{\partial}{\partial x}p_\alpha(t,x)]^2 + \beta p_\alpha{}^{\beta-1}(t,x) \, \Delta p_\alpha(t,x).$$

But

(A.18) $$\frac{\partial}{\partial x}p_2(t,x) = - x(2t)^{-1}p_2(t,x)$$

and

$$\Delta p_2(t,x) = \frac{\partial}{\partial t}p_2(t,x) = [-(2t)^{-1} + x^2(2t)^{-2}]p_2(t,x).$$

Then the first claimed inequality follows. By

$$p_2(t,x) = p_2(t,x/2)\exp[-3x^2/16t]$$

combined with (A.12), we also arrive at the second one. □

LEMMA A.19. *We have*

$$\left|\frac{\partial}{\partial t}p_\alpha(t,x)\right| \le \text{const } t^{-1}p_\alpha(t,x/2), \quad t>0, \quad x\in\mathbb{R}.$$

Proof. Because of Lemma A.15 we may restrict to $\alpha<2$. By the subordination (A.2), the self-similarity (A.0), and a substitution of integration variable,

$$p_\alpha(t,x) = \int_0^\infty ds\ q_{\alpha/2}(1,s)p_2(t^{2/\alpha}s,x). \tag{A.20}$$

Thus,

$$\left|\frac{\partial}{\partial t}p_\alpha(t,x)\right| \le \text{const} \int_0^\infty ds\ q_{\alpha/2}(1,s)t^{2/\alpha-1}s\left|\frac{\partial}{\partial t}p_2\right|(t^{2/\alpha}s,x).$$

Applying Lemma A.15 and again (A.20), we are done. □

LEMMA A.21. *Given* $k,K>0$, *for* $0<t\le K$ *and* $|x|\ge k$ *we have*

$$\left|\frac{\partial}{\partial x}p_\alpha(t,x)\right| \le \text{const } t^{-1/\alpha}p_\alpha(t,x/2)$$

and

$$|\Delta p_\alpha(t,x)| \le \text{const } t^{-2/\alpha}p_\alpha(t,x/2).$$

Proof. By (A.2) and Lemma A.15, for $\alpha<2$ we get

$$|\Delta p_\alpha(1,x)| \le \int_0^\infty ds\ q_{\alpha/2}(1,s)|\Delta p_2(s,x)|$$
$$\le \text{const} \int_0^\infty ds\ q_{\alpha/2}(1,s)s^{-1}[1+x^2/s]p_2(s,x).$$

But, for $|x|\ge k$,

$$p_2(s,x) \le p_2(s,x/2)\ \exp[-3x^2/32s]\exp[-3k^2/32s].$$

Then with (A.12) and (A.2) we arrive at the second inequality in the case $\alpha<2$ and $t=1$. The latter restriction can be removed using (A.1) and Lemma A.6, whereas the case $\alpha=2$ was contained in Lemma A.15.

The proof of the first inequality is quite analogous except we apply (A.18) instead of Lemma A.15. □

LEMMA A.22. *Given* $k,K>0$, *for* $0<t\le K$ *and* $|x|\ge k$ *we have*

$$\left|\frac{\partial}{\partial t}p_\alpha{}^\beta(t,x)\right| \le \text{const } t^{-1}p_\alpha{}^\beta(t,x/2)$$

$$\leq \text{ const } t^{\beta-1}|x|^{-\beta(1+\alpha)} \leq \text{ const } t^{\beta-1}.$$

Proof. It is enough to prove the first inequality. In fact, to get from this the second one use

$$p_\alpha(t,x/2) \leq \text{ const } t|x|^{-1-\alpha}, \quad t>0,\ x\neq 0$$

which follows from Lemma A.8 in the case $\alpha<2$ and is valid for $\alpha=2$, too.

By (A.16) and Lemma A.19, for $t>0$ and $x\in\mathbb{R}$ we get

$$\text{(A.23)} \qquad |\tfrac{\partial}{\partial t}p_\alpha{}^\beta(t,x)| \leq \text{ const } p_\alpha{}^{\beta-1}(t,x)t^{-1}p_\alpha(t,x/2).$$

Because of Lemma A.15, we may suppose that $\alpha<2$. Then from (A.1) (applied to $K=t^{-1/\alpha}$) and Lemma A.8 we recognize that

$$\text{(A.24)} \qquad p_\alpha(t,x/2) \leq \text{ const } p_\alpha(t,x), \quad 0<t\leq K,\ |x|\geq k.$$

If we combine this with (A.23), we are ready. □

LEMMA A.25. *Given* $k,K>0$, *for* $0<t\leq K$ *and* $|x|\geq k$ *we have*

$$|\Delta p_\alpha{}^\beta(t,x)| \leq \text{ const } t^{-2/\alpha}p_\alpha{}^\beta(t,x/2).$$

Proof. Because of Lemma A.15 we may suppose $\alpha<2$. Then apply (A.17), Lemma A.21, and (A.24). □

LEMMA A.26. *We have*

$$\|\Delta p_\alpha{}^\beta(t)\|_\infty < \infty, \quad t>0.$$

Proof. Because of Lemma A.15 we may suppose that $\alpha<2$ holds. For fixed $k,K>0$, by (A.9) and Lemma A.8, there is a positive constant const^+ such that

$$p_\alpha(t,k) = t^{-1/\alpha}p_\alpha(1,t^{-1/\alpha}k) \geq \text{const}^+t, \quad 0<t\leq K.$$

Hence (under $\alpha<2$)

$$\inf_{|x|\leq k} p_\alpha(t,x) \geq \text{const}^+t, \quad 0<t\leq K.$$

Therefore (A.17), Lemma A.14, and Lemma A.25 imply the claim. □

LEMMA A.27. *If* $\alpha<2$, *for* $\lambda\le 1$ *we have*

$$|p_\alpha{}^\beta(1,x) - S_\lambda^2 p_\alpha{}^\beta(1)(x)| \le \text{const}\ \lambda[1\wedge|x|^{-\beta(1+\alpha)}], \qquad x\in\mathbb{R}.$$

Proof. By a change of the integration variable,

$$S_\lambda^2 p_\alpha{}^\beta(1)(x) = \int dy\ p_2(1,y)p_\alpha{}^\beta(1,x+\lambda^{1/2}y).$$

We apply the Taylor formula:

$$p_\alpha{}^\beta(1,x+\lambda^{1/2}y) = p_\alpha{}^\beta(1,x) + \lambda^{1/2}y\ \frac{\partial}{\partial x}p_\alpha{}^\beta(1,x)$$
$$+\ 2^{-1}\lambda y^2\Delta p_\alpha{}^\beta(1)(x+\theta\lambda^{1/2}y),$$

where θ (depending on x,y,λ) satisfies $0\le|\theta|\le 1$. Since

$$\int dy\ p_2(1,y)\lambda^{1/2}y\ \frac{\partial}{\partial x}p_\alpha{}^\beta(1,x) \equiv 0,$$

we get

$$|p_\alpha{}^\beta(1,x) - S_\lambda^2 p_\alpha{}^\beta(1)(x)|$$
$$\le\ 2^{-1}\lambda\int dy\ p_2(1,y)y^2|\Delta p_\alpha{}^\beta(1)|(x+\theta\lambda^{1/2}\ y).$$

By the Lemmas A.26, A.25, and A.8 we have

$$|\Delta p_\alpha{}^\beta(1,z)| \le \text{const}\ \min\{|z|^{-\beta(1+\alpha)},1\}, \qquad z\ne 0.$$

Hence, for the integral restricted to $|x+\theta\lambda^{1/2}y| \ge |x|/2$ we are done. On the other hand, $|x+\theta\lambda^{1/2}y| < |x|/2$ implies $|x|/2 < |\theta\lambda^{1/2}y| \le |y|$, and

$$\int_{|y|>|x|/2} dy\ p_2(1,y)y^2 \le \text{const}\ |x|^{-\beta(1+\alpha)}$$

is obviously true. □

LEMMA A.28. *Let* $\alpha\le\alpha'$ *and* $k,K>0$ *be given. In the case* $\alpha<2$ *we additionally require that* $\beta(1+\alpha) > 1$ *holds. Then for* $0<t\le K$ *and* $|x|\ge k$ *we have*

$$|\Delta_{\alpha'}p_\alpha{}^\beta(t,x)| \le \text{const}\ t^{-\alpha'/\alpha}p_\alpha{}^\beta(t,x/2).$$

Proof. Because of Lemma A.28 we may assume that $\alpha'<2$. Also, (A.1) and Lemma A.6 allow us to reduce the problem to $t=1$. Then by Lemma A.8 it suffices to show that

$$|\Delta_{\alpha'}p_\alpha{}^\beta(1,x)| \le \text{const}\ |x|^{-\beta(1+\alpha)}, \qquad |x|\ge k$$

holds.

By Lemma A.4,

(A.29) $$|\Delta_{\alpha'} p_\alpha{}^\beta(1,x)| \le \text{const} \int_0^\infty d\lambda\ \lambda^{-1-\alpha'/2}\left|p_\alpha{}^\beta(1,x) - S_\lambda^2 p_\alpha{}^\beta(1)(x)\right|.$$

We distinguish between $\lambda\le 1$ and $\lambda>1$. In the first case, the previous lemma yields the desired result. Now we suppose $\lambda>1$. By Lemma A.8 we have

$$p_\alpha{}^\beta(1,x) \le \text{const}\ |x|^{-\beta(1+\alpha)}.$$

On the other hand, for the integral $\int dy\ p_2(\lambda,y-x)p_\alpha{}^\beta(1,y)$ restricted to $|x-y| > |x|/2$ we get $\le$ const $p_2(\lambda,x/2)$, where we used Lemma A.13. By Lemma A.7 we have

$$\lambda^{-1-\alpha'/2} \le \text{const}\ q_{\alpha'/2}(1,\lambda), \quad \lambda\ge 1.$$

Hence, by subordination (A.2), Lemma A.8, and $\alpha'\ge\alpha$,

$$\int_1^\infty d\lambda\ \lambda^{-1-\alpha'/2}p_2(\lambda,x/2) \le \text{const}\ p_{\alpha'}(1,x/2) \le \text{const}\ |x|^{-\beta(1+\alpha)}, \quad |x|\ge k.$$

In the opposite case $|x-y| \le |x|/2$, we have $|y|\ge|x|/2$, and again we may apply Lemma A.8. Summarizing, the integral in the formula line (A.29), restricted to $\lambda>1$, has the claimed estimate, too. □

LEMMA A.30. *Let be given* $\alpha\le\alpha'$, *with the restriction* $\beta(1+\alpha) > 1$ *if* $\alpha<2$. *Then, for* $t>0$ *(and finite constants)*

$$\|\tfrac{\partial}{\partial t}p_\alpha{}^\beta(t)\|_1 = \text{const}\ t^{-1+(1-\beta)/\alpha},$$

$$\|\tfrac{\partial}{\partial t}p_\alpha{}^\beta(t)\|_\infty = \text{const}\ t^{-1-\beta/\alpha},$$

$$\|\Delta_{\alpha'}p_\alpha{}^\beta(t)\|_1 = \text{const}\ t^{-\alpha'/\alpha + (1-\beta)/\alpha},$$

$$\|\Delta_{\alpha'}p_\alpha{}^\beta(t)\|_\infty = \text{const}\ t^{-\alpha'/\alpha - \beta/\alpha}.$$

Proof. The claimed dependence in t is a consequence of the self-similarities expressed in Lemma A.6, and we may assume that $t=1$ holds. In the expressions defining the two norms we will distinguish between $|x|\ge 1$ and the opposite. In the first case we use the Lemmas A.22, A.28, and A.13. It remains to show boundedness in $|x|<1$. Concerning the first two terms in the lemma, we use the estimate

$$p_\alpha^{\beta-1}(1,x) \le p_\alpha^{\beta-1}(1,1) = \text{const}, \quad |x|<1,$$

formula (A.16), and Lemma A.14. Concerning the other two terms, the case $\alpha'=2$ follows from Lemma A.26, whereas under $\alpha'<2$ in (A.29) we again distinguish between $\lambda\le 1$ and $\lambda>1$. In the first case we apply Lemma A.27, whereas the remaining case is obvious. □

REFERENCES

[1] P. BILLINGSLEY, "Convergence of Probability Measures", Wiley, New York, 1968.

[2] C. CUTLER, "Some Measure-theoretic and Topological Results for Measure-valued and Set-valued Stochastic Processes", Carleton Univ., Lab. Research Stat. Probab., Tech. Report No. 49, Ottawa, 1984.

[3] D.A. DAWSON and K. FLEISCHMANN, Diffusion and reaction caused by point catalysts, (revised manuscript, Carleton Univ. Ottawa 1990a).

[4] D.A. DAWSON and K. FLEISCHMANN, Critical branching in a highly fluctuating random medium, (revised manuscript, Carleton Univ. Ottawa 1990b).

[5] D.A. DAWSON, K. FLEISCHMANN, and S. ROELLY-COPPOLETTA, Absolute continuity of the measure states in a branching model with catalysts, Carleton Univ., Lab. Research Stat. Probab., Tech. Report No. 134, Ottawa, 1989.

[6] D.A. DAWSON and K.J. HOCHBERG, The carrying dimension of a stochastic measure diffusion, *Ann. Probab.* 7 (1979), 693-703.

[7] D.A. DAWSON and E.A.PERKINS, Historical processes, Carleton Univ., Lab. Research Stat. Probab., Tech. Report No. 142, Ottawa, 1990.

[8] N. DUNFORD and J.T. SCHWARTZ, "Linear Operators. Part I: General Theory", Interscience Publishers, New York, 1958.

[9] E.B. DYNKIN, Branching particle systems and superprocesses, (manuscript, Cornell Univ. Ithaca 1990).

[10] P.J. FITZSIMMONS, Construction and regularity of measure-valued Markov branching processes, *Israel J. Math.* 64 (1988), 337-361.

[11] P.J. FITZSIMMONS, Correction and addendum to: Construction and regularity of measure-valued Markov branching processes, *Israel J. Math.* (to appear 1990).

[12] K. FLEISCHMANN, Critical behavior of some measure-

valued processes, *Math. Nachr.* 135 (1988), 131-147.

[13] N. KONNO and T. SHIGA, Stochastic differential equations for some measure-valued diffusions, *Probab. Th. Rel. Fields* 79 (1988), 201-225

[14] R. MEIDAN, On the connection between ordinary and generalized stochastic processes. *J. Mat. Analysis Appl.* 76, 124-133 (1980).

[15] S. ROELLY-COPPOLETTA, A criterion of convergence of measure-valued processes: Application to measure branching processes, *Stochastics* 17 (1986), 43-65.

[16] R. TRIBE, Path properties of superprocesses. Ph.D thesis, UBC, Vancouver, 1989.

[17] A. WULFSOHN, Random creation and dispersion of mass, J. *Multivariate Anal.* 18 (1986), 274-286. 86.

[18] K. YOSIDA, "Functional Analysis", 5-th edition, Springer-Verlag, Berlin, 1978.

[19] V.M. ZOLOTARIEV, "One-dimensional Stable Distributions" (in Russian), Nauka, Moscow, 1983.

DONALD A. DAWSON

Department of Mathematics
and Statistics,
Carleton University,
Ottawa, Canada K1S 5B6

KLAUS FLEISCHMANN

Karl Weierstrass Institute
of Mathematics,
Box 1304,
Berlin, DDR-1086

SYLVIE ROELLY

Laboratoire de calcul
des Probabilités,
Université Paris 6,
4, place Jussieu, Tour 56,
75230 Paris cedex 05,
France

Martingales Associated with Finite Markov Chains

ROBERT J. ELLIOTT

1. Introduction.

In a recent paper, [1], Phillipe Biane introduced martingales M^k associated with the different jump 'sizes' of a time homogeneous, finite Markov chain and developed homogeneous chaos expansions. It has long been known that the Kolmogorov equation for the probability densities of a Markov chain gives rise to a canonical martingale M. The modest contributions of this note, are that working with a non-homogeneous chain, we relate Biane's martingales M^k to M, calculate the quadratic variation of M and thereby that of the M^k. In addition, square field identities are obtained for each jump size.

For $0 \le i \le N$ write $e_i = (0, 0, \dots, 1, \dots, 0)^*$ for the i-th unit (column) vector in R^{N+1}, (so $e_0 = (1, 0, \dots, 0)^*$ etc.). Consider the (non-homogeneous) Markov process $\{X_t\}$, $t \ge 0$, defined on a probability space (Ω, F, P), whose state space, without loss of generality, can be identified with the set $S = \{e_0, e_1, \dots, e_N\}$. Write $p_t^i = P(X_t = e_i)$, $0 \le i \le N$. We shall suppose that for some family of matrices A_t, $p_t = (p_t^0, \dots, p_t^N)^*$ satisfies the forward Kolmogorov equation

$$\frac{dp_t}{dt} = A_t p_t. \tag{1.1}$$

$A_t = (a_{ij}(t))$ is, therefore, the family of Q-matrices of the process.

It has long been known (see, for example, Liptser and Shiryayev [4], Elliott [2]) that the process

$$M_t = X_t - X_0 - \int_0^t A_r X_{r-} dr \tag{1.2}$$

is a martingale. (See Lemma 2.3 below.)

ACKNOWLEDGMENTS: Research partially supported by NSERC Grant A7964, the Air Force Office of Scientific Research United States Air Force, under contract AFOSR-86-0332, and the U.S. Army Research Office under contract DAAL03-87-0102.

Solving (1.2) by 'variation of constants' we can immediately write

$$X_t = \Phi(t,0)\Big(X_0 + \int_0^t \Phi(0,r)^{-1} dM_r\Big) \tag{1.3}$$

where Φ is the fundamental matrix of the generator A. Equation (1.3) is a martingale representation result which in turn gives a representation result in terms of the M^k. (By iterating this representation Biane's homogeneous chaos expansion can be obtained; this is quite explicit, in terms of matrices Φ and matrices associated with A.) Functions of the chain are just given by vectors in R^{N+1} and in Section 4 'square field' identities are obtained for each jump 'size'.

2. Markov Chains.

Consider a Markov chain $\{X_t\}$, $t \geq 0$, with state space $S = \{e_0, \dots, e_N\}$ and Q-matrix generators A_t. We shall make the following assumptions.

ASSUMPTIONS 2.1. (i) For all $0 \leq i, j \leq N$ and $t \geq 0$

$$|a_{ij}(t)| \leq B' \tag{2.1}$$

for some bound B'; write $B = B' + 1$.

(ii) For all $0 \leq i, j \leq N$ and $t \geq 0$, $a_{ij}(t) > 0$ if $i \neq j$ and, (because A_t is a Q-matrix),

$$a_{ii}(t) = -\sum_{j \neq i} a_{ji}(t). \tag{2.2}$$

The fundamental transition matrix associated with A will be denoted by $\Phi(t,s)$, so with I the $(N+1) \times (N+1)$ identity matrix,

$$\frac{d\Phi(t,s)}{dt} = A_t \Phi(t,s), \qquad \Phi(s,s) = I \tag{2.3}$$

and

$$\frac{d\Phi(t,s)}{ds} = -\Phi(t,s) A_s, \qquad \Phi(t,t) = I. \tag{2.4}$$

(If A_t is constant $\Phi(t,s) = \exp A(t-s)$.)

BOUNDS 2.2. For a matrix $C = (c_{ij})$ consider a norm $|C| = \max_{i,j} |c_{ij}|$. Then for all t, $|A_t| \leq B$. The columns of Φ are probability distributions so $|\Phi(t,s)| \leq 1$ for all t, s.

Consider the process in state $x \in S$ at time s and write $X_{s,t}(x)$ for its state at time $t \geq s$.

Then $E[X_{s,t}(x)] = E_{s,x}[X_t] = \Phi(t,s)x$. Write F_t^s for the right continuous complete filtration generated by $\sigma\{X_r : s \leq r \leq t\}$ and $F_t^0 = F_t$.

LEMMA 2.3. *The process* $M_t = X_t - X_0 - \int_0^t A_r X_{r-} dr$ *is an* $\{F_t\}$ *martingale.*

Proof. Suppose $0 \le s \le t$. Then

$$\begin{aligned} E[M_t - M_s \mid F_s] &= E\Big[X_t - X_s - \int_s^t A_r X_{r-} dr \mid F_s\Big] \\ &= E\Big[X_t - X_s - \int_s^t A_r X_r dr \mid X_s\Big] \\ &= E_{s,X_s}[X_t] - X_s - \int_s^t A_r E_{s,X_s}[X_r] dr \\ &= \Phi(t,s)X_s - X_s - \int_s^t A_r \Phi(r,s) X_s dr = 0 \qquad \text{by (2.3).} \end{aligned}$$

Therefore,

$$X_t = X_0 + \int_0^t A_r X_r dr + M_t = X_0 + \int_0^t A_r X_{r-} dr + M_t$$

where M is an $\{F_t\}$ martingale.

NOTATION 2.4. If $x = (x_0, x_1, \dots, x_N)^* \in R^{N+1}$ then diag x is the matrix

$$\begin{pmatrix} x_0 & & & 0 \\ & x_1 & & \\ & & \ddots & \\ 0 & & & x_N \end{pmatrix}.$$

LEMMA 2.5.

$$\langle M, M \rangle_t = \operatorname{diag} \int_0^t A_r X_{r-} dr - \int_0^t (\operatorname{diag} X_{r-}) A_r^* dr - \int_0^t A_r (\operatorname{diag} X_{r-}) dr.$$

Proof. Recall $X_t \in S$ is one of the unit vectors e_i. Therefore,

$$X_t \otimes X_t = \operatorname{diag} X_t. \tag{2.5}$$

Now by the differentiation rule

$$\begin{aligned} X_t \otimes X_t = X_0 \otimes X_0 &+ \int_0^t X_{r-} \otimes (A_r X_{r-}) dr \\ &+ \int_0^t X_{r-} \otimes dM_r + \int_0^t (A_r X_{r-}) \otimes X_{r-} dr \\ &+ \int_0^t dM_r \otimes X_{r-} + \langle M, M \rangle_t + N_t \end{aligned}$$

where N_t is the F_t martingale

$$[M,M]_t - \langle M,M\rangle_t.$$

However, a simple calculation shows

$$X_{r-} \otimes (A_r X_{r-}) = (\text{diag } X_{r-})A_r^*$$

and

$$(A_r X_{r-}) \otimes X_{r-} = A_r(\text{diag } X_{r-}).$$

Therefore,

$$X_t \otimes X_t = X_0 \otimes X_0 + \int_0^t (\text{diag } X_{r-})A_r^* dr$$
$$+ \int_0^t A_r(\text{diag } X_{r-})dr + \langle M,M\rangle_t + \text{ martingale.} \tag{2.6}$$

Also, from (2.5)

$$X_t \otimes X_t = \text{diag } X_t = \text{diag } X_0 + \text{diag} \int_0^t A_r X_{r-} dr + \text{diag } M_t. \tag{2.7}$$

The semimartingale decompositions (2.6) and (2.7) must be the same, so equating the predictable terms

$$\langle M,M\rangle_t = \text{diag} \int_0^t A_r X_{r-} dr - \int_0^t (\text{diag } X_{r-})A_r^* dr - \int_0^t A_r(\text{diag } X_{r-})dr.$$

We next note the following representation result:

LEMMA 2.6.

$$X_t = \Phi(t,0)\Big(X_0 + \int_0^t \Phi(r,0)^{-1} dM_r\Big). \tag{2.8}$$

Proof. This result follows immediately by 'variation of constants'.

REMARKS 2.7. A function of $X_t \in S$ can be represented by a vector

$$f(t) = (f_0(t),\dots,f_N(t))^* \in R^{N+1}$$

so that $f(t,X_t) = f(t)^* X_t = \langle f(t), X_t\rangle$ where $\langle\ ,\ \rangle$ denotes the inner product in R^{N+1}.

We, therefore, have the following differentiation rule and representation result:

LEMMA 2.8. *Suppose the components of $f(t)$ are differentiable in t. Then*

$$f(t, X_t) = f(0, X_0) + \int_0^t \langle f'(r), X_r \rangle dr + \int_0^t \langle f(r), A_r X_{r-} \rangle dr + \int_0^t \langle f(r), dM_r \rangle. \tag{2.9}$$

Here, $\int_0^t \langle f(r), dM_r \rangle$ is an F_t-martingale. Also,

$$f(t, X_t) = \langle f(t), \Phi(t,0) X_0 \rangle + \int_0^t \langle f(t), \Phi(t,r) dM_r \rangle. \tag{2.10}$$

This gives the martingale representation of $f(t, X_t)$.

REMARK 2.9. With an obvious abuse of notation, if the jump times of the chain are $T_1(w), T_2(w), \ldots$, we can write down a 'random measure' decomposition of X_t from (1.2) as

$$X_t = X_0 + \int_0^t \sum_i (e_i - X_{r-})(\sum_k \delta_{T_k(w)}(dr) \delta_{i_k(w)}(i) - a_{iX_{r-}} dr)$$

$$+ \int_0^t \sum_i (e_i - X_{r-}) a_{iX_{r-}} dr,$$

because $\sum_i (e_i - X_{r-}) a_{iX_{r-}} = A_{r-} X_{r-}$. Here, $\delta_{T_k(w)}(dr)$ is the unit mass at $T_k(w)$ and, with $X_{T_k(w)} = e_{i_k(w)}$, $\delta_{i_k(w)}(i)$ is 1 if $i = i_k(w)$ and 0 otherwise. That is,

$$M_t = \int_0^t \sum_i (e_i - X_{r-})(\sum_k \delta_{T_k(w)}(dr) \delta_{i_k(w)}(i) - a_{iX_{r-}} dr).$$

This representation would provide another means of calculating $\langle M, M \rangle_t$.

3. Shift Operators.

The formulae of Section 2, particularly the martingale representations (2.8) and (2.10), provide basic information about the Markov process X. However, if the 'size' of the jumps is considered some other expressions, including a homogeneous chaos expansion, were obtained recently by Biane [1]. We wish to indicate how the results of Biane relate to the above expressions. First we introduce some notation.

NOTATION 3.1. Write $i \oplus j$ for addition $\mod (N+1)$. For $X_s \in S = \{e_0, e_1, \ldots, e_N\}$, say $X_s = e_i$, and $k = 1, \ldots, N$, write

$$X_s^k = e_{i \oplus k}.$$

That is, $X_s \to X_s^k$ corresponds to a cyclic jump of size k in the index of the unit vector corresponding to the state.

Suppose $X_{s-} = e_i$ and $X_{s-}^k = e_j$, where $j = i \oplus k$, then clearly

$$(X_{s-}^k)^* A_s X_{s-} = a_{ji}(s). \tag{3.1}$$

We now wish to introduce some subsidiary matrices associated with $A_s = (a_{ij}(s))$. These can best be explained by first considering the 3×3 case. Suppose

$$A = \begin{pmatrix} a_{00} & a_{01} & a_{02} \\ a_{10} & a_{11} & a_{12} \\ a_{20} & a_{21} & a_{21} \end{pmatrix}.$$

Then

$$A^1 := \begin{pmatrix} -a_{10} & 0 & a_{02} \\ a_{10} & -a_{21} & 0 \\ 0 & a_{21} & -a_{02} \end{pmatrix},$$

$$A^2 := \begin{pmatrix} -a_{20} & a_{01} & 0 \\ 0 & -a_{01} & a_{12} \\ a_{20} & 0 & -a_{12} \end{pmatrix}.$$

Note that if A is a Q-matrix $a_{0i} + a_{1i} + a_{2i} = 0$, so $A^1 + A^2 = A$.

In general, if $A_s = (a_{ij}(s))$ is an $(N+1) \times (N+1)$ Q-matrix, A_s^k is obtained by forming a matrix from the k-th subdiagonal (continued as a superdiagonal), with the negative of the column entries on the diagonal and zeros elsewhere. By construction, A^k is a Q-matrix, and it is clearly related to those jumps of 'size' k. As above,

$$A_s = \sum_{k=1}^{N} A_s^k. \tag{3.2}$$

Also,

$$((X_{s-}^k)^* A_s X_{s-})(X_{s-}^k - X_{s-}) = A_s^k X_{s-}, \tag{3.3}$$

so

$$\sum_{k=1}^{N} ((X_{s-}^k)^* A_s X_{s-})(X_{s-}^k - X_{s-}) = A_s X_{s-}. \tag{3.4}$$

We also wish to introduce matrices $\widetilde{A}^k$, $k \neq 0$, whose off-diagonal entries are the (positive) square roots of those of A^k, and whose diagonal entries are the negative of that square root in the same column. That is, in the (3×3) case above:

$$\widetilde{A}^1 := \begin{pmatrix} -\sqrt{a_{10}} & 0 & \sqrt{a_{02}} \\ \sqrt{a_{10}} & -\sqrt{a_{21}} & 0 \\ 0 & \sqrt{a_{21}} & -\sqrt{a_{02}} \end{pmatrix}$$

$$\widetilde{A}^2 := \begin{pmatrix} -\sqrt{a_{20}} & \sqrt{a_{01}} & 0 \\ 0 & -\sqrt{a_{01}} & \sqrt{a_{12}} \\ \sqrt{a_{20}} & 0 & -\sqrt{a_{12}} \end{pmatrix}.$$

For $k = 1, \dots, N$ write

$$\Lambda_s^k = ((X_{s-}^k)^* A_s X_{s-})^{-1/2} (X_{s-}^k)^*,$$

so Λ_s^k is a predictable process.

Definition 3.2. In our notation the matrices M^k introduced by Biane [1] are, for $k = 1, \dots, N$

$$M_t^k = \sum_{0\le s\le t} ((X_{s-}^k)^* A_s X_{s-})^{-1/2} I(X_s = X_{s-}^k) - \int_0^t ((X_{s-}^k)^* A_s X_{s-})^{1/2} ds. \tag{3.5}$$

Lemma 3.3. *For $k = 1, \dots, N$,*

$$M_t^k = \int_0^t \Lambda_s^k \cdot dM_s.$$

Proof. First note

$$(X_{s-}^k)^* \cdot dX_s = (X_{s-}^k)^* \cdot \Delta X_s = (X_{s-}^k)^* \cdot (X_s - X_{s-}) = I(X_s = X_{s-}^k). \tag{3.6}$$

Also, $X_t = X_0 + \int_0^t A_s X_{s-} ds + M_t$, so

$$\begin{aligned} M_t^k &= \int_0^t \Lambda_s^k \cdot dX_s - \int_0^t \Lambda_s^k \cdot A_s X_{s-} ds \\ &= \int_0^t ((X_{s-}^k)^* A_s X_{s-})^{-1/2} (X_{s-}^k)^* \cdot dX_s \\ &\quad - \int_0^t ((X_{s-}^k)^* A_s X_{s-})^{-1/2} ((X_{s-}^k)^* A_s X_{s-}) ds, \end{aligned}$$

and the result follows from (3.6).

LEMMA 3.4. *For $k = 1,\dots,N$, $\langle M^k, M^k\rangle_t = t$.*

Proof. $M_t^k = \int_0^t \Lambda_s^k \cdot dM_s$, so

$$\langle M^k, M^k\rangle_t = \int_0^t \Lambda_s^k d\langle M,M\rangle_s (\Lambda_s^k)^*$$

$$= \int_0^t ((X_{s-}^k)^* A_s X_{s-})^{-1/2} (X_{s-}^k)^*$$

$$\cdot \left(\text{diag}\,(A_s X_{s-}) - (\text{diag}\, X_{s-})A_s^* - A_s(\text{diag}\, X_{s-})\right)$$

$$\cdot (X_{s-}^k)((X_{s-}^k)^* A_s X_{s-})^{-1/2} ds.$$

Now for $k \neq 0$:

$$(X_{s-}^k)^* \text{diag}\, X_{s-} = 0 = (\text{diag}\, X_{s-})(X_{s-}^k)$$

and

$$(X_{s-}^k)^* \cdot (\text{diag}\,(A_s X_{s-})) \cdot (X_{s-}^k) = (X_{s-}^k)^* A_s X_{s-}.$$

Therefore, $\langle M^k, M^k\rangle_t = \int_0^t ds = t.$ □

REMARKS 3.5. For $k \neq \ell$, M^k and M^ℓ have no common jumps, so $[M^k, M^\ell]_t = 0$ and $\langle M^k, M^\ell\rangle_t = 0$. Therefore, $M^1,\dots,M^N$ are a family of orthogonal martingales, each of which has predictable variation t.

Having expressed M^k in terms of M we now wish to express M in terms of the M^k.

THEOREM 3.6. *$M_t = \sum_{k=1}^N \int_0^t \widetilde{A}_s^k X_{s-} dM_s^k$, so the M^k form a basis.*

Proof. From (3.6) first note that

$$dX_s = \sum_{k=1}^N (X_{s-}^k - X_{s-})(X_{s-}^k)^* \cdot dX_s.$$

Therefore,

$$X_t - X_0 = \int_0^t A_s X_{s-} ds + M_t = \int_0^t dX_s \tag{3.7}$$

$$= \sum_{k=1}^N \int_0^t (X_{s-}^k - X_{s-})(X_{s-}^k)^* \cdot (A_s X_{s-} ds + dM_s). \tag{3.8}$$

By definition $dM_s^k = ((X_{s-}^k)^* A_s X_{s-})^{-1/2}(X_{s-}^k)^* \cdot dM_s$ so $(X_{s-}^k)^* \cdot dM_s = ((X_{s-}^k)^* A_s X_{s-})^{1/2} dM_s^k$. Substituting in (3.8)

$$X_t - X_0 = \int_0^t \sum_{k=1}^N (X_{s-}^k - X_{s-})((X_{s-}^k)^* A_s X_{s-})ds$$

$$+ \sum_{k=1}^N \int_0^t (X_{s-}^k - X_{s-})((X_{s-}^k)^* A_s X_{s-})^{1/2} dM_s^k.$$

From (3.3) and (3.4) this equals

$$= \int_0^t A_s X_{s-} ds + \sum_{k=1}^N \int_0^t \widetilde{A}_s^k X_{s-} dM_s^k. \tag{3.9}$$

Comparing (3.7) and (3.9) we see

$$M_t = \sum_{k=1}^N \int_0^t \widetilde{A}_s^k X_{s-} dM_s^k. \tag{3.10}$$

4. Discrete Derivatives for Different Jump Sizes.

Consider a function f on $S = \{e_i\}$. For simplicity we suppose f is constant in time. Then, as noted in Section 2, f is represented by a vector $f = (f_0, \dots, f_N)^*$ and

$$f(X_t) = \langle f, X_t \rangle = \langle f, X_0 \rangle + \int_0^t \langle f, A_r X_{r-} \rangle dr + \langle f, M_t \rangle$$

$$= \langle f, X_0 \rangle + \int_0^t \langle f, A_r X_{r-} \rangle dr + \sum_{k=1}^N \int_0^t \langle f, \widetilde{A}_r^k X_{r-} \rangle dM_r^k. \tag{4.1}$$

from (3.9), and this is

$$= \langle f, X_0 \rangle + \int_0^t \langle A_r^* f, X_{r-} \rangle dr + \sum_{k=1}^N \int_0^t \langle (\widetilde{A}_r^k)^* f, X_{r-} \rangle dM_r^k. \tag{4.2}$$

We now re-establish the 'square field' formula of Biane [1] by calculating $f(X_t)^2$ in two ways.

LEMMA 4.1. $A_r^* f^2 - 2f \cdot A_r^* f = \sum_{k=1}^N ((\widetilde{A}_r^k)^* f)^2$.

Proof. Function multiplication is pointwise in each coordinate, so f^2 corresponds to the vector $(f_0^2, \dots, f_N^2)^*$, and

$$f^2(X_t) = \langle f^2, X_0 \rangle + \int_0^t \langle A_r^*, f^2, X_{r-} \rangle dr + \sum_{k=1}^N \int_0^t \langle (\widetilde{A}_r^k)^* f^2, X_{r-} \rangle dM_r^k \tag{4.3}$$

$$= (f(X_t))^2.$$

Using the differentiation rule this also equals

$$= f(X_0)^2 + 2\int_0^t f(X_{r-})df(X_r) + [f(X), f(X)]_t$$

$$= f(X_0)^2 + 2\int_0^t \langle f, X_{r-}\rangle\langle A_r^* f, X_{r-}\rangle dr$$

$$+ 2\sum_{k=1}^{N}\int_0^t \langle f, X_{r-}\rangle\langle(\widetilde{A}_r^k)^* f, X_{r-}\rangle dM_r^k + [f(X), f(X)]_t. \tag{4.4}$$

Now

$$[f(X), f(X)]_t = \sum_{0\le r\le t} \Delta f(X_r)\Delta f(X_r)$$

$$= \sum_{k=1}^{N}\sum_{0\le r\le t} \langle(\widetilde{A}_r^k)^* f, X_{r-}\rangle^2(\Delta M_r^k)^2$$

$$= \sum_{k=1}^{N}\int_0^t \langle(\widetilde{A}_r^k)^* f, X_{r-}\rangle^2((X_{r-}^k)^* A_r X_{r-})^{-1/2} dM_r^k$$

$$+ \sum_{k=1}^{N}\int_0^t \langle(\widetilde{A}_r^k)^* f, X_{r-}\rangle^2 dr, \qquad \text{from (3.5).}$$

Substituting in (4.4)

$$f(X_t)^2 = f(X_0)^2 + 2\int_0^t \langle f, X_{r-}\rangle\langle A_r^* f, X_{r-}\rangle dr$$

$$+ 2\sum_{k=1}^{N}\int_0^t \langle f, X_{r-}\rangle\langle(\widetilde{A}_r^k)^* f, X_{r-}\rangle dM_r^k$$

$$+ \sum_{k=1}^{N}\int_0^t \langle(\widetilde{A}_r^k)^* f, X_{r-}\rangle^2((X_{r-}^k)^* A_r X_{r-})^{-1/2} dM_r^k$$

$$+ \sum_{k=1}^{N}\int_0^t \langle(\widetilde{A}_r^k)^* f, X_{r-}\rangle^2 dr. \tag{4.5}$$

The special semimartingales (4.3) and (4.5) are equal, so equating the bounded variation terms

$$\langle A_r^* f^2, X_{r-}\rangle = 2\langle f, X_{r-}\rangle\langle A_r^* f, X_{r-}\rangle + \sum_{k=1}^{N}\langle(\widetilde{A}_r^k)^* f, X_{r-}\rangle^2.$$

That is, as functions on S

$$\sum_{k=1}^{N}((\widetilde{A}_r^k)^*f)^2 = A_r^*f^2 - 2f\cdot A_r^*f.$$

□

$(\widetilde{A}_r^k)^*$ corresponds to a discrete derivative of 'amount', or in 'direction' k. However, the algebra suggests that $(\widetilde{A}_r^k)^2$ should be related to A_r^k.

A more specific relation is now obtained.

LEMMA 4.2. *For* $k = 1, \ldots, N$

$$((\widetilde{A}_r^k)^*f)^2 = (A_r^k)^*f^2 - 2f\cdot(A_r^k)^*f.$$

Proof. From the form of $\widetilde{A}^k$ and A^k, for any $f \in R^{N+1}$

$$\begin{aligned}(A^k)^*f &= (a_{k0}(-f_0+f_k), a_{k\oplus 1,1}(-f_1+f_{k\oplus 1}),\\ &\qquad \ldots, a_{k\oplus N,N}(-f_N+f_{k\oplus N})),\\ (A^k)^*f^2 &= (a_{k0}(-f_0^2+f_k^2), a_{k\oplus 1,1}(-f_1^2+f_{k\oplus 1}^2),\\ &\qquad \ldots, a_{k\oplus N,N}(-f_N^2+f_{k\oplus N}^2)),\\ (\widetilde{A}^k)^*f &= (\sqrt{a_{k0}}(-f_0+f_k), \sqrt{a_{k\oplus 1,1}}(-f_1+f_{k\oplus 1}),\\ &\qquad \ldots, \sqrt{a_{k\oplus N,N}}(-f_N+f_{k\oplus N})).\end{aligned}$$

Therefore, as function multiplication is pointwise, that is coordinatewise:

$$((\widetilde{A}^k)^*f)^2 = (a_{k0}(f_0^2 - 2f_0f_k + f_k^2), \ldots, a_{k\oplus N,N}(f_N^2 - 2f_Nf_{k\in N} + f_{k\oplus N}^2))$$

$$f((A^k)^*f) = (a_{k0}(-f_0^2 + f_0f_k), \ldots, a_{k\oplus N,N}(-f_N^2 + f_Nf_{k\oplus N})).$$

Operating coordinatewise, for example,

$$(-f_j^2 + f_{k\oplus j}^2) - 2(-f_j^2 + f_jf_{k\oplus j}) = f_j^2 - 2f_jf_{k\oplus j} + f_{k\oplus j}^2$$

and the result follows. □

Finally, we note that substituting (3.10) in (2.9) we have

$$X_t = \Phi(t,0)\Big(X_0 + \sum_{k=1}^{N}\int_0^t \Phi(r,0)^{-1}\widetilde{A}_r^k X_{r-}dM_r^k\Big). \tag{4.6}$$

Now X_{r-} is a.s. equal to X_r which equals

$$X_r = \Phi(r,0)\Big(X_0 + \sum_{k_2=1}^{N} \int_0^r \Phi(r_2,0)^{-1} \widetilde{A}_{r_2}^{k_2} X_{r_2 -} dM_{r_2}^{k_2}\Big).$$

Substituting in (5.1) we have

$$X_t = \Phi(t,0)X_0 + \sum_{k=1}^{N} \int_0^t \Phi(t,r)\widetilde{A}_r^k \Phi(r,0)X_0 dM_r^k$$

$$+ \sum_{k_1=1}^{N} \sum_{k_2=1}^{N} \int_0^t \int_0^{r_1} \Phi(t,r_1)\widetilde{A}_{r_1}^{k_1} \Phi(r_1,r_2)\widetilde{A}_{r_2}^{k_2} X_{r_2 -} dM_{r_2}^{k_2} dM_{r_2}^{k_1}.$$

Iterating this process we obtain the homogeneous chaos expansions of Biane [1], (see also Elliott and Kohlmann [3]), in terms of the non-homogeneous transition matrices Φ and the matrices $\widetilde{A}^k$.

REFERENCES

[1] P. Biane, Chaotic representation for finite Markov chains. *Stochastics and Stoch. Reports* 30 (1990), 61–68.

[2] R.J. Elliott, Smoothing for a finite state Markov process. *Springer Lecture Notes in Control and Info. Sciences*, Vol. 69, (1985), 199–206.

[3] R.J. Elliott and M. Kohlmann, Integration by parts, homogeneous chaos expansions and smooth densities. *Ann. of Prob.* 17 (1989), 194–207.

[4] R.S. Liptser and A.N. Shiryayev, *Statistics of Random Processes*, Vol. 1, Springer–Verlag, Berlin, Heidelberg, New York, 1977.

Robert J. Elliott
Department of Statistics and Applied Probability
University of Alberta
Edmonton, Alberta, Canada T6G 2G1.

Equivalence and Perpendicularity of Local Field Gaussian Measures

STEVEN N. EVANS*

1. **Introduction.**

One way of thinking about Gaussian measures is that they are the class of probability measures that naturally arise when we seek measures with properties that are intimately linked to the linearity and orthogonality structure of the spaces on which the measures are defined.

There are fields other than $\mathbb{R}$ or $\mathbb{C}$ for which there is a well-developed and interesting theory of orthogonality for the vector spaces over them. These fields are the so-called local fields. In Evans (1989) we worked from the above perspective and defined a suitable concept of a "Gaussian" measure on vector spaces over local fields. In many particulars the theory of such measures resembles the Euclidean prototype, but there are a number of interesting departures.

Here we continue this investigation with a consideration of various questions concerning equivalence, absolute continuity and perpendicularity of local field Gaussian measures.

We begin in §2 with some preliminaries regarding both the general theory of local fields and the particular properties of local field Gaussian measures.

In §3 we observe that, unlike the Gaussian case, one local field Gaussian measure can be absolutely continuous with respect to another without the two measures being equivalent and the only time two such measure will be equivalent is when they are equal.

Theorem 4.1 is a "Cameron-Martin"-type result on the effect of translating local field Gaussian measures. As a consequence, we show in Theorems 4.2 and 4.3 that the local field Gaussian measures on a Banach space are precisely the normalised Haar measures on compact additive subgroups which satisfy an extra "convexity" condition. This allows us to conclude that there is a "flat" local field Gaussian measure on a Banach space if and only if the space is finite dimensional

* Research supported in part by an NSF grant

(see Corollary 4.4).

In Theorem 5.1 we examine the effect of contracting or dilating a local field Gaussian measure, and observe that we get qualitatively different behaviour depending on whether the measure is "finite dimensional" or "infinite dimensional". In the Banach space case we show that the "dimension" of the measure shows up in the mass assigned to small balls around zero (see Theorem 5.2).

2. Preliminaries

We begin this section with a brief overview of some of the theory of local fields. We refer the reader to Taibleson (1975) or Schikhof (1984) for fuller accounts. Later we also recall some of the salient details from Evans (1989) regarding local field Gaussian measures.

Let K be a locally compact, non-discrete, topological field. If K is connected, then K is either $\mathbb{R}$ or $\mathbb{C}$. If K is disconnected, then K is totally disconnected and we say that K is a *local field.*

From now on, we let K be a fixed local field. There is a distinguished real-valued mapping on K which we denote by $x \mapsto |x|$ and call the *valuation map.* The set of values taken by the valuation is the set $\{q^k : k \in \mathbb{Z}\} \cup \{0\}$, where $q = p^c$ for some prime p and positive integer c. The valuation has the following properties:

$$|x| = 0 \iff x = 0;$$

$$|xy| = |x|\,|y|;$$

$$|x+y| \le |x| \vee |y|.$$

The last property is known as the *ultrametric inequality* and implies that if $|x| \ne |y|$ then $|x+y| = |x| \vee |y|$ — the so-called *isosceles triangle property.* The mapping $(x, y) \mapsto |x - y|$ on $K \times K$ is a metric which gives the topology of K.

There is a unique measure μ on K for which

$$\mu(x + A) = \mu(A), \quad x \in K,$$

$$\mu(xA) = |x|\mu(A), \quad x \in K,$$

$$\mu(\{x \in K : |x| \le 1\}) = 1.$$

The measure μ is Haar measure suitably normalised.

There is a character χ on the additive group of K with the properties

$$\chi(\{x : |x| \le 1\}) = \{1\}$$

and

$$\chi(\{x : |x| \le q\}) \ne \{1\}.$$

For $N = 1,2,\ldots$, the correspondence $\lambda \longleftrightarrow \chi_\lambda$, where $\chi_\lambda(x) = \chi(\lambda \cdot x)$ establishes an isomorphism between the additive group of K^N and its dual.

Let E be a vector space over K. A *norm* on E is a map $\| \quad \|_E : E \to [0, \infty[$ such that

$$\|x\|_E = 0 \iff x = 0;$$

$$\|\lambda x\|_E = |\lambda| \, \|x\|_E, \lambda \in K;$$

$$\|x + y\|_E \le \|x\|_E \vee \|y\|_E .$$

The last property is also called the ultrametric inequality and implies the obvious generalisation of the isosceles triangle property. We call the pair $(E, \| \quad \|_E)$ a *normed vector space* (over K).

If E is complete in the metric $(x,y) \mapsto \|x - y\|_E$ we say that E is a *Banach space*. For $N = 1,2,\ldots$ the space $(K^N, | \quad |)$, where

$$|(x_1, \ldots, x_N)| = |x_1| \vee \cdots \vee |x_N|,$$

is a Banach space. More generally, if (Ω, F, P) is a probability space and we let L^∞ be the set of measurable functions $f: \Omega \to K$ such that $\operatorname{ess\,sup}\{|f(\omega)| : \omega \in \Omega\} < \infty$ then L^∞ becomes a Banach space when we equip it with the norm $\| \quad \|_\infty$ defined by $\|f\|_\infty = \operatorname{ess\,sup}\{|f(\omega)| \mid \omega \in \Omega\}$ (we adopt the usual convention that we regard two functions to be equal if they are equal almost surely).

A subset C of a normal vector space E is said to be *orthogonal* if for every finite subset $\{x_1, \ldots, x_N\} \subset C$ and each $\lambda_1, \ldots, \lambda_N \in K$, we have

$$\|\sum_{i=1}^{N} \lambda_i x_i\|_E = \vee_{i=1}^{N} |\lambda_i| \, \|x_i\|_E .$$

If, moreover, $\|x\|_E = 1$ for all $x \in C$, then C is said to be *orthonormal.*

We now recall from Evans (1989) our general definition for the local field analogue of Gaussian measures. Let E be a measurable vector space (over K). Let E_1 and E_2 be two copies of E and write $X_i: E_1 \times E_2 \to E_i$, $i = 1,2$, for the two coordinate maps. A measure P on E is said to be K-*Gaussian* if for every pair of orthonormal vectors $(\alpha_{11}, \alpha_{12})$, $(\alpha_{21}, \alpha_{22}) \in K^2$ the law of $(\alpha_{11}X_1 + \alpha_{12}X_2, \alpha_{21}X_1 + \alpha_{22}X_2)$ under $P \times P$ is also $P \times P$.

Note: in future when we consider measure on a measurable vector space E we will always reserve the notation X for the identity map on E.

The K-Gaussian measures on $E = K$ are those measures P such that $X \in L^\infty(P)$ and $\int \chi(\xi x) P(dx) = \Phi(\|X\|_\infty |\xi|)$, where we let Φ denote the indicator function of the interval $[0,1]$. Thus, either $X = 0$ or

$$P(dx) = \|X\|_\infty^{-1} \Phi(\|X\|_\infty^{-1} |x|) \mu(dx).$$

The theory of K-Gaussian measures is particularly tractable when **B**, the σ-field on the measurable vector space E, is the σ-field generated by some collection, F, of linear functionals on E. In this case we say that the triple (E, F, **B**) satisfies the hypothesis (*) of Evans (1989). One example is the case when E is a separable Banach space, $F = E^*$ (the dual of E) and **B** is the Borel σ-field of E.

If (E, F, **B**) satisfies the hypothesis (*) then a measure P on E is K-Gaussian if and only if T(X) is a K-valued, K-Gaussian random variable for all $T \in \operatorname{span} F$. Moreover, P is then uniquely determined by the laws of the individual random variables T(X), $T \in \operatorname{span} F$, and hence by the set of numbers $\|T(X)\|_\infty$, $T \in \operatorname{span} F$.

3. Equivalence

A well-known feature of the Gaussian theory is, in a variety of general settings, that two Gaussian measures on the same space are either equivalent or perpendicular. We refer the reader to Kuo (1975) for a discussion of such results and some relevant references. An analogous theorem certainly doesn't hold for the K-Gaussian theory. For example, suppose that P and Q are two K-Gaussian measures on E = K such that $\|X\|_\infty = 1$ in $L^\infty(P)$ and $\|X\|_\infty = q$ in $L^\infty(Q)$. Then $P \ll Q$ but P and Q are not equivalent. As the following theorem shows, equivalence is a much more restrictive condition in the K-Gaussian case.

Theorem 3.1. Suppose that (E, F, **B**) satisfies the hypothesis (*). Let P and Q be two K-Gaussian measures on E. Then $P \sim Q$ if and only if $P = Q$.

Proof. Suppose that $P \sim Q$. Then $\|T(X)\|_\infty$ is the same in $L^\infty(P)$ and $L^\infty(Q)$ for all $T \in \operatorname{span} F$. As T(X) is K-valued, K-Gaussian for all such T we see that the distribution of T(X) under P is the same as that under Q, and hence $P = Q$.

The converse is obvious. □

4. Translation

Suppose that Z is a real-valued, centred, Gaussian process indexed by some set I. Let H be the corresponding reproducing kernel Hilbert space. If $z \in \mathbb{R}^I$ then it is known that the laws of Z and z + Z are either equivalent or perpendicular depending upon whether or not $z \in H$ (see, for example, Feldman (1958) or Hajek (1959)). In the K-Gaussian case we have the following analogue.

Theorem 4.1. Suppose that (E, F, **B**) satisfies the hypothesis (*). Let P be a K-Gaussian measure on E. Set

$$S = \{x \in E : |T(x)| \le \|T(X)\|_\infty \text{ all } T \in \operatorname{span} F\}.$$

If $x \in S$ then the law of x + X under P is P itself. Otherwise, the law of x + X under P is perpendicular to P.

Proof. Note that if Y is a K-valued, K-Gaussian random variable and $y \in K$ is such that $|y| \leq \|Y\|_\infty$ then, since the law of Y is Haar measure on the subgroup $\{z: |z| \leq \|Y\|_\infty\}$, we see that the law of $y + Y$ is that of Y. Hence, if $x \in S$ we have that the law of $T(x + X) = T(x) + T(X)$ under P is that of $T(X)$ under P for all $T \in \operatorname{span} F$. Thus the law of $x + X$ under P is that of X under P.

If $x \notin S$ then there exists $T \in \operatorname{span} F$ such that $|T(x)| > \|T(X)\|_\infty$. Then, by the isosceles triangle property, $|T(x + X)| = |T(x)|$ P-almost surely and hence the law of $x + X$ is perpendicular to P. □

We remark that S determines P uniquely. Also, S is an additive subgroup of E which is closed under multiplication by scalars $\alpha \in K$ such that $|\alpha| \leq 1$. If we combine parts (i) and (ii) of the following result with Theorem 4.1 we see when E is a separable Banach space that the group S supports P, S is compact and that P is just normalised Haar measure on S. Part (iii) extends Theorem 6.1 of Evans (1989), where it was shown that if P is a K-Gaussian measure on measurable vector space $(E, \mathbf{B})$ and M is a measurable subspace of E then either $P(M) = 0$ or 1. Here we see that if E is a separable Banach space then it is possible to give an analytic condition which determines what branch of the dichotomy holds. No comparable condition on the covariance structure seems to be known for the various Gaussian analogues of this zero-one law.

Theorem 4.2. Suppose that $(E, \|\ \|_E)$ is a separable Banach space with dual E^* and P is a K-Gaussian measure on E. Set

$$S = \{x \in E: |T(x)| \leq \|T(X)\|_\infty \text{ all } T \in E^*\}.$$

(i) The group S is the closed support of P.

(ii) The group S is compact.

(iii) If M is a measurable vector subspace of E then $P(M)$ is either 1 or 0, depending on whether or not $S \subset M$.

Proof. (i) It is clear that S is closed. Let $\{T_i\}_{i=1}^\infty$ be a countable dense subset of E^*. Then

$$P(S) = P\Big(\bigcap_{i=1}^{\infty} \{x \in E: |T_i(x)| \leq \|T_i(X)\|_\infty\}\Big) = 1.$$

Conversely, suppose that $x \in S$ and U is an open neighbourhood of x. Let $\{x_i\}_{i=1}^\infty$ be a countable dense subset of S. Then $\cup_{i=1}^\infty [x_i + (U - x)]$ covers S and hence $P(x_i + (U - x)) > 0$ for at least one i; but, by Theorem 4.1, $P(U) = P((x_i - x) + U)$, since $x_i - x \in S$.

(ii) As E is complete and separable, all probability measures on E are tight and so there exists a compact set $C \subset S$ such that $P(C) > 0$. Let G be the smallest closed additive group containing C. We claim that G is also compact. Given $\varepsilon > 0$, there exists a finite set $\{x_1^\varepsilon, \ldots, x_{n(\varepsilon)}^\varepsilon\} \subset C$ such that if $x \in C$ then $\|x - x_i^\varepsilon\|_E < \varepsilon$ for some x_i^ε. The smallest closed group containing $\{x_1^\varepsilon, \ldots, x_{n(\varepsilon)}^\varepsilon\}$ is $G^\varepsilon = (D \times x_1^\varepsilon) + \cdots + (D \times x_{n(\varepsilon)}^\varepsilon)$ where D is the ring of integers in K, that is, $D = \{k \in K : |k| \le 1\}$. Clearly, G^ε is compact. Moreover, from the ultrametric inequality it is clear that if $x \in G$, then there exists $y \in G^\varepsilon$ such that $\|x - y\|_E < \varepsilon$. Thus G is totally bounded and hence compact.

Part (ii) will now follow if G has only finitely many distinct cosets in S; but this must be the case, since otherwise we could find infinitely many disjoint cosets $G_1, G_2, \ldots$ for which, by Theorem 4.1, $P(G_i) = P(G) > 0$.

(iii) From Theorem 6.1 of Evans (1989) we know that $P(M)$ is either 0 or 1. If $S \subset M$ then it follows from (i) that $P(M) = 1$. Conversely, suppose that $P(M) = 1$. If there exists $x \in S$ such that $x \notin M$ then M and $x + M$ are disjoint; but this is impossible, since $P(x + M) = P(M) = 1$ by Theorem 4.1. □

The converse to Theorem 4.2 holds.

Theorem 4.3. Suppose that $(E, \|\ \|_E)$ is a separable Banach space and that G is a compact additive subgroup of E such that $\alpha G \subset G$ when $\alpha \in K$ with $|\alpha| \le 1$. Let P be normalised Haar measure on G. Then P is a K-Gaussian measure for which $S = G$.

Proof. It follows from Corollary 7.4 of Evans (1989) that P is K-Gaussian. Part (i) of Theorem 4.2 shows that $S = G$. □

If $(H, \langle \cdot, \cdot \rangle)$ is a real, separable Hilbert space then it is well-known that there exists a probability measure P on H such that $\int e^{i\langle x,y\rangle} P(dy) = e^{-\langle x,x\rangle/2}$ for all $x \in H$ if and only if H is finite dimensional. The corresponding result in our setting is the following.

Corollary 4.4. Let $(E, \|\ \|_E)$ be a separable Banach space. There exists a probability measure P on E such that $\int \chi(Ty) P(dy) = \Phi(\|T\|_{E^*})$ for all $T \in E^*$ if and only if E is finite dimensional.

Proof. Suppose that E is infinite dimensional and such a probability measure P exists. It is clear that P is K-Gaussian and $\|T(X)\|_\infty = \|T\|_{E^*}$ for all $T \in E^*$. Therefore we have

$$S = \{x : |T(x)| \le \|T\|_{E^*} \text{ all } T \in E^*\}$$

and so $S \supset \{x : \|x\|_E \le 1\}$ (in fact, we have equality). Since the unit ball of E is certainly not compact, this contradicts part (ii) of Theorem 4.2.

Suppose, on the other hand, that E has finite dimension n. Let $e_1, \ldots, e_n$ be an orthogonal basis for E (see Theorem 50.8 of Schikhof (1984) for the existence of such a basis). By rescaling, we may assume that $q^{-1} < \|e_i\|_E \le 1$, $1 \le i \le n$, so that $\{x \in E : \|x\|_E \le 1\} = \{\Sigma_i \alpha_i e_i : \vee_{i=1}^n |\alpha_i| \le 1\}$ and hence $\|T\|_{E^*} = \vee_{i=1}^n |Te_i|$ for each $T \in E^*$. Let $X_1, \ldots, X_n$ be independent, K-valued, K-Gaussian random variables for which $\|X_i\|_\infty = 1$, so that $X_1, \ldots, X_n$ are orthonormal in L^∞ (see Theorem 7.5 of Evans (1989)). Set $X = \Sigma_i X_i e_i$. Then X is K-Gaussian and $\|T(X)\|_\infty = \vee_{i=1}^n |Te_i|$, so the law of X is the probability measure we are seeking. □

5. Contraction

Suppose that $\{Z(i) : i \in I\}$ is a real-valued, centred, Gaussian process on some index set I. Using, for example, Theorem II. 4.3 in Kuo (1975) it is not difficult to see that if $\alpha \in \mathbb{R} \setminus \{-1, 0, 1\}$ then the law of αZ is either equivalent or perpendicular to the law of Z depending on whether the subspace of L^2 spanned by $\{Z(i) : i \in I\}$ is finite dimensional or infinite dimensional. The corresponding result holds in our setting.

Theorem 5.1. Suppose that $(E, F, \mathbf{B})$ satisfies the hypothesis (*) and P is a K-Gaussian measure on E. Given $\alpha \in K$ with $0 < |\alpha| < 1$, let Q be the law of αX. Then either $Q \ll P$ or $Q \perp P$, depending on whether the subspace of $L^\infty(P)$ spanned by $\{T(X) : T \in F\}$ is finite dimensional or infinite dimensional.

Proof. Suppose that the dimension of the subspace of $L^\infty(P)$ spanned by $\{T(X) : T \in F\}$ is $m < \infty$. Then there exists $T_1^*, \ldots, T_m^* \in \operatorname{span} F$ such that $\operatorname{span}\{T_1^*(X), \ldots, T_m^*(X)\} = \operatorname{span}\{T(X) : T \in F\}$ and $T_1^*(X), \ldots, T_m^*(X)$ are orthonormal in $L^\infty(P)$ (see Theorem 50.8 of Schikhof (1984)). From Theorem 7.5 of Evans (1989), $T_1^*(X), \ldots, T_m^*(X)$ are independent. If g is a **B**-measurable function then g is of the form $g(x) = G(T_1(x), T_2(x), \ldots)$ for some measurable function G on $K^{\mathbb{N}}$ and some sequence $\{T_i\} \subset F$. We may find coefficients $\beta_{ij} \in K$, $1 \le i < \infty$, $1 \le j \le m$, such that $T_i(X) = \Sigma_j \beta_{ij} T_j^*(X)$, $1 \le i < \infty$. Define $G^* : K^m \to \mathbb{R}$ by $G^*(t_1, \ldots, t_m) = G((\Sigma_j \beta_{ij} t_j)_{i=1}^\infty)$. A straightforward calculation of the density of the law of $(T_1^*(X), \ldots, T_m^*(X))$ with respect to μ^m shows that

$$\begin{aligned} Q[g(X)] &= P[g(\alpha X)] \\ &= P[G(\alpha T_i(X)_{i=1}^\infty)] \\ &= P[G^*(\alpha T_1^*(X), \ldots, \alpha T_m^*(X))] \\ &\le |\alpha|^{-m} P[G^*(T_1^*(X), \ldots, T_m^*(X))] \\ &= P[g(X)], \end{aligned}$$

and so $Q \ll P$.

Suppose now that the dimension of the subspace of $L^\infty(P)$ spanned by $\{T(X) : T \in F\}$ is infinite dimensional. Then we may find a sequence $\{T_i\} \subset \operatorname{span} F$ such that $T_1(X), T_2(X), \ldots$ are orthonormal in $L^\infty(P)$ and hence independent (see Theorem 7.5 of Evans (1989)). The event $\{|T_i(X)| \le |\alpha|,\ 1 \le i < \infty\}$ has probability zero under P and probability one under Q, so that $P \perp Q$. □

When E is a Banach space the dimension appearing in Theorem 5.1 shows up in the probability of small balls.

Theorem 5.2. Suppose that $(E, \|\ \|_E)$ is a separable Banach space and P is a K-Gaussian measure on E. Let

$$m = \dim\{T(X) : T \in E^*\}.$$

Then

$$m = -\lim_{n\to\infty} \frac{\log_q P(\|X\|_E \le q^{-n})}{n}.$$

Proof. The result is obvious if $m = 0$, since in that case $X = 0$ almost surely. Suppose next that $1 \le m < \infty$. We may find $T_1, \ldots, T_m \in E^*$ such that $\{T_1(X), \ldots, T_m(X)\}$ forms an orthonormal basis for $\{T(X) : T \in E^*\}$ in $L^\infty(P)$. Consequently, $\{T_1(X), \ldots, T_m(X)\}$ are independent random variables. Let $(e_i)_{i=1}^\infty$ be a basis for E. If $\pi_i : E \to K$, $1 \le i < \infty$, is the ith coordinate map then $\pi_i \in E^*$ and so we have

$$\begin{aligned} X &= \Sigma_{i=1}^\infty \pi_i(X) e_i \\ &= \Sigma_{i=1}^\infty (\Sigma_{j=1}^m \alpha_{ij} T_j(X)) e_i \\ &= \Sigma_{j=1}^m T_j(X) (\Sigma_{i=1}^\infty \alpha_{ij} e_i) \\ &= \Sigma_{j=1}^m T_j(X) f_j, \end{aligned}$$

say, for some choice of coefficients (α_{ij}). It is easy to see that $\{f_1, \ldots, f_m\}$ must be linearly independent, otherwise we would have a contradiction to the linear independence of $\{T_1(X), \ldots, T_m(X)\}$. Thus $(x_1, \ldots, x_m) \mapsto \|\Sigma_{j=1}^m x_j f_j\|_E$ is a norm on K^m. Since all norms on K^m are equivalent (see Theorem 13.3 of Schikhof (1984)) there exists a constant $c > 1$ such that

$$c^{-1} \vee_{j=1}^m |T_j(X)| \le \|X\|_E \le c \vee_{j=1}^m |T_j(X)|$$

and the result now follows easily.

Suppose finally that $m = \infty$. We may then find a sequence $(T_i)_{i=1}^\infty \subset E^*$ such that $(T_i(X))_{i=1}^\infty$ are orthonormal in $L^\infty(P)$ and hence independent. Then

$$P(\|X\|_E \le q^{-n}) \le P(\cap_{i=1}^\infty \{|T_i(X)| \le \|T_i\|_{E^*} q^{-n}\})$$

$$= \prod_{i=1}^{\infty} P(|T_i(X)| \le \|T_i\|_{E^*} q^{-n}),$$

and so $-\lim_{n\to\infty} \log_q P(\|X\|_E \le q^{-n})/n = \infty$, as required. □

REFERENCES

[1] Evans, S.N. (1989). Local field Gaussian measures. In *Seminar on Stochastic Processes 1988* (E. Cinlar, K.L. Chung, R.K. Getoor eds.) pp.121-160. Birkhäuser.

[2] Feldman, J. (1958). Equivalence and perpendicularity of Gaussian processes. *Pacific J. Math.* **4**, 699-708.

[3] Hájek, J. (1959). On a simple linear model in Gaussian processes. In *Trans. Second Prague Conf. Information Theory*, pp.185-197.

[4] Kuo, H.-H. (1975). *Gaussian measures in Banach Spaces*. Lecture Notes in Mathematics 463. Springer.

[5] Schikhof, W.H. (1984). *Ultrametric Calculus*. Cambridge University Press.

[6] Taibleson, M.H. (1975). *Fourier Analysis on Local Fields*. Princeton University Press.

Department of Statistics
University of California
367 Evans Hall
Berkeley, CA 94720
U.S.A.

Skorokhod Embedding by Randomized Hitting Times

P. J. FITZSIMMONS*

1. Introduction.

The "Skorokhod embedding" problem was solved for general strong Markov processes by Rost [**R70,R71**]: given such a process $X = (X_t;\, t \geq 0)$, an initial law μ with σ-finite potential, and a target law ν, there is a randomized stopping time T such that

$$X_T \sim \nu \quad \text{when} \quad X_0 \sim \mu \tag{1.1}$$

if and only if the potential of μ dominates that of ν. Subsequently various authors have shown that under additional hypotheses on X one can take T to be nonrandomized, i.e. a stopping time of the natural filtration of X. For recent work on this subject see [**C85**] and [**FF90**]; see also [**Fa81,Fa83**] which contain references for the earlier literature.

Our object in this note is different. We shall deal with a general right Markov process X, but we shall show that a randomized stopping time T achieving the embedding (1.1) can be chosen from the reasonably narrow class of "randomized hitting times." More precisely, we show that if the potential of μ is σ-finite and dominates that of ν, then there is a monotone family of sets $\{B(r); 0 \leq r \leq 1\}$ such that if T is the first entry time of $B(R)$, where R is independent of X and uniformly distributed over $[0,1]$, then (1.1) holds. The reader will recall that this

* Research supported in part by NSF grant DMS 8721347.

is the same sort of stopping time constructed by Skorokhod in his original work [**Sk65**] on embedding mean zero random variables in Brownian motion.

Our main result, Theorem (2.1) is stated and proved in the next section. The proof is based on a result of Meyer [**Me71**], and on the version of Rost's theorem found in [**Fi88**]. This latter result relies on a technique due to Mokobodzki which was used by Heath [**H74**] to prove what amounts to Theorem (2.1) in the special case of Brownian motion in three or more dimensions. In Sect. 3 we provide a new proof of the result of Meyer mentioned above. This is included since it yields an explicit description of the family $\{B(r); 0 \leq r \leq 1\}$ involved in the main result.

2. Main Result.

Let $X = (\Omega, \mathcal{F}, \mathcal{F}_t, X_t, \theta_t, P^x)$ be a right process in the sense of Sharpe [**Sh88**]. Thus X is a strong Markov process with right continuous paths, along which the α-excessive functions are almost surely right continuous. The state space E of X is homeomorphic to a universally measurable subset of some compact metric space. The Borel σ-field in E is denoted $\mathcal{E}$, and $\mathcal{E}^*$ is the universal completion of $\mathcal{E}$. The transition probabilities $(P_t\,; t \geq 0)$ form a semigroup of subMarkovian kernels on $(E, \mathcal{E}^*)$. In particular, a cemetery point $\Delta \notin E$ is adjoined to E as an isolated point and the lifetime $\zeta := \inf\{t : X_t = \Delta\}$ may be finite. The potential kernel U is defined by

$$Uf(x) := \int_0^\infty P_t f(x)\, dt = P^x\left(\int_0^\zeta f(X_t)\, dt\right).$$

Recall that $\mathcal{E}^e (\supset \mathcal{E})$ denotes the σ-field on E generated by the 1-excessive functions of X. If $B \in \mathcal{E}^e$ then the entry time (or début) of B

$$D_B := \inf\{t \geq 0 : X_t \in B\}$$

is a stopping time of the natural filtration $(\mathcal{F}_t)$. We write

$$H_B f(x) := P^x(f(X_{D_B}); D_B < \zeta)$$

for the associated hitting operator.

Here is the main result of the paper.

(2.1) Theorem. *Let μ and ν be measures on $(E, \mathcal{E})$ such that μU is σ-finite and $\nu U \le \mu U$. Then there is a decreasing family $\{B(r); 0 \le r \le 1\}$ of finely closed $\mathcal{E}^e$-measurable sets such that*

$$\nu = \int_0^1 \mu H_{B(r)}\, dr. \tag{2.2}$$

If $\{A(r); 0 \le r \le 1\}$ is a second such family, then $P^\mu(D_{B(r)} \ne D_{A(r)}) = 0$ for a.e. $r \in [0,1]$.

Remarks. (a) According to a result of Mokobodzki [**Mo71**], if μU is σ-finite, then the extreme points of the convex set $A_\mu := \{\nu : \nu U \le \mu U\}$ are precisely the measures μH_B, $B \in \mathcal{E}^e$. One could use Mokobodzki's theorem and an abstract integral representation theorem of Arsove and Leutwiler [**AL75**] to prove the existence part of Theorem (2.1). We shall give a more direct probabilistic proof. Of course, Mokobodzki's theorem is an immediate corollary of Theorem (2.1).

(b) The measures μ and ν in Theorem (2.1) need not be finite but they are σ-finite: if $f > 0$ and $\mu U(f) < \infty$, then $Uf > 0$ and $\nu(Uf) \le \mu(Uf) < \infty$. The σ-finiteness of μU amounts to a transience hypothesis. Indeed with f as before, X restricted to the absorbing set $\{Uf < \infty\}$ is transient, and each of the measures μ, ν, $\mu H_{B(r)}$, is carried by $\{Uf < \infty\}$.

(c) The probabilistic interpretation of (2.2) is as noted in Sect. 1: if R is chosen independently of X and uniformly distributed over the interval $[0,1]$, and if D is the début of $B(R)$, then X_D has law ν when X_0 has law μ.

For the proof of (2.1) we require two lemmas. The first of these is taken from Sect. 3 of [**Fi88**] and was proved there under the hypotheses of Borel measurability. However the argument is valid in the general case considered here; cf. [**G90**, (5.23)]. The second lemma is due to Meyer [**Me71**, Prop. 8] and, independently, to Mokobodzki.

Recall that an *excessive measure* of X is a σ-finite measure ξ on $(E, \mathcal{E})$ such that $\xi P_t \le \xi$ for all $t > 0$. For example, any potential λU is excessive provided it is σ-finite. If ξ and η are excessive measures then the réduite $R(\xi - \eta)$ is the smallest excessive measure ρ such that $\rho + \eta$ dominates ξ. If ξ is a potential, then so is $R(\xi - \eta)$. For a stopping T the kernel P_T is defined by $P_T f(x) := P^x(f(X_T); T < \zeta)$.

(2.3) Lemma. *Let μU and νU be σ-finite potentials with $\nu U \leq \mu U$. Then there is a family $\{T(r); 0 \leq r \leq 1\}$ of $(\mathcal{F}_t)$ stopping times, with $r \mapsto T(r,\omega)$ increasing and right continuous for each $\omega \in \Omega$, such that*

$$\nu = \int_0^1 \mu P_{T(r)}\, dr \tag{2.4}$$

and

$$R(\nu U - r \cdot \mu U) = \int_r^1 \mu P_{T(s)} U\, ds, \quad \forall r \in [0,1]. \tag{2.5}$$

(2.6) Lemma. *Let ν and λ be measures on $(E, \mathcal{E})$ such that the potentials νU and λU are σ-finite. Then there exists a finely closed $\mathcal{E}^e$-measurable set B such that*

$$R(\nu U - \lambda U) = (\nu - \lambda) H_B U.$$

Proof of Theorem (2.1). Fix $r \in]0,1[$. By (2.5), Lemma (2.6) (with $\lambda = r \cdot \mu$), and the uniqueness of charges [**G90**, (2.12)], there is a finely closed set $B(r) \in \mathcal{E}^e$ such that

$$\int_r^1 \mu P_{T(s)}\, ds = (\nu - r \cdot \mu) H_{B(r)}. \tag{2.7}$$

Since $B(r)$ is finely closed, the measure on the R.H.S. of (2.7) is carried by $B(r)$; the same is therefore true of the L.H.S. It follows that $X_{T(s)} \in B(r)$ a.s. P^μ on $\{T(s) < \infty\}$ for a.e. $s \in [r,1]$. Consequently $T(s) \geq D_{B(r)}$ a.s. P^μ for a.e. $s \in [r,1]$. Invoking Fubini's theorem and the right continuity of $s \mapsto T(s,\omega)$, we conclude that

$$T(r) \geq D_{B(r)} \quad \text{a.s. } P^\mu. \tag{2.8}$$

On the other hand if we apply $H_{B(r)}$ to both sides of (2.7), then by (2.4) and the identity $H_{B(r)} = H_{B(r)}H_{B(r)}$,

$$\mu H_{B(r)} = r^{-1} \int_0^r \mu P_{T(s)} H_{B(r)}\, ds. \tag{2.9}$$

But $D_{B(r)} \leq D(s) := T(s) + D_{B(r)} \circ \theta_{T(s)}$ on $\{T(s) < \infty\}$. Since μU is σ-finite, we can choose $f > 0$ such that $\mu U f < \infty$, and then by (2.9) and the strong Markov property

$$P^\mu \left(\int_{D_{B(r)}}^\infty f(X_t)\, dt \right) = r^{-1} \int_0^r ds\, P^\mu \left(\int_{D(s)}^\infty f(X_t)\, dt \right) < \infty,$$

so

$$D_{B(r)} = T(s) + D_{B(r)} \circ \theta_{T(s)} \geq T(s) \quad \text{a.s. } P^\mu \tag{2.10}$$

for a.e. $s \in [0, r]$. By (2.8), (2.10), and the monotonicity of $s \mapsto T(s,\omega)$ we therefore have

$$T(r-) \leq D_{B(r)} \leq T(r) \quad \text{a.s. } P^\mu. \tag{2.11}$$

Since $r \in]0,1[$ was arbitrary and $T(\cdot,\omega)$ has only countably many discontinuities, formula (2.2) now follows easily from (2.4) and (2.11). The sets $B(r)$ just constructed need not be monotone in r; to remedy this replace $B(r)$ by the fine closure of $\cup\{B(s) : r < s < 1,\ s \text{ rational}\}$ (taking $B(1) = \emptyset$). In view of (2.11) and the monotonicity of $T(\cdot,\omega)$, this change does not disturb the validity of (2.2).

It remains to prove the uniqueness. Let $\{A(r); 0 \leq r \leq 1\}$ be a second family of sets with the properties of $\{B(r); 0 \leq r \leq 1\}$. Then

$$\nu U = \int_0^1 \mu H_{A(s)} U\, ds \leq r \cdot \mu U + \int_r^1 \mu H_{A(s)} U\, ds,$$

from which it follows that

$$\int_r^1 \mu H_{B(s)} U\, ds = R(\nu U - r \cdot \mu U) \leq \int_r^1 \mu H_{A(s)} U\, ds.$$

Consequently, since the $A(r)$'s decrease,

$$\int_0^r \mu H_{B(s)} U\, ds \geq \int_0^r \mu H_{A(s)} U\, ds \geq r \cdot \mu H_{A(r)} U.$$

Thus, by a lemma of Rost [**R74**, p.201],

$$T(s) = D_{B(s)} \leq D_{A(r)} \quad \text{a.s. } P^\mu, \text{ for a.e. } s \in [0, r],$$

hence $T(r-) \leq D_{A(r)}$ a.s. P^μ. Since $\int_0^1 \mu H_{A(r)} U\, dr = \nu U = \int_0^1 \mu H_{B(r)} U\, dr$, the argument used earlier yields $P^\mu(D_{B(r)} \neq D_{A(r)}) = 0$ for a.e. $r \in [0,1]$, as required. □

Remark. The proof of Lemma (2.6) given in Sect. 3 reveals the following recipe for the sets $B(r)$ of Theorem (2.1). For $r \in [0,1]$, the excessive measures $R(\nu U - r \cdot \mu U)$ and νU are both dominated by μU; let their "fine" densities (Lemma (3.1)) be denoted γ_r and ψ respectively. Then $B(r)$ can be taken to be the fine closure of $\{x \in E : \gamma_r(x) \leq \psi(x) - r\}$. (Note that $\{x \in E : \gamma_r < \psi - r\}$ is μU-null.) With a little care one can arrange that $r \mapsto \gamma_r(x)$ is decreasing and convex for each x; this being done, $r \mapsto \{\gamma_r \leq \psi - r\}$ is decreasing, and so is $r \mapsto B(r)$.

3. Proof of Lemma (2.6)

The proof of Lemma (2.6) rests on a domination principle, which is based on the choice of precise versions of certain Radon-Nikodym derivatives. We fix a σ-finite potential $m = \rho U$. A set $B \in \mathcal{E}^e$ is *ρ-evanescent* provided $P^\rho(X_t \in B \text{ for some } t \geq 0) = 0$. The following two lemmas sharpen results in [**Fi87, Fi89**] by taking advantage of the fact that the excessive measure m is a potential. For a complete discussion of these and related results see [**FG90**].

(3.1) Lemma. *Let νU be a σ-finite potential dominated by a multiple of m. Then there is a bounded $\mathcal{E}^e$-measurable version ψ of $d(\nu U)/dm$ and a set $A \in \mathcal{E}^e$ such that*

(i) A is absorbing for X and $E \setminus A$ is ρ-evanescent;

(ii) $\psi|_A$ *is finely continuous.*

The density ψ *is uniquely determined modulo a* ρ*-evanescent set.*

In the sequel we shall write ψ_ν for the "fine" version of $d(\nu U)/dm$ provided by Lemma (3.1). If νU and μU are both dominated by a multiple of m and $\nu U \le \mu U$, then both ψ_ν and $\psi_\nu \wedge \psi_\mu$ are fine versions of $d(\nu U)/dm$. Thus we can (and do) assume that $\psi_\nu \le \psi_\mu$ when $\nu U \le \mu U$. Also, note that if $\nu U \le \mu U$ then $\nu H_B U \le \mu H_B U$ for any $B \in \mathcal{E}^e$. In particular, if μU is dominated by a multiple of m, then μ charges no ρ-evanescent set; cf. [**Fa83**, Lemma 3]. These facts in hand the proof of [**Fi89**, (2.13)] can be adapted in the obvious way (replace "m-polar" by "ρ-evanescent") to yield the following domination principle.

(3.2) Lemma. *Let* μU *and* νU *be* σ*-finite potentials dominated by a multiple of* m*. If* $\psi_\nu \le \psi_\mu$ a.e. ν*, then* $\psi_\nu \le \psi_\mu$ *off a* ρ*-evanescent set, hence* $\nu U \le \mu U$*.*

Proof of Lemma (2.6). Given potentials νU and λU, the réduite $R(\nu U - \lambda U)$, being dominated by νU, is also a potential, say $\nu_1 U$. Moreover, $\nu_1 U$ is strongly dominated by νU in that there is a potential $\nu_2 U$ such that $\nu_1 U + \nu_2 U = \nu U$, and then $\nu_1 + \nu_2 = \nu$ by the uniqueness of charges. (The reader can consult Sect. 5 of [**G90**] for proofs of these well-known facts.) Since $\nu_1 U + \lambda U = R(\nu U - \lambda U) + \lambda U \ge \nu U$, we have

$$\nu_2 U \le \lambda U. \tag{3.3}$$

We take $\rho = \nu + \lambda$, and in the subsequent discussion all fine densities (ψ_ν, ψ_μ, etc.) are taken relative to $m = \rho U$. By a previous remark we can assume that

$$\psi_{\nu_1} + \psi_{\nu_2} = \psi_\nu, \quad \psi_{\nu_2} \le \psi_\lambda.$$

Let B denote the fine closure of $\{\psi_{\nu_2} = \psi_\lambda\}$. Clearly B is $\mathcal{E}^e$-measurable and $B \setminus \{\psi_{\nu_2} = \psi_\lambda\}$ is ρ-evanescent. We will show that

$$\nu_1 U = (\nu - \lambda) H_B U. \tag{3.4}$$

For $\epsilon \in]0,1[$, set $B(\epsilon) = \{\epsilon\psi_{\nu_1} < \psi_\nu - \psi_\lambda\}$, so that $\cap_n B(1-1/n) = B$ up to a ρ-evanescent set. By a lemma of Mokobodzki (see [**G90**, (5.6)]), and [**Fi88**, (2.17)]

$$\nu_1 H_{B(\epsilon)} U = R(1_{B(\epsilon)} \cdot \nu_1 U) = \nu_1 U$$

since $B(\epsilon)$ differs from it fine interior by a ρ-evanescent set. By the uniqueness of charges, $\nu_1 H_{B(\epsilon)} = \nu_1$, so ν_1 is carried by the fine closure of $B(\epsilon)$. But if $0 < \epsilon' < \epsilon < 1$, then $B(\epsilon')$ contains the fine closure of $B(\epsilon)$ up to a ρ-evanescent set not charged by ν_1. It follows that ν_1 is carried by B, hence $\nu_1 = \nu_1 H_B$. To finish the proof of (3.4) we must therefore establish

$$\nu_2 H_B U = \lambda H_B U. \tag{3.5}$$

On the one hand $\lambda U \geq \nu_2 U$, so $\lambda H_B U \geq \nu_2 H_B U$. On the other hand, the inequality $\lambda H_B U \leq \lambda U$ implies that $\{\psi_{\lambda H_B} > \psi_\lambda\}$ is ρ-evanescent. Thus

$$\psi_{\lambda H_B} \leq \psi_\lambda = \psi_{\nu_2} \quad \text{on} \quad B \cap \{\psi_{\lambda H_B} \leq \psi_\lambda\},$$

which carries λH_B. Lemma (3.2) allows us to conclude that $\lambda H_B U \leq \nu_2 U$, hence $\lambda H_B U = \lambda H_B H_B U \leq \nu_2 H_B U$, and (3.5) follows. □

References

[AL75] M. ARSOVE and H. LEUTWILER. Infinitesimal generators and quasi-units in potential theory. Proc. Nat. Acad. Sci. **72** (1975) 2498–2500.

[C85] P. CHACON. The filling scheme and barrier stopping times. Ph. D. Thesis, Univ. Washington, 1985.

[Fa81] N. FALKNER. The distribution of Brownian motion in $\mathbf{R}^n$ at a natural stopping time. Adv. Math. **40** (1981) 97–127.

[Fa83] N. FALKNER. Stopped distributions for Markov processes in duality. Z. Wahrscheinlichkeitstheor. verw. Geb. **62** (1983) 43–51.

[FF90] N. FALKNER and P. J. FITZSIMMONS. Stopping distributions for right processes. Submitted to Probab. Th. Rel. Fields.

[Fi87] P. J. FITZSIMMONS. Homogeneous random measures and a weak order for the excessive measures of a Markov process. Trans. Am. Math. Soc. **303** (1987) 431–478.

[Fi88] P. J. FITZSIMMONS. Penetration times and Skorohod stopping. *Sém. de Probabilités XXII*, pp. 166–174. Lecture Notes in Math. **1321**, Springer,Berlin, 1988.

[Fi89] P. J. FITZSIMMONS. On the equivalence of three potential principles for right Markov processes. Probab. Th. Rel. Fields **84** (1990) 251–265.

[FG90] P. J. FITZSIMMONS and R. K. GETOOR. A fine domination principle for excessive measures. To appear in Math. Z.

[G90] R. K. GETOOR. *Excessive Measures*. Birkhäuser, Boston, 1990.

[H74] D. HEATH. Skorohod stopping via potential theory. *Sém. de Probabilités VIII*, pp. 150–154. Lecture Notes in Math. **381**, Springer, Berlin, 1974.

[Me71] P.-A. MEYER. Le schéma de remplissage en temps continu. *Sém. de Probabilités VI*,pp.130–150. Lecture Notes in Math. **258**, Springer, Berlin, 1971.

[Mo71] G. MOKOBODZKI. Eléments extrémaux pour le balayage. *Séminaire Brelot-Choquet-Deny (Théorie du potentiel)*, 13e année, 1969/70, no.5, Paris, 1971.

[R70] H. ROST. Die Stoppverteilungen eines Markoff-Processes mit lokalendlichem Potential. Manuscripta Math. **3** (1970) 321–329.

[R71] H. ROST. The stopping distributions of a Markov process. Z. Wahrscheinlichkeitstheor. verw. Geb. **14** (1971) 1–16.

[R74] H. ROST. Skorokhod stopping times of minimal variance. *Sém. de Probabilités X*, pp. 194–208. Lecture Notes in Math. **511**, Springer, Berlin, 1974.

[Sh88] M. J. SHARPE. *General Theory of Markov Processes*. Academic Press, San Diego, 1988.

[Sk65] A. V. SKOROKHOD. *Studies in the Theory of Random Processes*. Addison-Wesley, Reading, Mass., 1965.

P. J. FITZSIMMONS
Department of Mathematics, C-012
University of California, San Diego
La Jolla, California 92093

Multiplicative Symmetry Groups of Markov Processes

JOSEPH GLOVER*
RENMING SONG

1. Introduction.

In [**5**], Glover and Mitro formulated a group G consisting of symmetries of the cone $\mathcal{S}$ of excessive functions of a transient Markov process X_t. Roughly speaking, G is defined to be the collection of all bimeasurable bijections φ of the state space E of X_t onto itself such that $\mathcal{S} = \{f \circ \varphi : f \in \mathcal{S}\}$. This group G can also be characterized as the collection of all bimeasurable bijections $\varphi : E \longrightarrow E$ with the following properties: i) $\varphi(X)$ is a transient Markov process; and ii) there is a continuous additive functional A_t^φ of X_t which is strictly increasing and finite on $[0,\zeta)$ with right continuous inverse $\tau(\varphi,t)$ such that $(\varphi(X_t), P^{\varphi^{-1}(x)})$ and $(X_{\tau(\varphi,t)}, P^x)$ are identical in law. Because of this, we call the group G the additive symmetry group of X_t. From each subgroup H of G, Glover and Mitro constructed a new state space F and a surjection $\Phi : E \longrightarrow F$. They showed that, under some mild topological hypotheses, there is a time change $\tau(t)$ of X_t

*Research supported in part by NSA and NSF by grant MDA904-89-H-2037

such that $\Phi(X_{\tau(t)})$ is a strong Markov process. Following this, Glover [3] used appropriate transitive subgroups of G to introduce a group structure on the state space E and showed that, under appropriate conditions, X_t is a Lévy process in this new group structure.

There are at least two important classes of functionals in the theory of Markov processes: one is the class of additive functionals mentioned above, and the other is the class of multiplicative functionals. It is therefore natural to ask if we can formulate a multiplicative symmetry group by using multiplicative functionals and develop a theory similar to that of the additive case.

By using a "diagonal principle", Glover [4] proved results similar to those of [3] for multiplicative symmetry groups when the underlying process X_t is a regular step process. The argument in [4] depends heavily on the special properties of regular step processes, and it seems that his method cannot be extended easily to more general processes.

In this paper, we are going to assume that X_t is a general Markov process but that H is a subgroup of the multiplicative symmetry group with a special property: we shall assume that H has a finite left–invariant measure. The contents of this paper are organized as follows. Section 2 serves as a preparation: the basic framework is set up in this section and a preliminary result is proven. A result similar to that of [3] is given in Section 3.

2. Preparation.

Let E be a Lusin space and let $\mathcal{B}(E)$ be the Borel field of E. Adjoin a cemetery point Δ to E and denote the extended space and Borel field by E_Δ and $\mathcal{B}(E_\Delta)$. Let $X = (\Omega, \mathcal{F}, \mathcal{F}_t, X_t, \theta_t, P^x)$ be a right process on $(E, \mathcal{B}(E))$. For convenience, we shall assume that Ω is the space of all maps $\omega : [0, \infty) \longrightarrow E_\Delta$ which are right continuous and such that $\omega(t) = \Delta$ if and only if $\omega(t+s) = \Delta$ for every $s > 0$. Set $X_t(\omega) = \omega(t)$, and let $\mathcal{F}_t$ and $\mathcal{F}$ be the appropriate completions of $\mathcal{F}_t^0 = \sigma\{X_s : s \le t\}$ and $\mathcal{F}^0 = \sigma\{X_s : s \ge 0\}$. For each $t \ge 0$, $\theta_t : \Omega \longrightarrow \Omega$ is the shift operator characterized by $X_s \circ \theta_t = X_{s+t}$. Under the measure P^x, X_t is a time homogeneous strong Markov process with $X_0 = x$ a.s. P^x. In general, if $\mathcal{E}$ is a σ-algebra, we write $b\mathcal{E}$(resp. $p\mathcal{E}$) to denote the collection of bounded (resp. positive) $\mathcal{E}$-measurable functions.

Let P_t denote the semigroup of X. We assume throughout this article that X is a Borel right process, by which we mean P_t maps Borel functions into Borel functions.

Let G be the collection of bijections $\varphi : E \longrightarrow E$ satisfying the following properties:

(1) φ and φ^{-1} are $\mathcal{B}(E)$ measurable.

(2) φ and φ^{-1} are finely continuous.

PROPOSITION. *If $\varphi \in G$, then $Y_t = \{\varphi(X_t), P^{\varphi^{-1}(x)}\}$ is also a Borel right process.*

PROOF: Let $P_\varphi^x = P^{\varphi^{-1}(x)}$. We must check first that Y_t is a right continuous strong Markov process on E. If g is any continuous function on E, then $g \circ \varphi$ is finely continuous, so $g(Y_t)$ is right continuous a.s. P_φ^x for every x in E. Therefore Y_t is right continuous a.s. P_φ^x for every $x \in E$. Since φ is a measurable bijection, Y_t inherits the strong Markov property from X_t.

Second, we need to check that $g(Y_t)$ is right continuous whenever g is excessive for Y_t. But g is excessive for Y_t if and only if $g \circ \varphi$ is excessive for X_t, so $g(Y_t)$ is right continuous a.s. P_φ^x for every x. Let U_φ^α be the resolvent of Y_t. For $x \in E$, we have

$$\begin{aligned}\epsilon_x U_\varphi^\alpha(f) &= P^{\varphi^{-1}(x)} \int e^{-\alpha t} f(Y_t)\, dt \\ &= \epsilon_{\varphi^{-1}(x)} U^\alpha(f \circ \varphi) \\ &= \varphi(\epsilon_{\varphi^{-1}(x)} U^\alpha)(f)\end{aligned}$$

for every bounded positive function $f \in \mathcal{B}(E)$. Therefore, for every $x \in E$,

$$\epsilon_x U_\varphi^\alpha = \varphi(\epsilon_{\varphi^{-1}(x)} U^\alpha).$$

Since φ and φ^{-1} are both $\mathcal{B}(E)$-measurable, (U_φ^α) is a Borel resolvent and we have proved that Y_t is a right process. ∎

DEFINITION. *A family $M = \{M_t;\ 0 \le t < \infty\}$ of positive real-valued random variables on $(\Omega, \mathcal{F})$ is called a multiplicative functional of X provided:*

(1) $M_t \in \mathcal{F}_t$ *for each* $t \ge 0$.

(2) $M_{t+s} = M_t(M_s \circ \theta_t)$ a.s. *for each* $t, s \ge 0$.

A multiplicative functional M of X is called nonvanishing if $M_t > 0$ a.s. P^x on $\{t < \zeta\}$ for every $t > 0$ and every $x \in E$.

A multiplicative functional M of X is said to be a strong multiplicative functional provided that

$$M_{t+T} = M_T M_t \circ \theta_T$$

a.s. P^x *on $\{T < \infty\}$ for every $x \in E$, every $t \ge 0$ and every stopping time T.*

It follows from [**1**] that any right continuous multiplicative functional of X is a strong multiplicative functional

Given two multiplicative functionals M and N of X, we say that N is a version of M if $P^x[M_t \ne N_t;\ t < \zeta] = 0$ for all t and x.

DEFINITION. *$\varphi \in G$ is called a multiplicative symmetry of X if there is a right continuous nonvanishing multiplicative functional M_t^φ such that*

$$P^{\varphi^{-1}(x)}[f \circ \varphi(X_t)] = P^x[f(X_t)M_t^\varphi] \tag{1}$$

for every $f \in p\mathcal{B}(E)$, for every t. We let GM denote the collection of all multiplicative symmetries of X.

It is easy to see that for every $\varphi \in GM$, M_t^φ is a supermartingale multiplicative functional.

PROPOSITION. *GM is a subgroup of G.*

PROOF: First, we must show that if $\varphi \in GM$, then $\varphi^{-1} \in GM$. Define a map $\Gamma^\varphi : \Omega \longrightarrow \Omega$ by $\Gamma^\varphi(\omega) = \varphi(X_t(\omega))$ (where $\varphi(\Delta) = \Delta$); then (1) implies

$$\Gamma^\varphi(P^{\varphi^{-1}(x)})[F; t < \zeta] = P^x[F \cdot M_t^\varphi]$$

for every $F \in p\mathcal{F}_t$. In particular, if we let $F = f(X_t)/M_t^\varphi$, we have

$$\begin{aligned} P^x[f(X_t)] &= \Gamma^\varphi(P^{\varphi^{-1}(x)})[f(X_t)/M_t^\varphi] \\ &= P^{\varphi^{-1}(x)}[f \circ \varphi(X_t) \cdot 1/(M_t^\varphi \circ \Gamma^\varphi)] \end{aligned}$$

If we replace $f \circ \varphi$ with g and $\varphi^{-1}(x)$ with z, we see that

$$P^{\varphi(z)}[g \circ \varphi^{-1}(X_t)] = P^z[g(X_t) \cdot 1/(M_t^\varphi \circ \Gamma^\varphi)].$$

Since $1/(M_t^\varphi \circ \Gamma^\varphi)$ is a right continuous, nonvanishing multiplicative functional of X, $\varphi^{-1} \in GM$.

Second, we must show that if φ and ψ are in GM, then $\varphi \circ \psi \in GM$. To do this, we compute

$$\begin{aligned} P^{(\varphi\circ\psi)^{-1}(x)}[f \circ \varphi \circ \psi(X_t)] &= P^{\varphi^{-1}(x)}[f \circ \varphi(X_t) M_t^\psi] \\ &= P^x[f(X_t) M_t^\varphi \cdot (M_t^\psi \circ \Gamma^{\varphi^{-1}})] \end{aligned}$$

Since $M^\varphi \cdot (M_t^\psi \circ \Gamma^{\varphi^{-1}})$ is also a right continuous, nonvanishing multiplicative functional of X, $\varphi \circ \psi \in GM$ and we conclude that GM is a group. ∎

From the proof above, we can see that we have the following important

COROLLARY. *For any $\varphi, \psi \in GM$, we have*

(1) $1/M_t^\varphi \circ \Gamma^\varphi$ *is a version of* $M_t^{\varphi^{-1}}$

(2) $M_t^\varphi \cdot (M_t^\psi \circ \Gamma^{\varphi^{-1}})$ *is a version of* $M_t^{\varphi\circ\psi}$

3. Lévy processes.

Take a subgroup H of GM. In this article we are going to assume:

HYPOTHESIS. *H is transitive, i.e., for each pair of points x and y in E, there is a map $\varphi \in H$ such that $\varphi(x) = y$.*

Let us fix, once and for all, a point $e \in E$ to serve as a reference point in E and let

$$H_e = \{\varphi \in H;\ \varphi(e) = e\}.$$

This is a subgroup of H, and we let $\mathcal{T} = H/H_e$ be the collection of left cosets

$$\varphi H_e = \{\varphi \circ \psi;\ \psi \in H_e\}.$$

From [3] we know that

$$\varphi H_e = \{\psi \in H;\ \psi(e) = \varphi(e)\}.$$

Because of this, we can define a map Ψ from E to $\mathcal{T}$ as follows:

$$\Psi(x) = \{\varphi \in H;\ \varphi(e) = x\}.$$

In fact, it is easy to show that Ψ is a bijection from E to $\mathcal{T}$ (see [3]).

The bijection $\Psi : E \longrightarrow \mathcal{T}$ allows us to identify E with $\mathcal{T}$; we thereby endow E with the structure of a coset space.

Now we are going to assume the following:

HYPOTHESIS. *H_e is trivial; i.e., H_e consists only of the identity map.*

Under this hypothesis, $\mathcal{T}$ and H are isomorphic, and Ψ is a bijection from E to H. We use Ψ to identify E and H and in particular, Ψ endows E with the group structure of H given by the product

$$xy = \Psi^{-1}(\Psi(x) \circ \Psi(y))$$

whenever $x, y \in E$.

The group product notation above is useful, but we also find it convenient to use the product in H (which is composition $\circ$) by identifying the point $x \in E$ with the map $\varphi_x = \Psi(x) \in H$.

HYPOTHESIS. *$(x, y) \longrightarrow xy$ and $(x, y) \longrightarrow x^{-1}y$ are $\mathcal{B}(E) \times \mathcal{B}(E)$-measurable.*

DEFINITION. *If μ is a measure on $(E, \mathcal{B}(E))$ and $x \in E$, μ^x is the measure on $(E, \mathcal{B}(E))$ defined by $\mu^x(A) = \mu(xA)$ for every $A \in \mathcal{B}(E)$. A σ-finite measure μ on $(E, \mathcal{B}(E))$ is said to be left quasi-invariant if $\mu^x << \mu$ for every $x \in E$. A σ-finite measure m on $(E, \mathcal{B}(E))$ is said to be a left Haar measure if $m^x = m$ for every $x \in E$.*

In this paper we assume the following

HYPOTHESIS. *There is a σ–finite left quasi-invariant measure μ on $(E, \mathcal{B}(E))$.*

By the Mackey-Weil theorem (see [3]) we know that this hypothesis implies there is a topology on E making E into a locally compact second countable metric group such that:

(1) the Borel σ-algebra of the topology is $\mathcal{B}(E)$;

(2) there is a left Haar measure n, and μ and n have the same null sets.

We are going to call this topology the Mackey-Weil topology, and we set

$$m(A) = n(\Psi^{-1}(A))$$

so m is a measure on $(H, \mathcal{B}(H))$, where $\mathcal{B}(H) = \{\Psi^{-1}(A) : A \in \mathcal{B}(E)\}$.

In this article we are going to **assume m is a finite measure**. Without loss of generality we can assume that $m(H) = 1$.

The purpose of this article is to use $\{M_t^\varphi; \varphi \in H\}$ to produce a nice multiplicative functional M_t so that (X_t, M_t) is a Lévy process. In order to proceed, we need to know that M_t^φ can be made jointly measurable.

PROPOSITION. *There is a process N_t^φ such that*

(1) *for each φ, N^φ is a version of M^φ;*

(2) *$(t, x, \omega) \longrightarrow N_t^{\Psi(x)}$ is $\mathcal{B}(R^+) \times \mathcal{B}(E) \times \mathcal{F}^0$-measurable.*

PROOF: First we fix a $t > 0$. For each pair $(x, \varphi) \in E \times H$, define a measure $L_t((x, \varphi), d\omega)$ by setting $L_t((x, \varphi), F) = P^x[M_t^\varphi \cdot F]$ for every $F \in p\mathcal{F}_t^0$. Assume for the moment that we have shown that $(x, z) \longrightarrow L_t((x, \varphi_z), F)$ is $\mathcal{B}(E) \times \mathcal{B}(E)$-measurable. Doob's lemma then yields a density $C_t(x, z, \omega) \in \mathcal{B}(E) \times \mathcal{B}(E) \times \mathcal{F}_t^0$ such that

$$L_t((x, \varphi_z), F) = P^x[C_t(x, z, \cdot)F(\cdot)]$$

for every $F \in p\mathcal{F}_t^0$. If we set $C_t^z(\omega) = \dot{C}_t(X_0(\omega), z, \omega)$, then $C_t^z(\omega)$ is $\mathcal{B}(E) \times \mathcal{F}_t^0$-measurable and $C_t^z(\omega) = M_t^{\Psi(z)}$ a.s. P^x for every x.

Now we define

$$N_t^{\Psi(z)}(\omega) = \liminf_{s \downarrow\downarrow t, s \in Q} C_s^z(\omega).$$

Then $t \longrightarrow N_t^{\Psi(z)}$ is right continuous a.s., $N_t^{\Psi(z)}$ and $M_t^{\Psi(z)}$ are indistinguishable, $(t, x, \omega) \longrightarrow N_t^{\Psi(x)}$ is $\mathcal{B}(R^+) \times \mathcal{B}(E) \times \mathcal{F}^0$- measurable and $N_t^{\Psi(x)}$ is $\mathcal{F}_t$-measurable for every t.

So all that remains to complete the proof of this proposition is to verify that $(x, z) \longrightarrow P^x[M_t^{\Psi(z)} \cdot F]$ is $\mathcal{B}(E) \times \mathcal{B}(E)$-measurable whenever $F \in p\mathcal{F}_t^0$. Since $\mathcal{F}_t^0$ is generated by random variables of the form

$$f_1(X(t_1))f_2(X(t_2)) \cdots f_n(X(t_n))$$

with $t_1 < t_2 < \cdots < t_n \le t$ and $n = 1, 2, \cdots$, it suffices to prove that

$$(x, z) \longrightarrow P^x[M_t^{\Psi(z)} f_1(X(t_1)) \cdots f_n(X(t_n))]$$

is $\mathcal{B}(E) \times \mathcal{B}(E)$-measurable for every n and all $t_1 < \cdots < t_n \le t$.

We proceed by induction on n. When $n = 1$

$$\begin{aligned} P^x[M_t^{\Psi(z)} f_1(X(t_1))] &= P^x[M_{t_1}^{\Psi(z)} M_{t-t_1}^{\Psi(z)} \circ \theta_{t_1} f_1(X(t_1))] \\ &= P^x[M_{t_1}^{\Psi(z)} f_1(X(t_1)) P^{X(t_1)}[M_{t-t_1}^{\Psi(z)}]] \\ &= P^{\varphi_z^{-1}(x)}[f_1 \circ \varphi_z(X(t_1)) P^{\varphi_z(X(t_1))}[M_{t-t_1}^{\Psi(z)}]] \\ &= P^{\varphi_z^{-1}(x)}[f_1 \circ \varphi_z(X(t_1)) P^{X(t_1)}[t - t_1 < \zeta]] \end{aligned}$$

Since $(z, x) \to \varphi_z^{-1}(x)$ and $(z, x) \to \varphi_z(x)$ are jointly measurable, and X is a Borel right process, we immediately obtain the desired measurability.

Now we assume that for any $f_1, \cdots, f_{n-1} \in \mathcal{B}(E)$ and any $t_1 < \cdots < t_{n-1} \le t$

$$(x, z) \longrightarrow P^x[M_t^{\Psi(z)} f_1(X(t_1)) \cdots f_{n-1}(X(t_{n-1}))]$$

is $\mathcal{B}(E) \times \mathcal{B}(E)$-measurable. Then for any $f_1, \cdots, f_n \in \mathcal{B}(E)$ and any $t_1 < \cdots < t_n \le t$,

$$\begin{aligned} &P^x[M_t^{\Psi(z)} f_1(X(t_1)) \cdots f_n(X(t_n))] \\ &= P^x[M_{t_1}^{\Psi(z)} \cdot M_{t-t_1}^{\Psi(z)} \circ \theta_{t_1} f_1(X(t_1))[f_2(X(t_2 - t_1)) \cdots f_n(X(t_n - t_1))] \circ \theta_{t_1}] \\ &= P^x[M_{t_1}^{\Psi(z)} f_1(X(t_1)) P^{X(t_1)}[M_{t-t_1}^{\Psi(z)} f_2(X(t_2 - t_1)) \cdots f_n(X(t_n - t_1))]]. \end{aligned}$$

By the induction assumption we know that

$$(x,z) \longrightarrow P^x[M_{t-t_1}^{\Psi(z)} f_2(X(t_2-t_1)) \cdots f_n(X(t_n-t_1))]$$

is $\mathcal{B}(E) \times \mathcal{B}(E)$-measurable. It follows that

$$(x,z) \longrightarrow P^x[M_t^{\Psi(z)} f_1(X(t_1)) \cdots f_n(X(t_n))]$$

is $\mathcal{B}(E) \times \mathcal{B}(E)$-measurable. ∎

Now let us put

$$A_t^\varphi = \ln N_t^\varphi.$$

Aside from our generic assumptions about the structure of H, which have appeared before in [**3**] and [**4**], we have made the special assumption that $m(H) = 1$. We need one other special hypothesis, without which our proposed method cannot work.

HYPOTHESIS. *There is a null set N such that for any $t > 0$ and for any $\omega \in \Omega - (N \cup \{t > \zeta\})$,*

$$(A_t^\varphi)^- \in L^1(dm)$$

Under this hypothesis, we can define, for any $t > 0$, $A_t(\omega) = \int_H A_t^\varphi \, m(d\varphi)$ when $t < \zeta(\omega)$ and $A_t(\omega) = -\infty$ when $t > \zeta(\omega)$. Since $\int (A_t^\varphi)^- \, m(d\varphi)$ and $\int (A_s^\varphi)^- \, m(d\varphi)$ are finite on $\Omega - (N \cup \{t+s > \zeta\})$, the following identities are true almost surely on $\{t+s < \zeta\}$:

$$\begin{aligned} A_{t+s} &= \int (A_t^\varphi + A_s^\varphi \circ \theta_t) \, m(d\varphi) \\ &= \int A_t^\varphi \, m(d\varphi) + \int A_s^\varphi \circ \theta_t \, m(d\varphi) \\ &= A_t + A_s \circ \theta_t \end{aligned}$$

Thus the A_t defined above is an additive functional. In fact, the same argument yields the fact that for every $x \in E$, every $t > 0$ and every stopping time T,

$$A_{t+T} = A_T + A_t \circ \theta_T$$

a.s. P^x on $\{T < \infty\}$.

Put

$$M_t = e^{A_t}$$

Then Jensen's inequality implies

$$\begin{aligned} P^x[M_t] &= P^x[\exp\{\int A_t^{\varphi} m(d\varphi)\}] \\ &\leq P^x[\int \exp\{A_t^{\varphi}\} m(d\varphi)] \\ &\leq 1. \end{aligned}$$

So M_t is finite almost surely and furthermore, the inequality above shows that M_t is a supermartingale strong multiplicative functional. Our hypothesis insures that M_t is nonvanishing.

As we mentioned above, we need the hypothesis to insure that M_t does not vanish. To see what can happen without this hypothesis, let $E = [0, 2\pi)$ be the circle group, and let Y_t be the Lévy process on E which sits for an exponential length of time at its starting point x, after which it jumps to the point $x + \pi$ $(mod\ 2\pi)$, where it sits for an exponential length of time, etc. Let $c > 0$, define

$$\begin{aligned} B_t &= \int_0^t c(Y_s) ds \\ R_t &= \exp(-B_t) \end{aligned}$$

and let X_t be the process Y_t killed by R_t. Let H be the group of rotations on E: $H = \{\varphi_a : \varphi_a(x) = x + a \ (mod\ 2\pi)\}$. Then H is isomorphic to E, and H has a finite left–invariant measure, namely, normalized Lebesgue measure. If $\varphi = \varphi_a$, then $M_t^{\varphi} = \exp(B_t^{\varphi})$, where

$$B_t^{\varphi} = \int_0^t [c(X_s) - c(X_s - a)] ds$$

We see that

$$\int B_t^{\varphi} m(d\varphi) = \int \int_0^t [c(X_s) - c(X_s - a)] ds \, da$$

is finite only when $c \in L^1(da)$, so the hypothesis above is a necessary assumption.

PROPOSITION. *M_t has a right continuous version.*

PROOF: In the proof of this proposition we are going to use the original topology on E. In this topology X is a Borel right process with lifetime ζ. Define a kernel $Q^x(d\omega)$ from $(E_\Delta, \mathcal{B}(E_\Delta))$ to $(\Omega, \mathcal{F})$ by

$$Q^x f(X(t)) = P^x[M_t f(X(t))].$$

Then clearly

(1) For $x \in E$, $Q^x(X_0 = x) = 1$.

(2) For every $t \geq 0$, every $f \in b\mathcal{B}(E_\Delta)$ and every optional time T over $\mathcal{F}^0_{t+} \subset \mathcal{F}_t$,

$$Q^x\{f(X(T+t))|\mathcal{F}^0_{T+}\} = Q^{X(T)} f(X(t)).$$

(3) For every $x \in E$, the trace $Q^x|_{\mathcal{F}^0_t}$ on $\{t < \zeta\}$ is absolutely continuous relative to the trace of $P^x|_{\mathcal{F}^0_t}$ on $\{t < \zeta\}$.

Thus by Theorem 62.26 of [8] we know that there exists a right continuous supermartingale multiplicative functional $\overline{M}$ such that for every stopping time T over $\mathcal{F}_t$ and every $H \in b\mathcal{F}_T$,

$$Q^x(H 1_{\{T<\zeta\}}) = P^x(H \overline{M}_T).$$

In other words, M has a right continuous version. ∎

Because of this proposition we are going to assume, in the sequel, that M is actually right continuous.

DEFINITION. *X is called H-translation invariant if the processes $(\varphi(X_t), P^{\varphi^{-1}(x)})$ and (X_t, P^x) are identical in law for every $x \in E$ and every $\varphi \in H$.*

THEOREM. *If $\varphi \in H$, then we have*

$$P^{\varphi^{-1}(x)}[f \circ \varphi(X_t) M_t] = P^x[f(X_t) M_t]$$

for any $f \in p\mathcal{B}(E_\Delta)$, *any* $x \in E$ *and any* t.

PROOF: By the definition of M_t, we have

$$\begin{aligned} P^{\varphi^{-1}(x)}[f \circ \varphi(X_t)M_t] &= P^{\varphi^{-1}(x)}[f \circ \varphi(X_t)e^{\int A_t^\psi m(d\psi)}] \\ &= P^x[f(X_t)M_t^\varphi \exp(\int A_t^\psi \circ \Gamma^{\varphi^{-1}} m(d\psi))] \\ &= P^x[f(X_t)\exp(\int [A^\varphi + A_t^\psi \circ \Gamma^{\varphi^{-1}}] m(d\psi))] \\ &= P^x[f(X_t)\exp(\int [A^{\varphi\circ\psi}] m(d\psi)) \end{aligned}$$

by the corollary at the end of section 2. Since m is left–invariant, this last expression is $P^x[f(X_t)M_t]$. ∎

From this theorem, we can immediately get the following:

COROLLARY. $\hat{X} = (\Omega, \mathcal{F}, \mathcal{F}_t, X_t, \theta_t, Q^x)$ *is* H*-translation invariant.*

Now let f be a bounded positive continuous function on E, and let F be a positive $\mathcal{F}_s$-measurable random variable. Then

$$\begin{aligned} Q^x[f(X_s^{-1}X_{t+s})1_{\{t<\zeta\circ\theta_s\}}F1_{\{s<\zeta\}}] &= Q^x[Q^{X(s)}[f(X_s^{-1}X_t); t < \zeta]F; s < \zeta] \\ &= Q^x[Q^e[f(X_t); t < \zeta]F; s < \zeta] \\ &= Q^e[f(X_t); t < \zeta]Q^x[F; s < \zeta]. \end{aligned}$$

In particular, if we let $f = F = 1$ in the above, we get

$$Q^x[t + s < \zeta] = Q^e[t < \zeta]Q^x[s < \zeta].$$

This together with the H-translation invariance implies that $Q_t 1 = e^{-\alpha t}$ for some $\alpha \geq 0$. Define

$$N_t = e^{\alpha t}1_{[0,\zeta)}(t).$$

Then N_t is a right continuous, nonvanishing martingale multiplicative functional of $\hat{X}$. Now let $\tilde{X} = (\Omega, \mathcal{F}, \mathcal{F}_t, X_t, \theta_t, \tilde{P}^x)$ be the subprocess of $\hat{X}$, constructed in Theorem 62.19 of [8], corresponding to N. Then $\tilde{X}$ is again a Borel right process.

Summarizing the above, we get our final result.

THEOREM. *With the group structure on* E *given by* H, $\tilde{X}$ *is a Lévy process.*

REFERENCES

1. R. M. Blumenthal and R. K. Getoor, "Markov Processes and Potential Theory," Academic Press, New York, 1968.
2. C. Dellacherie and P. A. Meyer, "Probability and Potentials," North-Holland, Amsterdam, 1982.
3. J. Glover, *Symmetry groups and translation invariant representations of Markov processes*, Annals of Probab. to appear.
4. J. Glover, *Symmetry groups of Markov processes and the diagonal principle*, to appear.
5. J. Glover and J. Mitro, *Symmetries and functions of Markov processes*, Annals of Prob. **18** (1990), 655–668.
6. H. Heyer, "Probability Measures on Locally Compact Groups," Interscience, New York, 1977.
7. D. Montegomery and L. Zippin, "Topological Transformation Groups," Interscience, New York, 1955.
8. M. J. Sharpe, "General Theory of Markov Processes," Academic Press, San Diego, 1988.

Department of Mathematics, University of Florida, Gainesville, FL32611

On the Existence of Occupation Densitites of Stochastic Integral Processes via Operator Theory

PETER IMKELLER

INTRODUCTION

Fourier analysis provides one of the well known methods by which local behaviour of Gaussian processes, especially their occupation densities, can be investigated. Berman [3] initiated on approach which proved to be rather successful also in the more general area of Gaussian random fields and random fields with independent increments (see Geman, Horowitz [6] for a survey, Ehm [4]). The observation basic to this approach is comprised in the statement: the Fourier transform of the occupation measure of a real valued function is square integrable if and only if it posesses a square integrable density which then serves as a "local time" or "occupation density". It is therefore, at least in principle, quite general.

Random fields of a different source have recently been studied intensively. They originate for example from stochastic differential equations, involving the Wiener process, with boundary conditions (e.g. periodic ones) destroying the adaptedness of their solutions with respect to some filtration. They can therefore only be described by stochastic integrals and their associated processes able to integrate non-adapted data (see Ocone, Pardoux [16], Nualart, Pardoux [14],[15]).

Combining Skorohod's [21] original construction of an appropriate stochastic integral with ideas of Malliavin's calculus on Wiener space, and taking into account the surprising fact that Skorohod's integral has a simple interpretation as the adjoint operator of Malliavin's derivative, Nualart, Pardoux [13] presented a stochastic calculus fulfilling these requirements. They were able to explain some fine structure properties of the random fields described by Skorohod's integral, as for example the existence of a non-trivial quadratic variation. Yet their calculus provided no answer to questions about existence and properties of occupation densities.

In [7], we took up Berman's Fourier analytic approach on only a small portion of Wiener space, the second chaos, on which Skorohod's integral produces, so to speak, the simplest non-Gaussian fields in this setting. They are mainly described by a generally infinite-dimensional interaction matrix T of pairwise orthogonal Gaussian components. We wound up with translating sample properties into purely analytic terms and this way obtained a necessary and sufficient integral condition for the existence of occupation densities involving only T and Hilbert-Schmidt operators derived from it. At that time, however, the theory of operators and integral equations we fell upon after performing this translation procedure, was rather new to us. So for example it came just as a surprise and was puzzling for some time that our integral condition seemed to be necessary and sufficient only in case T is a trace class operator. Meanwhile, after becoming just a little better acquainted with the relevant literature (a look at the books of Smithies [22], Jörgens [9] and in particular Simon [20] proved to be very profitable), the problem found its natural solution. We mainly learned that our integral condition could be nicely put into the terms of Fredholm's theory of integral equations, developed already in the first half of this century.

This is what our translation of the problem into operator theoretic language ultimately lead to: a necessary and sufficient integral condition in terms of "Fredholm determinants" and "minors", if T is of trace class, and regularized Fredholm determinants and minors of the second order, if T is not of trace class. Its different versions, along with the "computable" descriptions of there objects we could find in the literature, will be presented in section 1. They still look rather complex and formidable. One reason for this might be our ignorance of a highly developed and sophisticated area of mathematics, leading to possibly awkward formulations. Another reason might well be the delicacy of the problem of the existence of occupation densities for complex objects as the ones considered, which might call for some stochastically intuitive notions, at least in a less abstract setting of fields described as solutions of particular stochastic differential equations, for example.

In section 2, we consider Skorohod integral processes defined by non-necessarily symmetric finite-dimensional operators T. Put stochastically, only finitely many orthogonal Gaussian components are allowed to interact. In solving the problem of the existence of square integrable occupation densities in this innocently looking context, again the complexity of the analysis to be invested came as a surprise to us. The easiest and simplest way we could think of was looking at a two-parameter family of finite-dimensional matrices, which form the essential building block of the integral condition to be confirmed, in the coordinates of their major axes. This lead to considering the smoothness of the associated families of eigenvalues and orthogonal matrices, a non-trivial problem which could be formulated in the framework of the perturbation theory of linear operators as presented in Kato's [10] book. A major role is played by the variational description of eigenvalues, as described in the min-max principle of Courant-Fischer. Along the way, for technical reasons, we lost track of the dependence of the upper bound we ultimately turn up with, on the interaction matrix T. We therefore can only conjecture that the method developed will have some bearing also if T takes into account infinitely many Gaussian components. Our main result most likely can be carried over to a non-compact parameter space and a "locally finite" interaction, i.e. each point in parameter space has a neighborhood in which only finitely many of the Gaussian components considered are "alive".

0. NOTATIONS AND CONVENTIONS

We will be dealing with the Wiener process W indexed by [0,1], defined on some fixed probability space $(\Omega,\mathcal{F},P)$, and its stochastic integrals in the "second chaos". More precisely, if for $g,h\in L^2([0,1])$ the tensor product of g and h is denoted by $g\otimes h(s,t) = g(s)h(t)$, and $\int_0^{\cdot} h\,dW$ is the usual stochastic integral of a deterministic function with respect to a Gaussian process, and if a kernel $f\in L^2([0,1]^2)$ is described in terms of an orthonormal basis $(h_n)_{n\in\mathbf{N}}$ of $L^2([0,1])$ by

$$f = \sum_{i,j=1}^{\infty} a_{ij}\, h_i \otimes h_j,$$

we consider the integrand

$$u_t = \sum_{i,j=1}^{\infty} a_{ij}\, h_i(t) \int_0^1 h_j\,dW,$$

and its "Skorohod integral process"

$$U_t = \sum_{i,j=1} a_{ij}\left(\int_0^t h_i\,dW \int_0^1 h_j\,dW - \int_0^t h_i h_j\,d\lambda\right), \quad t\in[0,1].$$

Apart from this simple definition, we will essentially not need results of the theory of Skorohod's integral based on Malliavin's calculus. But, of course, it will always be present in the background. We refer to Nualart, Pardoux [13], Nualart [11] or Watanabe [23]. For a system Γ of subsets of Ω, $\sigma(\Gamma)$ denotes the σ-algebra generated by Γ.

In terms of linear operators on the Hilbert space $L^2([0,1])$, the integral kernel f defines a Hilbert-Schmidt operator T. By T^* we denote its adjoint which is associated with the kernel $f^*(s,t) = f(t,s)$, by tr(T) its trace, if it exists, by I the identity on $L^2([0,1])$. If f,g are L^2-kernels,

$$fg(s,t) = f(s,\cdot)g(\cdot,t) = \int_0^1 f(s,u)g(u,t)du, \quad s,t\in[0,1],$$

is their product kernel. If f=g, it induces the operator T^2. The scalar product on $L^2([0,1])$ is denoted by $\langle\cdot,\cdot\rangle$, the norm by $\|\cdot\|_2$.

Especially in section 2, we will mostly be working in finite-dimensional spaces, say of dimension n and use a matrix description. In this context, $I = (\delta_{ij}:1\le i,j\le n)$ is the unit matrix. The scalar product in $\mathbb{R}^n$ is written x^*y, $x,y\in\mathbb{R}^n$. A vector of functions $h_1,\ldots,h_n \in L^2([0,1])$ will be denoted by h, the vector of their Gaussian integrals $\int_0^1 h_1dW,\ldots,\int_0^1 h_ndW$ occasionally by $\int_0^1 hdW$. The Lebesgue measure on the Borel subsets of any measurable subspace of $\mathbb{R}^n$ is sometimes written λ, regardless of the dimension.

1. A CRITERION FOR EXISTENCE IN TERMS OF FREDHOLM'S THEORY

In [7], we gave a necessary and sufficient condition for the existence of occupation densities of Skorohod integral processes in the second Wiener chaos. It was essentially described by an integral condition featuring the term exp(-i tr(H)) det (I+iH) where H is a Hilbert-Schmidt operator closely related to the one which determines the stochastic integral process considered. Written in the above way, the condition of course only makes sense if H is a trace class operator. This, in turn, restricted the validity of the criterion to integral processes based on a trace class operator themselves. For some time, this strange circumstance proved to be puzzling. Only when we tried to reinterpret the integral condition in the light of the theory of integral equations named after Fredholm and developed already in the first decades of this century, the problem dissolved completely. Formally, the two components of the term mentioned above can be taken together as a "regularized" Fredholm determinant of second order, which is defined and behaves smoothly on the whole space of Hilbert-Schmidt operators and requires no condition on the trace. Consequently, after being put into these and related terms, our existence criterion for occupation densities generalized completely and naturally. Therefore, we reformulate essential parts of [7] using Fredhom's theory as developed by Carleman, Schmidt, Hilbert, Smithies, Plemelj a.o., hereby using Smithies' [22] and Jörgens' [9] books as guidelines, but mainly the more modern presentation of Simon [20] for references. For simplicity, the results will not be stated in the most flexible form of [7], using an arbitrary subspace of $L^2([0,1])$ containing the range of the basic Hilbert-Schmidt operator as "universe", but $L^2([0,1])$ itself. On the other hand, we will choose a slightly more general setting, including non-symmetric kernels as well. We first recall the main general result. If $f \in L^2([0,1]^2)$ is a not necessarily symmetric kernel, T the Hilbert-Schmidt operator associated with it,

$$u_t = \int_0^1 f(t,s)dW_s, \text{ and}$$

$$U_t = \delta(1_{[0,t]}u),\ t\in[0,1],$$ its Skorohod integral process, we set

$$A(s,t) = \operatorname{sgn}(t-s)1_{[s\wedge t,s\vee t]},$$

$$H(s,t,x) = -x(T^* A(s,t) + A(s,t)T),$$

$$F(s,t,x) = I + iH(s,t,x), \quad s,t\in[0,1],\ x\in\mathbf{R},$$

we obtain that, provided T is of trace class, U posesses a square integrable occupation density iff (the attribute "balanced" of [7] is omitted here, since this is the only kind discussed in this paper)

$$(1) \qquad \int_{\mathbf{R}}\int_0^1\int_0^1 \exp(1/2\ \operatorname{tr}(H(s,t,x)))\ (\det F(s,t,x))^{-1/2} [f(s,\cdot)\ F(s,t,x)^{-1}\ f^*(\cdot,s)\ f(t,\cdot)\ F(s,t,x)^{-1}\ f^*(\cdot,t) + 2(f(s,\cdot)\ F(s,t,x)^{-1}\ f^*(\cdot,t))^2]\ ds\ dt\ dx < \infty.$$

To obtain (1) from theorem 2.1 of [7], we took care of the non-symmetry of T, f and translated a basis dependent description into a basis independent one using integral kernels instead. To express the main ingredients of (1) in Fredholm's theory, we have to introduce the following objects. Assume S is a Hilbert-Schmidt operator. Then the operator

$$R_2(S) = (I+S)\exp(-S) - I$$

is a trace class operator (see Simon [20], p. 106). It therefore makes sense to define

$$\det_2(I+S) = \det(I+R_2(S)),$$

the "regularized Fredholm determinant", and

$$D_2(S) = -S(I+R_2(S))^{-1}\det_2(I+S)\exp(-S),$$

the "regularized Fredholm minor" (see Simon [20], p. 107).
In case S is a trace class operator, we may reverse the regularization and wind up with the familiar formula

$$(2) \qquad \det_2(I+S) = \det(I+S)\exp(-\operatorname{tr}(S)).$$

In this case we need not regularize to get "Fredholm minors"

$$D_1(S) = -S(I+S)^{-1} \det(I+S)$$

(see Simon [20], p. 67). For the cases we are interested in, we will give alternative and more transparent descriptions of these quantities below. Now determinants and resolvents in (1) can be given a new shape. This leads to the following integral condition.

THEOREM 1: U possesses a square integrable occupation density iff

$$\int_{\mathbb{R}} \int_0^1 \int_0^1 (\det_2 F(s,t,x))^{-5/2}$$
$$\{f(s,\cdot)[\det_2(F(s,t,x))I + D_2(iH(s,t,x))]\ f^*(\cdot,s)$$
$$f(t,\cdot)[\det_2(F(s,t,x))\ I + D_2(iH(s,t,x))]f^*(\cdot,t)$$
$$+ 2(f(s,\cdot)[\det_2(F(s,t,x))I + D_2(iH(s,t,x,))]\ f^*(\cdot,t))^2\}\ ds\ dt\ dx < \infty.$$

PROOF: If T is of trace class, (1) gives the necessary and sufficient condition. Now apply (2) to $S = iH(s,t,x)$ and use the well known formula for the resolvent

$$F(s,t,x)^{-1} = I + \det_2(F(s,t,x))^{-1} D_2(iH(s,t,x))$$

(see Simon [20], pp. 107, 108 and Smithies [22], pp. 96-99). The resulting integral condition now makes sense for arbitrary Hilbert-Schmidt operators. Hence an approximation argument as contained in propositions 2.8, 2.9 of [7] completes the proof. ∎

In case T is of trace class, we can use non-regularized determinants and minors.

THEOREM 2: Assume T is a trace class operator. Then U posesses a square integrable occupation density iff

$$\int_{\mathbb{R}} \int_0^1 \int_0^1 \exp(i/2\, \mathrm{tr}(H(s,t,x)))\, \det(F(s,t,x))^{-5/2}$$

$$\{f(s,\cdot)[\det(F(s,t,x))I + D_1(iH(s,t,x))]\, f^*(\cdot,s)$$

$$f(t,\cdot)[\det(F(s,t,x))I + D_1(iH(s,t,x))]\, f^*(\cdot,t)$$

$$+ 2(f(s,\cdot)[\det(F(s,t,x))I + D_1(iH(s,t,x))]\, f^*(\cdot,t))^2\}\, ds\, dt\, dx < \infty.$$

PROOF: Proceed as in the proof of the preceding theorem and use the alternative equation for the resolvent

$$F(s,t,x)^{-1} = I + \det F(s,t,x)^{-1}\, D_1(iH(s,t,x))$$

(see Simon [20], p. 67). ∎

So far we have gained some generality. But we have only replaced the complicated resolvents $F(s,t,x)^{-1}$ be another set of complex objects. Now determinants and minors can be developed in power series featuring new expressions which look a little more easily accessible. This interpretation is due to the work of Fredholm, Plemelj and Smithies in the case of trace class operators. For general HS-operators, Hilbert, Plemelj and Smithies deduced the formulas we will now be using. This time, we start by looking at trace class operators.

PROPOSITION 1: Let $g \in L^2([0,1])^2)$ induce the trace class operator G. For $n \in \mathbf{N}$, $x_1,\ldots,x_n,y_1,\ldots,y_n \in [0,1]$ let

$$G\begin{pmatrix} x_1 \ldots x_n \\ y_1 \ldots y_n \end{pmatrix} = \det(g(x_i,y_j) : 1 \le i,j \le n),$$

$$\alpha_n(G) = \int_0^1 \ldots \int_0^1 G\begin{pmatrix} x_1 \ldots x_n \\ x_1 \ldots x_n \end{pmatrix} dx_1 \ldots dx_n,\quad \alpha_0(G) = 1,$$

$$G_n(x,y) = \int_0^1 \int_0^1 G\begin{pmatrix} x & x_1 \ldots x_n \\ y & x_1 \ldots x_n \end{pmatrix} dx_1 \ldots dx_n, \quad x,y \in [0,1],$$

$\beta_n(G)$ the operator induced by the kernel G_n, $\beta_0(G) = I$. Then for any $\lambda \in \mathbf{C}$

$$\det(I+\lambda G) = \sum_{n=0}^{\infty} \frac{\lambda^n}{n!} \alpha_n(G),$$

$$D_1(\lambda G) = \sum_{n=0}^{\infty} \frac{\lambda^{n+1}}{n!} \beta_n(G).$$

<u>PROOF</u>: See Simon [20], pp. 51,69. ∎

The formulas of proposition 1 are due to Fredholm [5]. Alternatively, we can use formulas developed by Plemelj-Smithies.

<u>PROPOSITION 2</u>: Let $g \in L^2([0,1]^2)$ induce the trace class operator G. For $n \in \mathbf{N}$ let $\sigma_n = \mathrm{tr}(G^n)$ and

$$\gamma_n(G) = \det \begin{pmatrix} \sigma_1 & n-1 & 0 & \cdots & 0 \\ \sigma_2 & \sigma_1 & n-2 & & \vdots \\ \vdots & \vdots & \sigma_1 & & 0 \\ \vdots & \vdots & \vdots & \ddots & 1 \\ G^{n+1} & \sigma_n & \sigma_{n-1} & \cdots & \sigma_1 \end{pmatrix}, \quad \gamma_0(G) = 1,$$

$$\delta_n(G) = \det \begin{pmatrix} G & n & 0 & \cdots & 0 \\ G^2 & \sigma_1 & n-1 & & \vdots \\ \vdots & \vdots & \sigma_1 & & 0 \\ \vdots & \vdots & \vdots & & 1 \\ G^{n+1} & \sigma_n & \sigma_{n-1} & \cdots & \sigma_1 \end{pmatrix}, \quad \delta_0(G) = I$$

(HS-operators!).

Then for any $\lambda \in \mathbf{C}$

$$\det(I+\lambda G) = \sum_{n=0}^{\infty} \frac{\lambda^n}{n!} \gamma_n(G),$$

$$D_1(\lambda G) = \sum_{n=0}^{\infty} \frac{\lambda^{n+1}}{n!} \delta_n(G).$$

PROOF: See Simon [20], pp. 68,69. ∎

If G is not of trace class, we know already that determinants and minors have to be regularized. In terms of the matrices used in their power series description, this simply amounts to removing the diagonal.

PROPOSITION 3: Let $g \in L^2([0,1]^2)$ induce the Hilbert-Schmidt operator G. For $n \in \mathbf{N}$, $x_1,\ldots,x_n,y_1,\ldots,y_n \in [0,1]$ let

$$\tilde{G}\begin{pmatrix} x_1 \ldots x_n \\ y_1 \ldots y_n \end{pmatrix} = \det(g(x_i,y_j)(1-\delta_{ij}) \ : \ 1 \le i,j \le n),$$

$$\tilde{\tilde{G}}\begin{pmatrix} x_1 \ldots x_n \\ y_1 \ldots y_n \end{pmatrix} = \det(g(x_i,y_j)(1-\delta_{ij}+\delta_{1j}) \ : \ 1 \le i,j \le n),$$

$$\tilde{\alpha}_n(G) = \int_0^1 \cdot\cdot \int_0^1 \tilde{G}\begin{pmatrix} x_1 \ldots x_n \\ x_1 \ldots x_n \end{pmatrix} dx_1 \ldots dx_n, \quad \tilde{\alpha}_0(G) = 1,$$

$$\tilde{G}_n(x,y) = \int_0^1 \cdot\cdot \int_0^1 \tilde{\tilde{G}}\begin{pmatrix} x \ x_1 \ldots x_n \\ y \ x_1 \ldots x_n \end{pmatrix} dx_1 \ldots dx_n,$$

$\tilde{\beta}_n(G)$ the operator induced by the kernel $\tilde{G}_n$, $\tilde{\beta}_0(G) = I$.

Then for any $\lambda \in \mathbf{C}$

$$\det{}_2(I+\lambda G) = \sum_{n=0}^{\infty} \frac{\lambda^n}{n!} \tilde{\alpha}_n(G),$$

$$D_2(\lambda G) = \sum_{n=0}^{\infty} \frac{\lambda^{n+1}}{n!} \tilde{\beta}_n(G).$$

PROOF: See Simon [20], p. 108, and Smithies [22], p. 99. ∎

Plemelj-Smithies' formulas possess the following regularizations.

PROPOSITION 4: Let $g \in L^2([0,1]^2)$ induce the Hilbert-Schmidt operator G.

For $n \in \mathbf{N}$ let $\sigma_n = tr(G^n)$ and

$$\tilde{\gamma}_n(G) = \det \begin{pmatrix} 0 & n-1 & \cdots & 0 \\ \sigma_2 & \sigma_2 & & \vdots \\ \vdots & \vdots & & 0 \\ \vdots & \vdots & & 1 \\ \sigma_n & \sigma_{n-1} & \cdots & 0 \end{pmatrix}, \quad \tilde{\gamma}_0(G) = 1,$$

$$\tilde{\delta}_n(G) = \det \begin{pmatrix} G & n & 0 & \cdots & 0 \\ G^2 & 0 & n-1 & & \vdots \\ \vdots & \sigma_2 & 0 & & \vdots \\ \vdots & \vdots & \sigma_2 & & 0 \\ \vdots & \vdots & \vdots & & 1 \\ G^{n+1} & \sigma_n & \sigma_{n-1} & & 0 \end{pmatrix}, \quad \tilde{\delta}_0(G) = I$$

(HS-operators!).

Then for any $\lambda \in \mathbf{C}$

$$\det{}_2(I+\lambda G) = \sum_{n=0}^{\infty} \frac{\lambda^n}{n!} \tilde{\gamma}_n(G), \quad D_2(\lambda G) = \sum_{n=0}^{\infty} \frac{\lambda^{n+1}}{n!} \tilde{\delta}_n(G).$$

<u>PROOF</u>: See Simon [20], p. 108, and Smithies [22], p. 94. ∎

The preceding proposition now allows us to put the conditions of theorems 1 and 2 into more readily accessible, yet rather complex, forms.

<u>THEOREM 3</u>: For $s,t \in [0,1]$ let

$$H = T^*A(s,t) + A(s,t)T.$$

U possesses a square integrable occupation density iff

(3) $$\int_{\mathbf{R}} \int_0^1 \int_0^1 \left(\sum_{n=0}^{\infty} \frac{(-ix)^n}{n!} \tilde{\alpha}_n(H) \right)^{-5/2}$$

$$\left\{ \left[ff^*(s,s) + \sum_{n=1}^{\infty} \frac{(-ix)^n}{n!} \left(ff^*(s,s)\tilde{\alpha}_n(H) + n\, f\tilde{\beta}_n(H) f^*(s,s) \right) \right] \cdot \right.$$

$$\cdot \left[ff^*(t,t) + \sum_{n=1}^{\infty} \frac{(-ix)^n}{n!} \left(ff^*(t,t)\tilde{\alpha}_n(H) + n\, f\tilde{\beta}_n(H) f^*(t,t) \right) \right]$$

$$+ 2[ff^*(s,t) + \sum_{n=1}^{\infty} \frac{(-ix)^n}{n!} (ff^*(s,t)\tilde{\alpha}_n(H) + n\, f\tilde{\beta}_n(H)f^*(s,t))]^2\}$$

$$ds\, dt\, dx < \infty.$$

Alternatively, $\tilde{\alpha}_n(H)$ resp. $\tilde{\beta}_n(H)$ may be replaced by $\tilde{\gamma}_n(H)$ resp. $\tilde{\delta}_n(H)$.

PROOF: Combine propositions 3,4 with theorem 1 and compare the resulting power series term by term. ∎

In the trace class case, we can again replace the "~"-coefficients in the integral criterion of theorem 3 with their counterparts.

THEOREM 4: Assume T is a trace class operator. Then U possesses a square integrable occupation density iff the analogue of (3) holds with $\tilde{\alpha}_n(H)$ replaced by $\alpha_n(H)$, $\tilde{\beta}_n(H)$ by $\beta_n(H)$. Alternatively, $\alpha_n(H)$ resp. $\beta_n(H)$ may be replaced by $\gamma_n(H)$ resp. $\delta_n(H)$.

PROOF: This time we have to combine propositions 1, 2 and theorem 2, and compare the power series appearing term by term. ∎

REMARK: Though the constituents of (3) are computable and there are relatively simple recursive formulas for the coefficients $\alpha_n(H)$, $\beta_n(H)$ etc. (see Smithies [22], pp. 74,88), the criteria of theorem 3 or theorem 4 seem to be hard to verify. In particular, the series in (3) seem to simplify further in only rather special cases. Therefore, so far we have just been able to use the integral conditions directly in some particular cases. Other cases, for example the one considered in the subsequent section, seem to favor the more flexible criterion of theorem 2.1. of [7] in which the analysis is restricted to a subspace of $L^2([0,1])$.

2. OCCUPATION DENSITIES IN THE FINITE DIMENSIONAL CASE

We will now look at Skorohod integral processes in the second chaos described by only finitely many interacting orthogonal Gaussian components. The main result of this section is that they always possess square integrable occupation densities. The nature of the problem makes it more convenient to work with a form of the integral conditions discussed in section 1, the operators of which live on the finite dimensional range of T. Criteria of this form were presented in the first two theorems of section 2 of [7]. Choosing $N = R(T)$ there (cf. p. 14 of [7]) makes appear a nontrivial real part of $F(s,t,x)$, namely

$$G(s,t,x) = P + x^2 T^* A(s,t)B(s,t)T, \text{ where } B(s,t) = P - A(s,t),$$

P the orthogonal projection on N. Instead of $F(s,t,x)$, we will be able to work with $G(s,t,x)$ alone. To verify the resulting integral condition, we look at $G(s,t,x)$ in its diagonal form. This amounts to following the major axes of $A(s,t)$ all along the way as s,t run through [0,1]. The main problem we have to face hereby consists in keeping track of the eigenvalues $\lambda(s,t)$ and orthogonal projections $0(s,t)$. As long as $A(s,t)$ itself varies analytically, the perturbation theory of linear operators based upon the variational description of its eigenvalues in the Courant-Schmidt min-max principle yields nice results about the connection between $\lambda(s,t)$ and $0(s,t)$ and the analyticity of these functions. This enables us to solve our problem for analytic data first. We then approximate general data by analytic ones to carry the result over to any finite dimensional operator T. Finally, a very simple example will be given to underline that our results are out of reach of the usual techniques of enlargment of filtrations hooked up with martingale theory, as developed in Jeulin, Yor [8].

To be more precise now, assume $(h_1,...,h_n)$ is an orthonormal family in $L^2([0,1])$ and

$$f = \sum_{1\le i,j\le n} a_{ij}\, h_i \otimes h_j$$

with some real matrix $(a_{ij})_{1\le i,j\le n}$, T the operator associated with f. Moreover, let

$$N = \text{span } \{h_i : 1\le i\le n\}$$

and

P the orthogonal projection on N.

We tacitly assume, finally, that n, via the orthogonal family, is chosen "minimal", i.e. that T is invertible. In particular, from now on, n is supposed to be fixed and will not get a special mention in propositions and theorems.

Before we start analyzing the Fourier analytic criterion for the existence of occupation densities for the Skorohod integral process assoicated with T, we present an inequality for the inverses of an ordered pair of symmetric, positive definite matrices which will prove to be very useful along the way.

PROPOSITION 1: Let A,B be n-dimensional real symmetric non-negative definite matrices. Suppose that

$$A \ge B > 0.$$

Then

$$A^{-1} \le B^{-1}.$$

PROOF: We found the following nice argument in the book of Bellman [2], p. 93. Consider the function

$$f: \mathbf{R}^n \to \mathbf{R},$$

$$y \to 2x^* y - y^* Ay.$$

To determine the extrema of f we may, by an orthogonal transformation of the coordinates, assume that A is in diagonal form, i.e.

$$A = \begin{pmatrix} a_1 & & 0 \\ & \ddots & \\ 0 & & a_n \end{pmatrix} \quad \text{with } a_1, \dots, a_n > 0.$$

Here we used the assumption that the considered matrices are symmetric and positive definite. Now

$$Df(y) = 2x - 2Ay, \quad D^2 f(y) = -A, \quad y \in \mathbf{R}^n.$$

Since $-A$ is negative definite, f has a maximum at $x = Ay$, i.e. $y = A^{-1}x$. We therefore obtain

$$x^* A^{-1} x = \max \{2x^* y - y^* Ay : y \in \mathbf{R}^n\}.$$

The same equation being true for B, we obtain

$$\begin{aligned} x^* A^{-1} x &= \max \{2x^* y - y^* Ay : y \in \mathbf{R}^n\} \\ &\le \max \{2x^* y - y^* By : y \in \mathbf{R}^n\} \qquad (A \ge B) \\ &= x^* B^{-1} x. \end{aligned}$$

Since this inequality holds for any $x \in \mathbf{R}^n$, we are done. ∎

With the aid of proposition 1, we gain the following sufficient condition from theorem 2.2 of [7].

<u>PROPOSITION 2</u>: For $s,t \in [0,1]$, $x \in \mathbf{R}$ let

$$A(s,t) = \operatorname{sgn}(t-s) \cdot P\, 1_{[s\wedge t, s\vee t]} P = \int_s^t h\, h^* d\lambda,$$

$$B(s,t) = I - A(s,t),$$

$$G(s,t,x) = I + x^2 T^* A(s,t) B(s,t) T.$$

Assume that

$$\int_{\mathbb{R}} \int_0^1 \int_0^1 \det G(s,t,x)^{-1/2} [h^*(s) T G(s,t,x)^{-1} T^* h(s)$$

$$h^*(t) T G(s,t,x)^{-1} T^* h(t)] ds\, dt\, dx < \infty.$$

Then U possesses a square integrable occupation density.

PROOF: Since T is a trace operator, a slight extension of theorem 2.2 of [7] to non-symmetric operators (see section 1) tells us that U possesses a square integrable occupation density if

$$(4) \quad \int_{\mathbb{R}} \int_0^1 \int_0^1 \det G(s,t,x)^{-1/2} \det C(s,t,x)^{-1/4}$$

$$\{h^*(s) T G(s,t,x)^{-1/2} C(s,t,x)^{-1/2} G(s,t,x)^{-1/2} T^* h(s)$$

$$h^*(t) T G(s,t,x)^{-1/2} C(s,t,x)^{-1/2} G(s,t,x)^{-1/2} T^* h(t)\} ds\, dt\, dx < \infty.$$

Here

$$C(s,t,x) = P + G(s,t,x)^{-1/2} H(s,t,x) G(s,t,x)^{-1} H(s,t,x) G(s,t,x)^{-1/2},$$

$$H(s,t,x) = -x(T^* A(s,t) + A(s,t) T).$$

Now let

$$J = G(s,t,x)^{-1/2} H(s,t,x) G(s,t,x)^{-1/2}.$$

Then

$$C(s,t,x) = P + J^2 \geq P > 0.$$

Hence by proposition 1

$$C(s,t,x)^{-1} \leq P^{-1}$$

and so, since P is represented by the unit matrix on N,

$$C(s,t,x)^{-1/2} \leq P^{-1/2}.$$

Also,

$$\det C(s,t,x)^{-1/4} \leq 1.$$

Hence (4) follows from the integral condition in the statement of the proposition and the proof is finished. ∎

REMARK: It is worth noting that we actually got a little more than what the statement of proposition 2 says. We have

$$\int_{\mathbb{R}} \int_0^1 \int_0^1 E(\exp(ix(U_t-U_s))u_s^2u_t^2)ds\ dt\ dx$$

$$\le \int_{\mathbb{R}} \int_0^1 \int_0^1 \det G(s,t,x)^{-1/2}[h^*(s)TG(s,t,x)^{-1}\ T^*h(s)$$

$$h^*(t)T\ G(s,t,x)^{-1}T^*h(t)]\ ds\ dt\ dx.$$

To see this, look at proposition 2.10 of [7].

Next, to establish the integral condition figuring in proposition 2, we eliminate the influence of the "interaction amplitudes" described in the coefficients of T. This is done to avoid some technical problems.

PROPOSITION 3: For $s,t\in[0,1]$, $x\in\mathbb{R}$ let

$$K(s,t,x) = I + x^2A(s,t)\ B(s,t).$$

There is a constant c_1 which only depends on $(a_{ij}:1\le i,j\le n)$ such that for any $s,t\in[0,1]$, $x\in\mathbb{R}$

$$\det G(s,t,x)^{-1/2}\ \cdot$$

$$[h^*(s)T\ G(s,t,x)^{-1}T^*\ h(s)\ h^*(t)T\ G(s,t,x)^{-1}T^*\ h(t)]$$

$$\le c_1 \cdot \det K(s,t,x)^{-1/2}\ \cdot$$

$$[h^*(s)\ K(s,t,x)^{-1}\ h(s)\ h^*(t)K(s,t,x)^{-1}\ h(t)].$$

PROOF: We have to show that there is a constant $c_2 > 0$ only depending on $(a_{ij}:1\le i,j\le n)$ such that

$$T\ G(s,t,x)^{-1}\ T^* \le (c_2I+x^2A(s,t)B(s,t))^{-1},\ s,t\in[0,1],\ x\in\mathbb{R}.$$

Now by definition

$$G(s,t,x) = T^*((TT^*)^{-1} + x^2(A(s,t)B(s,t))T.$$

So we have to show

$$((TT^*)^{-1} + x^2A(s,t)B(s,t))^{-1} \le (c_2I+x^2A(s,t)B(s,t))^{-1}.$$

But proposition 1 reduces this inequality further to

$$(TT^*)^{-1} \geq c_2 I.$$

This inequality obviously holds, if we let c_2 be the smallest eigenvalue of $(TT^*)^{-1}$. This quantity, due to the fact that TT^* is symmetric and positive definite, is obviously positive. ∎

To treat the integrand figuring on the right hand side of the inequality of proposition 3 further, we look at the A(s,t) along their major axes. This involves working in moving coordinate systems and with moving eigenvalues. Since we want to have some smoothness in s,t for both objects, we face a problem usually encountered in the perturbation theory of finite dimensional linear operators. Its main theorems state that analytic behaviour of one-parameter families of linear operators is inherited by both eigenvalues and projections on the eigenspaces. Continuity or differentiability alone is inherited by just the eigenvalues, whereas eigenspaces may behave rather badly (see Kato [10], p. 111, for an example of Rellich [19]). Of course, since $h_1,\ldots,h_n$ are just square integrable functions in general, A(s,t) is no more than continuously differentiable in s,t. To make things even worse, it is a two-parameter family of matrices. And in this situation, perturbation theory becomes more complicated. Not even analyticity is inherited by the eigenvalues (see Rellich [19], p. 37, Baumgärtel [1]). We circumvent these problems in the following way. First of all, we fix either s or t and consider the one-parameter families of matrices as the respective other paramater varies. In addition, we assume that $h_1,\ldots,h_n$ are analytic (for example polynomials) first and come back to the general situation later using a global approximation argument.

<u>PROPOSITION 4</u>: Let $h_1,\ldots,h_n$ be analytic functions, $s \in [0,1]$. Then

there exist families $(\lambda_i(s,t):s\le t\le 1, 1\le i\le n)$ of real numbers and $(o_i(s,t):s\le t\le 1, 1\le i\le n)$ of vectors in $\mathbb{R}^n$ such that

(i) $t \to \lambda_i(s,t)$, $t \to o_i(s,t)$ is analytic except at finitely many points,

(ii) $t \to \lambda_i(s,t)$ is increasing,

(iii) $0 \le \lambda_i(s,t) \le 1$, $\lambda_i(s,s) = 0$, $\lambda_i(0,1) = 1$, $\lambda_i(s,t) \ge \lambda_{i+1}(s,t)$, $t\in[s,1]$, $1\le i\le n$,

(iv) for $t\in[s,1]$ the matrix $O(s,t) = (o_1(s,t),\dots,o_n(s,t))$ is orthogonal and

$$O^*(s,t)\ A(s,t)\ O(s,t) = \begin{pmatrix} \lambda_1(s,t) & & 0 \\ & \ddots & \\ 0 & & \lambda_n(s,t) \end{pmatrix}.$$

A similar statement holds with respect to $s\in[o,t]$ for t fixed.

<u>PROOF</u>: Since $h_1,\dots,h_n$ are analytic, so is the family of matrices $(A(s,t):s\le t\le 1)$. Hence there is an integer $m \le n$, a family $(\mu_1(s,t),\dots,\mu_m(s,t):s\le t\le 1)$ (eigenvalues), integers $p_1,\dots,p_m$ (their multiplicities) such that $\sum_{j=1}^{m} p_j = n$, and a family $(P_1(s,t),\dots,P_m(s,t):s\le t\le 1)$ of orthogonal projections such that for any $1\le j\le m$

$P_j(s,t)$ is the orthogonal projection on the eigenspace of $\mu_j(s,t)$, $s\le t\le 1$, and such that the functions

$t \to \lambda_j(s,t)$, $t \to P_j(s,t)$ are analytic, $1\le j\le m$

(see Kato [10], pp. 63-65, 120). Now fix $1\le j\le m$. Using an analytic family of unitary transformations (see Kato [10], pp. 104-106, 121, 122), we can construct analytic families of orthonormal vectors, say

$$(e_j^1(s,t),\dots,e_j^{p_j}(s,t):s\le t\le 1),$$

a smoothly moving basis of the subspaces of $\mathbf{R}^n (P_j(s,t): s \le t \le 1)$ project on. Next, we take multiplicities into account. For

$$p_1+\ldots+p_{j-1}+1 \le i \le p_1+\ldots+p_j$$

let

$$\nu_i(s,t) = \mu_j(s,t),$$

$$e_i(s,t) = e_j^{i-p_1-\ldots-p_{j-1}}(s,t), \quad s \le t \le 1.$$

Then the eigenvectors $e_i(s,t)$ correspond to the eigenvalues $\nu_i(s,t)$, $1 \le i \le n$. But still, $\nu_i(s,t) < \nu_{i+1}(s,t)$ is possible. We therefore have to arrange the eigenvalues to make (iii) valid. For $s \le t \le 1$ fixed we therefore define a permutation σ of $\{1,\ldots,m\}$ such that

$$\mu_{\sigma(1)}(s,t) \ge \ldots \ge \mu_{\sigma(m)}(s,t).$$

Due to continuity, we obtain the same permutations on whole subintervals of $[s,1]$. Analyticity and compactness imply that we need only finitely many permutations on the whole of $[s,1]$. If we perform these permutations on the $\nu_i(s,t)$ and $e_i(s,t)$, $1 \le i \le n$, $s \le t \le 1$, we obtain the desired families

$$(\lambda_i(s,t): s \le t \le 1, 1 \le i \le n) \text{ and } o_i(s,t): s \le t \le n, 1 \le i \le n).$$

By construction, they are analytic except at finitely many points of $[s,1]$. We have therefore proved (i) and (iv). To prove (ii) and the rest of (iii), let us look a little more closely at $A(s,t)$. Observe that for $y \in \mathbf{R}^n$

$$y^* A(s,t) y = \int_s^t (y^* h(u))^2 du.$$

Therefore the family $(A(s,t): s \le t \le 1)$ of nonnegative definite matrices possesses the properties

$$0 \le A(s,t) \le I, \; A(s,s) = 0, \; A(0,1) = I,$$

and

$t \to A(s,t)$ is increasing on $[s,1]$ with respect to the usual ordering of non-negative definite symmetric matrices. These facts together with the Courant-Fischer min-max principle, expressed for example in Kato [10], pp. 60,61, yield the desired inequalities. ∎

Proposition 4 allows a further reduction of the integral condition we have to establish. Due to the problems alluded to above, we will have to be careful with two-parameter families and symmetrically fix s for one part of the integrand, t for the other.

<u>PROPOSITION 5</u>: Let $h_1,\dots,h_n$ be analytic functions, $s,t\in[0,1]$, $s\le t$. Assume $(\lambda_i(s,v): s\le v\le 1, 1\le i\le n)$ and $(o_i(s,v): s\le v\le 1, 1\le i\le n)$ resp. $(\mu_i(u,t): 0\le u\le t, 1\le i\le n)$ and $(p_i(u,t): 0\le u\le t, 1\le i\le n)$ are given according to proposition 4 for s resp. t fixed. Let $O(s,v) = (o_1(s,v),\dots,o_n(s,v))$, $P(u,t) = (p_1(u,t),\dots,p_n(u,t))$ and

$$k(t) = O^*(s,t)\ h(t), \qquad \ell(s) = P^*(s,t)\ h(s).$$

Moreover, let

$$c_2 = \max\{\int_{\mathbf{R}} (1+x^4)^{-3/2}\,dx,\ \int_{\mathbf{R}} (1+x^2)^{-5/2}\,dx\}.$$

Then

(i) $\lambda_i(u,v) = \mu_i(u,v)$ for all $s\le u\le v\le t$, $1\le i\le n$,

(ii) $$\int_{\mathbf{R}} \det K(s,t,x)^{-1/2}\ [h^*(s)K(s,t,x)^{-1}h(s)\ h^*(t)K(s,t,x)^{-1}h(t)]\,dx$$
$$\le c_2 \sum_{i=1}^n \ell_i^2(s)\ [\lambda_i(s,t)(1-\lambda_i(s,t))]^{-1/4}$$
$$\cdot \sum_{j=1}^n k_j^2(t)\,[\lambda_j(s,t)(1-\lambda_j(s,t))]^{-1/4}.$$

<u>PROOF</u>: Though the procedure of arranging the eigenvalues in descending order in the proof of proposition 4 may destroy their overall analyticity, it preserves continuity. This obviously implies (i). To prove (ii), first observe that, due to the choice of $\lambda_i(s,t)$, $O(s,t)$,

$P(s,t)$, for any $x \in \mathbb{R}$

$$K(s,t,x) = O(s,t) \begin{pmatrix} 1+x^2\lambda_1(s,t)(1-\lambda_1(s,t)) & & 0 \\ & \ddots & \\ 0 & & 1+x^2\lambda_n(s,t)(1-\lambda_n(s,t)) \end{pmatrix} O^*(s,t)$$

and a similar equation with $P(s,t)$ in place of $O(s,t)$. Hence

$$h^*(t)\ K(s,t,x)^{-1}\ h(t) = \sum_{j=1}^{n} k_j^2(t)\ [1+x^2\lambda_j(s,t)(1-\lambda_j(s,t))]^{-1},$$

$$h^*(s)\ K(s,t,x)^{-1}\ h(s) = \sum_{i=1}^{n} \ell_i^2(s)\ [1+x^2\lambda_i(s,t)(1-\lambda_i(s,t))]^{-1},$$

and therefore

$$(5)\quad \det K(s,t,x)^{-1/2}\ [h^*(s)K(s,t,x)^{-1}h(s)\ h^*(t)K(s,t,x)^{-1}h(t)]$$

$$= \prod_{k=1}^{n} (1+x^2\lambda_k(s,t)(1-\lambda_k(s,t)))^{-1/2}.$$

$$\sum_{i=1}^{n} \ell_i^2(s)\ [1+x^2\lambda_i(s,t)(1-\lambda_i(s,t))]^{-1}$$

$$\cdot \sum_{j=1}^{n} k_j^2(t)\ [1+x^2\lambda_j(s,t)(1-\lambda_j(s,t))]^{-1}$$

$$\le \sum_{\substack{i,j=1 \\ i\neq j}}^{n} \ell_i^2(s)k_j^2(t)\ \{[1+x^2\lambda_i(s,t)(1-\lambda_i(s,t))]$$

$$[1+x^2\lambda_j(s,t)(1-\lambda_j(s,t))]\}^{-3/2}$$

$$+ \sum_{i=1}^{n} \ell_i^2(s)k_i^2(t)\quad [1+x^2\lambda_i(s,t)(1-\lambda_i(s,t))]^{-5/2}$$

$$\le \sum_{\substack{i,j=1 \\ i\neq j}} \ell_i^2(s)k_j^2(t)\ [1+x^4\lambda_i(s,t)(1-\lambda_i(s,t))\lambda_j(s,t)(1-\lambda_j(s,t))]^{-3/2}$$

$$+ \sum_{i=1}^{n} \ell_i^2(s)k_j^2(t)\ [1+x^2\lambda_i(s,t)(1-\lambda_i(s,t))]^{-5/2}.$$

Now observe that for $b_1, b_2 \geq 0$ we have

$$(6) \qquad \int_{\mathbb{R}} (1+x^4 b_1)^{-3/2}\, dx = b_1^{-1/4} \int_{\mathbb{R}} (1+x^4)^{-3/2}\, dx \leq c_2\, b_1^{-1/4},$$

$$\int_{\mathbb{R}} (1+x^2 b_2)^{-5/2}\, dx = b_2^{-1/2} \int_{\mathbb{R}} (1+x^2)^{-5/2}\, dx \leq c_2\, b_2^{-1/2}.$$

Applying (6) term by term to the right hand side of (5) yields the desired inequality. ∎

From this point on it is relatively obvious what has to be done to prove the integral condition of proposition 2. We ultimately have to integrate the rhs of (ii) in proposition 5 in s and t. The key observation we will exploit in doing this rests upon the extremal properties of the eigenvalues as expressed in the principle of Courant-Fischer. Intuitively, this can be most easily understood in the two-dimensional case. Assume the notation of proposition 5. Fix $s<t$ and suppose $\lambda_1(s,t) > \lambda_2(s,t)$. Then the principle of Courant-Fischer states

$$(7) \qquad \lambda_1(s,t) = \max_{0 \neq x \in \mathbb{R}^n} \frac{x^* A(s,t) x}{x^* x}, \quad \lambda_2(s,t) = \min_{0 \neq x \in \mathbb{R}^n} \frac{x^* A(s,t) x}{x^* x}.$$

Since $o_1(s,t)$, $o_2(s,t)$ are unit eigenvectors of $\lambda_1(s,t), \lambda_2(s,t)$, we also have

$$\lambda_1(s,t) = o_1^*(s,t) A(s,t) o_1(s,t), \quad \lambda_2(s,t) = o_2^*(s,t) A(s,t) o_2(s,t).$$

Now consider the functions

$$f_1(h) = o_1^*(s,t+h) A(s,t) o_1(s,t+h), \quad f_2(h) = o_2^*(s,t+h) A(s,t) o_2(s,t+h),$$

defined in some small neighborhood of t. If, as we may do, assume that t is not one of the exceptional points of proposition 4, f_1, f_2 are differentiable at 0. Moreover, (7) forces them to take their maximum resp. minimum there. Hence

$$f_1'(0) = f_2'(0) = 0$$

and we obtain the formulas

$$\begin{aligned} \frac{d}{dt}\lambda_1(s,t) &= o_1^*(s,t)\,\frac{d}{dt}A(s,t)o_1(s,t) \\ &= o_1^*(s,t)\begin{pmatrix} h_1^2(t) & h_1h_2(t) \\ h_1h_2(t) & h_2^2(t)\end{pmatrix} o_1(s,t) \qquad (8) \\ &= (o_1^*(s,t)h(t))^2 = k_1^2(t), \\ \frac{d}{dt}\lambda_2(s,t) &= k_2^2(t) \text{ correspondingly.}\end{aligned}$$

(8) enables us, while integrating the rhs of the inequality (ii) of proposition 5, to do a simple substitution of variables and the rest is "smooth sailing". As it turns out, (8) is true far more generally. The reasons, as given in Kato [10], pp. 77-81, are not as intuitive as the ones given above in the simplest case one can think of, yet rest upon the same observations. We are therefore led to the following proposition.

<u>PROPOSITION 6</u>: Let $h_1,\dots,h_n$ be analytic functions, $s,t\in[0,1]$, $s\le t$. In the notation of proposition 5, for $1\le i\le n$ let

$$I_i = \{j:\ 1\le j\le n,\ \lambda_j(u,v) = \lambda_i(u,v) \text{ for } 0\le u\le v\le 1\} \quad \text{(analyticity!)}.$$

Set

$$c_3 = \int_0^1 [u(1-u)]^{-1/2}\,du.$$

Then

$$\int_s^1 \sum_{j\in I_i} k_j^2(v)\ [\lambda_i(s,v)(1-\lambda_i(s,v))]^{-1/2}\,dv \le c_3\ |I_i|,$$

$$\int_0^t \sum_{j\in I_i} \ell_j^2(u)\ [\lambda_i(u,t)(1-\lambda_i(u,t))]^{-1/2}\ du \le c_3\ |I_i|.$$

<u>PROOF</u>: Since the asserted inequalities are symmetric, we may concentrate on the first one. Proposition 4 allows us to differentiate the function $v \to \lambda_i(s,v)$ at all but finitely many $v\in[s,1]$. We obtain

$$\frac{d}{dv}\lambda_i(s,v) = \frac{1}{|I_i|} \sum_{j\in I_i} o_j^*(s,v)\ \frac{d}{dv} A(s,v)\ o_j(s,v)$$

(Kato [10], p. 80)

$$= \frac{1}{|I_i|} \sum_{j\in I_i} o_j^*(s,v)h(v)h^*(v)o_j(s,v)$$

$$= \frac{1}{|I_i|} \sum_{j\in I_i} k_j^2(v).$$

We may therefore substitute

$$w = \lambda_i(s,v)$$

to get, observing proposition 4, (iii),

$$\int_s^1 \sum_{j\in I_i} k_j^2(v)\ [\lambda_i(s,v)(1-\lambda_i(s,v))]^{-1/2}\ dv \le \int_0^1 |I_i|(w(1-w))^{-1/2}\ dw$$

$$= c_3|I_i|,$$

which completes the proof. ∎

We are now ready to prove the integral condition of proposition 2.

<u>PROPOSITION 7</u>: Let $h_1,\ldots,h_n$ be analytic functions. Then

$$\int_{\mathbb{R}}\int_0^1\int_0^1 \det G(s,t,x)^{-1/2}\ [h^*(s)T\ G(s,t,x)^{-1}T^*h(s)$$
$$h^*(t)T\ G(s,t,x)^{-1}T^*h(t)]ds\ dt\ dx \le 2\ c_1c_2c_3\ n^2,$$

where c_1,c_2,c_3 are the constants of propositions 3, 5 and 6.

PROOF: We adopt the notations introduced in proposition 5. Fix $s,t\in[0,1]$ for a moment, $s\le t$. The inequality of Cauchy-Schwarz and the orthogonality of $O(s,t),P(s,t)$ allow us to estimate

$$(9)\quad \int_{\mathbb{R}} \det K(s,t,x)^{-1/2}\,[h^*(s)K(s,t,x)^{-1}h(s)h^*(t)K(s,t,x)^{-1}h(t)]\,dx$$

$$\le c_2 \sum_{i,j=1}^{n} \ell_i^2(s)k_j^2(t)\,[\lambda_i(s,t)(1-\lambda_i(s,t))\lambda_j(s,t)(1-\lambda_j(s,t))]^{-1/4}$$

(proposition 5)

$$\le c_2 \left[\sum_{i,j=1}^{n} \ell_i^2(s)k_j^2(t)\,[\lambda_i(s,t)(1-\lambda_i(s,t))]^{-1/2}\right]^{1/2} \cdot$$

$$\cdot \left[\sum_{i,j=1}^{n} \ell_i^2(s)k_j^2(t)\,[\lambda_j(s,t)(1-\lambda_j(s,t))]^{-1/2}\right]^{1/2}$$

$$= c_2 \left[\sum_{i=1}^{n} \ell_i^2(s)\,[\lambda_i(s,t)(1-\lambda_i(s,t))]^{-1/2}|h(t)|^2\right]^{1/2} \cdot$$

$$\cdot \left[\sum_{i=1}^{n} k_j^2(t)\,[\lambda_j(s,t)(1-\lambda_j(s,t))]^{-1/2}|h(s)|^2\right]^{1/2}.$$

To integrate both sides of (9) over s,t, $s\le t$, we may and do assume that k and ℓ have measurable versions in both variables. We then obtain

$$\int_0^1\int_0^1\int_{\mathbb{R}}\limits_{\{s\le t\}} \det K(s,t,x)^{-1/2}[h^*(s)K(s,t,x)^{-1}h(s)$$

$$h^*(t)K(s,t,x)^{-1}h(t)]\,dx\,ds\,dt$$

$$\le c_2 \left[\int_0^1\int_0^1\limits_{\{s\le t\}} \sum_{i=1}^{n} \ell_i^2(s)\,[\lambda_i(s,t)(1-\lambda_i(s,t))]^{-1/2}|h(t)|^2\,ds\,dt\right]^{1/2} \cdot$$

$$\cdot \left[\int_0^1\int_0^1\limits_{\{s\le t\}} \sum_{j=1}^{n} k_j^2(t)\,[\lambda_j(s,t)(1-\lambda_j(s,t))]^{-1/2}|h(s)|^2\,dt\,ds\right]^{1/2}$$

$$\le c_2c_3[n\int_0^1 |h(t)|^2\, dt]^{1/2}\ [n\int_0^1 |h(s)|^2\, ds]^{1/2} \qquad \text{(proposition 6)}$$

$$= c_2c_3\ n^2 \qquad (h_1,\dots,h_n \text{ orthonormal}).$$

It remains to apply proposition 3. Splitting $[0,1]^2$ into $\{s\le t\}$ and $\{t\le s\}$ leads to the factor 2 in the asserted inequality. This completes the proof. ∎

For analytic data we have therefore achieved our aim.

<u>PROPOSITION 8</u>: Let $h_1,\dots,h_n$ be analytic functions. Then U possesses a square integrable occupation density.

<u>PROOF</u>: Combine propositions 2 and 7. ∎

To generalize proposition 8 to non-analytic $h_1,\dots,h_n$, we first remark that indeed we have proved a little more.

<u>REMARK</u>: Let $h_1,\dots,h_n$ be analytic functions. Then

$$\int_{\mathbb{R}}\int_0^1\int_0^1 E(\exp(ix(U_t-U_s))u_s^2u_t^2)\ ds\ dt\ dx$$
$$\le 2\ c_1c_2c_3\ n^2.$$

This follows immediately from the remark to proposition 2 and proposition 7. An estimate like this with a dimension dependent bound makes one wonder whether the inequalities we have been using were too rough to carry over to the infinite dimensional case. Indeed, in proposition 3, when getting rid of the influence of the interaction T, our arguments were susceptible to some improvement. We suspect that the bound $c_1\ n^2$ can be replaced by a smaller constant depending only on T. But it is hard to say in which way this constant depends on n.

Our second step to generalize proposition 8 consists in approximating an orthonormal family $(h_1,\dots,h_n)$ by an orthonormal family of analytic functions.

<u>PROPOSITION 9</u>: Let $(h_1,\dots,h_n)$ be an orthonormal family in $L^2([0,1])$, $\delta>0$. Then there exists an orthonormal family $(g_1,\dots,g_n)$ consisting of analytic functions such that

$$\|g_i-h_i\|_2 \le \delta \quad \text{for} \quad 1\le i\le n.$$

<u>PROOF</u>: Choose $\Theta>0$ such that $3\Theta + 3n\Theta(1+3\Theta) < 1$. Using standard theorems of real analysis we obtain a family $(k_1,\dots,k_n)$ of polynomials on $[0,1]$ such that

$$\|h_i-k_i\|_2 \le \Theta \quad \text{for } 1\le i\le n.$$

To $(k_1,\dots,k_n)$ we apply the Gram-Schmidt orthogonalization procedure.

$$\text{Let } g_1 = \frac{1}{\|k_1\|_2} k_1,$$

$$g_i = [\|k_i - \sum_{j=1}^{i-1} \langle k_j,k_i\rangle k_j\|_2]^{-1} \cdot [k_i - \sum_{j=1}^{i-1} \langle k_j,k_i\rangle k_j],\ 2\le i\le n.$$

Note that for $i\neq j$

$$\langle k_i,k_j\rangle = \langle k_i-h_i,k_j-h_j\rangle + \langle k_i-h_i,h_j\rangle + \langle h_i,k_j-h_j\rangle,$$

due to orthogonality of h_i,h_j. Therefore, since h_i,h_j are unit vectors,

$$\begin{aligned}|\langle k_i,k_j\rangle| &\le \|k_i-h_i\|_2\, \|k_j-h_j\|_2 + \|k_i-h_i\|_2 + \|k_j-h_j\|_2 \\ &\le \Theta^2 + 2\Theta \le 3\Theta \qquad (\Theta<1).\end{aligned}$$

In the same way for $1\le i\le n$

$$|\langle k_i,k_i\rangle-1| = |\langle k_i,k_i\rangle - \langle h_i,h_i\rangle| \le 3\Theta.$$

Moreover, for $1\le i\le n$

$$\begin{aligned}\|k_i - \sum_{j=1}^{i-1} \langle k_j,k_i\rangle k_j\|_2 &\le \|k_i\|_2 + \sum_{j=1}^{i-1} |\langle k_j,k_i\rangle|\|k_j\|_2 \\ &\le 1 + 3\Theta + n \cdot 3\Theta(1+3\Theta),\end{aligned}$$

$$\begin{aligned}\|k_i - \sum_{j=1}^{i-1} \langle k_j,k_i\rangle k_j\|_2 &\ge \|k_i\|_2 - \sum_{j=1}^{i-1} |\langle k_j,k_i\rangle|\|k_j\|_2 \\ &\ge 1 - 3\Theta - n \cdot 3\Theta(1+3\Theta) > 0.\end{aligned}$$

Hence for $1 \le i \le n$

$$\|k_i - g_i\|_2 \le \left| [\|k_i - \sum_{j=1}^{i-1} \langle k_j, k_i\rangle k_j\|_2]^{-1} - 1 \right| \|k_i\|_2 + [\|k_i - \sum_{j=1}^{i-1} \langle k_j, k_i\rangle k_j\|_2]^{-1} \sum_{j=1}^{i-1} |\langle k_j, k_i\rangle| \|k_j\|_2$$
$$\le \frac{3\Theta + 3n\Theta(1+3\Theta)}{1-3\Theta-3n\Theta(1+3\Theta)} (1+3\Theta) + \frac{3n\Theta(1+3\Theta)}{1-3\Theta-3n\Theta(1+3\Theta)} .$$

Finally, we may make Θ small enough to keep both $\|h_i - k_i\|_2$ and $\|k_i - g_i\|_2$ below $\delta/2$. This completes the proof. ∎

We now can prove the main result of this section.

THEOREM 1: U possesses a square integrable occupation density.

PROOF: Using proposition 8, choose sequences $(g_1^m)_{m\in\mathbf{N}}, \ldots, (g_n^m)_{m\in\mathbf{N}}$ of analytic functions such that for any $m\in\mathbf{N}$

$$(g_1^m, \ldots, g_n^m) \text{ is an orthonormal family}$$

and

$$\lim_{m\to\infty} \|g_i^m - h_i\|_2 = 0, \quad 1 \le i \le n.$$

For $m\in\mathbf{N}$ let

$$T^m = \sum_{i,j=1}^{n} a_{ij}\, g_i^m \otimes g_j^m,$$

u^m, U^m the respective integrand and Skorohod integral process associated with T^m. Remember

$$T = \sum_{i,j=1}^{n} a_{ij}\, h_i \otimes h_j .$$

Now the remark following proposition 8 tells us that

$$\sup_{m\in\mathbf{N}} \int_{\mathbf{R}} \int_0^1 \int_0^1 E(\exp(ix(U_t^m - U_s^m))(u_s^m)^2 (u_t^m)^2)\, ds\, dt\, dx < \infty.$$

Moreover, by choice of the approximately sequence $u^m \to u$, $U^m \to U$, both in $L^2(\Omega\times[0,1])$ as $m \to \infty$. By selecting a subsequence, if necessary, we may assume that this convergence is $P \times \lambda$ - a.s. Hence the lemma of Fatou allows us to deduce from (10)

$$\int_{\mathbb{R}} \int_0^1 \int_0^1 E(\exp(ix(U_t - U_s))u_s^2 u_t^2)\, ds\, dt\, dx < \infty.$$

But by proposition 1.1 of [7], this implies that U possesses a square integrable occupation density. ∎

We will now illustrate by an example that the result of theorem 1, as simple and easy as it may seem, cannot be deduced from the results of the theory of Gaussian enlargements of the Wiener filtration. Indeed, it will turn out that it is enough to take two orthogonal interacting Gaussian components.

<u>EXAMPLE</u>: Let

$$f(s) = \log 2/2\ (\sqrt{s}\ \log s)^{-1},\ 0 \le s \le 1/2,$$

$$h_1(s) = f(s)\ 1_{[0,1/2]}(s) + f(1-s) 1_{[1/2,1]}(s),$$

$$h_2(s) = 1_{[0,1/2]}(s) - 1_{[1/2,1]}(s).$$

Using the transformation $t = -\log s$, it is easy to see that

$$\int_0^{1/2} f^2(s)\, ds = 1/2,$$

and therefore that $\|h_1\|_2 = 1 = \|h_2\|_2$. It is obvious from the definition that $\langle h_1, h_2\rangle = 0$. So (h_1, h_2) is an orthonormal pair of functions. Now let

$$u_t = h_1(t) \int_0^1 h_2\, dW,\ t\in[0,1].$$

The Skorohod integral process of u is given by

$$U_t = \int_0^t h_1 dW \int_0^1 h_2 dW - \int_0^t h_1 h_2\, d\lambda,\ t\in[0,1].$$

Theorem 1 shows that U possesses a square integrable occupation density. Let us show that U is not a semimartingale with respect to the enlarged Wiener filtrations to be used in this context (see Juelin,

Yor [8]). For $t\in[0,1]$ let

$$\mathcal{G}_t^1 = \sigma(W_s: s\le t) \vee \sigma(\int_0^1 h_2 dW),$$

completed w.r.t. to P so that the "usual hypotheses" of martingale theory are valid. Abbreviate $G^1 = (\mathcal{G}_t^1: 0\le t\le 1)$. Since h_1 is deterministic, théoreme I.1.1 of the paper of Chaleyat-Maurel, Jeulin in Jeulin, Yor [8], p. 64, is applicable and states that

$$\int_0^{\cdot} h_1 dW \text{ is a } G^1\text{-semimartingale iff } \int_0^t |h_1(s)| \frac{|h_2(s)|}{(\int_s^1 h_2^2(u)du)^{1/2}} ds < \infty$$

for all $t\in[0,1]$. Now

$$\begin{aligned}\int_0^1 |h_1(s)| \frac{|h_2(s)|}{(\int_s^1 h_2^2(u)du)^{1/2}} ds &\ge \log 2/2 \int_{1/2}^1 [(1-s)|\log(1-s)|]^{-1} ds \\ &= \log 2/2 \int_0^{1/2} [s|\log s|]^{-1} ds \\ &= \log 2/2 \int_{\log 2}^{\infty} \frac{1}{t} dt \\ &= \infty.\end{aligned}$$

For the equation of lines 2 and 3 of this inequality chain we have used the substitution $t = -\log s$ again. Hence

$$\int_0^1 h_1 dW \text{ is not a } G^1\text{-semimartingale.}$$

Since $\int_{1/2}^1 h_1 h_2 \, d\lambda < \infty$, also U is not a G^1-semimartingale. Of course, if we enlarge further, for example to

$$\mathcal{G}_t^2 = \mathcal{G}_t^1 \vee \sigma(\int_0^1 h_1 dW), \quad t\in[0,1],$$

this statement is true a fortiori.

<u>REMARK</u>: The question, whether the process U just constructed is a

semimartingale with respect to its natural filtration, remains open. It is hard to imagine how it could be approached.

REFERENCES

[1] Baumgärtel, H. Analytic perturbation theory for matrices and operators. Birkhäuser: Basel, Boston (1985).

[2] Bellman, R. Introduction to matrix analysis. McGraw-Hill: New York (1970).

[3] Berman, S.M. Local times and sample function properties of stationary Gaussian processes. Trans. Amer. Math. Soc. 137 (1969), 277-300.

[4] Ehm, W. Sample function properties of multi-parameter stable processes. Z. Wahrscheinlichkeitstheorie verw. Geb. 56 (1981), 195-228.

[5] Fredholm, I. Sur une classe d'équations fonctionnelles. Acta Math. 27 (1903), 365-390.

[6] Geman, D., Horowitz, J. Occupation densities. Ann. Probab. 8 (1980), 1-67.

[7] Imkeller, P. Occupation densities for stochastic integral processes in the second Wiener chaos. Preprint, Univ. of B.C. (1990).

[8] Jeulin, Th., Yor, M. (eds). Grossissement de filtrations: exemples et applications. Séminaire de Calcul Stochastique, Paris 1982/83. LNM 1118. Springer: Berlin, Heidelberg, New York (1985).

[9] Jörgens, K. Linear integral operators. Pitman: Boston, London (1982).

[10] Kato, T. Perturbation theory for linear operators. Springer: Berlin, Heidelberg, New York (1966).

[11] Nualart, D. Noncausal stochastic integrals and calculus. LNM 1516. Springer: Berlin, Heidelberg, New York (1988).

[12] Nualart, D., Pardoux, E. Stochastic calculus with anticipating integrands. Probab. Th. Rel. Fields 78 (1988), 535-581.

[13] Nualart, D., Pardoux, E. Boundary value problems for stochastic

differential equation. Preprint (1990).

[14] Nualart, D., Pardoux, E. Second order stochastic differential equations with Dirichlet boundary conditions. Preprint (1990).

[15] Nualart, D., Zakai, M. Generalized stochastic integrals and the Malliavin calculus. Probab. Th. Rel. Fields 73 (1986), 255-280.

[16] Ocone, D., Pardoux, E. Linear stochastic differential equations with boundary conditions. Probab. Th. Rel. Fields, to appear (1990).

[17] Pietsch, A. Eigenvalues and s-numbers. Cambridge University Press: Cambridge, London (1987).

[18] Reed, M., Simon, B. Methods of modern mathematical physics. IV: Analysis of operators. Acad. Press: New York (1978).

[19] Rellich, F. Perturbation theory of eigenvalue problems. Gordon, Breach: New York, London (1969).

[20] Simon, B. Trace ideals and their applications. London Math. Soc. Lecture Notes Series 35. Cambridge University Press: Cambridge, London (1979).

[21] Skorohod, A.V. On a generalization of a stochastic integral. Theor. Prob. Appl. 20 (1975), 219-233.

[22] Smithies, F. Integral equations. Cambridge University Press: Cambridge, London (1965).

[23] Watanabe, S. Lectures on stochastic differential equations and Malliavin calculus. Tata Institute of Fundamental Research. Springer: Berlin, Heidelberg, New York (1984).

[24] Zakai, M. The Malliavin calculus. Acta Appl. Math. 3 (1985), 175-207.

Peter Imkeller
Department of Mathematics
University of British Columbia
121 - 1984 Mathematics Road
Vancouver, B.C. V6T 1Y4
Canada

Calculating the Compensator: Method and Example

BY FRANK B. KNIGHT

1. METHOD: Let X_t, $t \geq 0$ be a real-valued stochastic process on a complete probability space $(\Omega, \mathcal{F}, P)$, adapted to a right-continuous filtration $\mathcal{F}_t$ containing all P-null sets. We recall from [**2**, VII, 23] that X_t is an $\mathcal{F}_t$-semimartingale if it can be expressed $X_t = X_0 + M_t + A_t$, where M_t is a local martingale of $\mathcal{F}_t$, $M_0 = 0$, and A_t is a right-continuous adapted process with paths of finite variation on finite time intervals. Moreover ([**2**, ibid]) X_t is called "special" if there is such a representation with A_t previsible, and then the previsible A_t is unique. In this case we will call A_t the "compensator" (or dual previsible projection) of X_t. Note that this terminology differs considerably from that of [**2**, VI], which seems not to give any general name to such A_t. The process A_t can be obtained from X_t more or less explicitely, at least in theory. Indeed, there exist stopping times $T_n \uparrow \infty$ such that $A_{t\wedge T_n}$ are of bounded variation ([**2**, VI, 2, (52.1)]). Then $A_{t\wedge T_n}$ may be constructed from $X_{t\wedge T_n}$ by the approximations of P. A. Meyer [**8**, VII, T29] or M. Rao ([**10**] and [**2**, VII; 1, 21]).

In the present work, by contrast, we do not assume that X_t is a semimartingale, but instead propose a method of checking that it is one, and of simultaneously obtaining the compensator A_t. We do not, however, have any results that evaluate how general this method is. Instead, we only wish to apply it to an example which seems to be of independent interest.

To describe the method in general terms, we assume that X_t is right-continuous in L^1, and that, for $\lambda > 0$, $E\int_0^\infty e^{-\lambda t}|X_t|dt < \infty$ (since it suffices to construct A_t in finite intervals $(0, K]$ it would be permissible to redefine $X_t \equiv 0$ for all $t \geq K$ in order to achieve the last hypothesis). In this case, it is known ([**6**] and [**8**)] that for $\lambda > 0$ the following expression is an $\mathcal{F}_t$-martingale

$$(1.1) \qquad M_\lambda(t) = R_\lambda(X_t) - R_\lambda(X_0) + \int_0^t X_s - \lambda R_\lambda(X_s)ds,$$

where $R_\lambda(X_t) = E\left(\int_0^\infty e^{-\lambda s} X_{t+s} ds | \mathcal{F}_t\right)$ is chosen to be right–continuous in t.

REMARK: The notation R_λ derives from the fact that if we represent $X_t = \varphi(Z_t)$, where Z_t is the "prediction process" of X_t, then $R_\lambda(X_t) = R_\lambda^Z \varphi(X_t)$, where$R_\lambda^Z$ is the resolvent of Z_t, (see for example [7]).

The method which we use may be spelled out as follows

PROPOSITION 1.1: If X_t has right–continuous paths, and the limits

$$A_t = \lim_{\lambda \to \infty} \lambda \int_0^t (X_u - \lambda R_\lambda X_u) du, \qquad 0 < t, \tag{1.2}$$

exists both pathwise a.s. and in L^1, and are of finite variation in finite time intervals, then X_t is a special semimartingale and A_t is its compensator.

PROOF: Since X_t is right–continuous in L_1, $\lim_{\lambda \to \infty} \lambda R_\lambda(X_t) = X_t$ in L^1. Thus if the limit A_t exists we have $\lim_{\lambda \to \infty} \lambda M_\lambda(t) = X_t + A_t$ in L_1. Hence $X_t + A_t \doteq X_0 + M_t$ is a martingale, and since $\lambda \int_0^t (X_u - \lambda R_\lambda X_u) du$ is continuous in t, A_t is previsible. Therefore A_t is the required compensator.

2. AN EXAMPLE: While simple to state, the above method (and probably any other method of finding the compensator as well) can lead to difficult calculations when put into practice. We have chosen to work an example in the form $X_t = B_{t \wedge Q}$, where B_t is a Brownian motion starting at 0 and Q is measurable over $G_\infty (\doteq \sigma(B_s, s < \infty))$, but Q is not a stopping time of $G_t (\doteq \sigma(B_s, s \le t))$. The general class of such processes might be called "arrested Brownian motions," and they behave rather differently from stopped Brownian motions, as is to be expected. What was not entirely anticipated is the degree of difficulty inherent in calculating A_t for such X_t, even for the simplest cases of Q. Indeed, we still do not know whether all such X_t are even semimartingales relative to $\mathcal{F}_t (\equiv \sigma(X_s, s \le t+))$. Our aim, however, was not to investigate this question, but to calculate A_t in the following special case (proposed by Professor Bruce Hajek).

PROPOSITION 2.1. *For $c > 0$, let $S_c = \max_{0 \le x \le c} B_s$, so that $B(Q_c) = S_c$ (it is well-known that Q_c is unique, P–a.s.) Then $X_t = B(t \wedge Q_c)$ is an $\mathcal{F}_t$–special semimartingale, with compensator $A_{t\wedge Q_c} = \int_0^{t\wedge Q_c} H_v(u, S(u) - X(u))du$, where $H(u,v) = \ell n \int_v^\infty \exp(-y^2/2(c-u))dy$, and $H_v = \frac{\partial}{\partial v}H$. Moreover, $X_t + A_t$ is a stopped Brownian motion, in the sense that $(X_{t\wedge Q_c} + A_{t\wedge Q_c})^2 - (t \wedge Q_c)$ is also a martingale.*

Before commencing the proof, it may be amusing to give an "economic" interpretation. Suppose that a certain stock market index (with appropriate scaling) performs a Brownian motion, but that there is an oracle who, given a time $c > 0$, can announce at its arrival the time when the market reaches its maximum in $0 \le t \le c$. The question is, how should a stock owner, who would otherwise have no inside knowledge, be fairly paid in lieu of using the oracle (and thus selling at the maximum). Thus, if he promises to give up the oracle until a time $t < c$ he (or his agent) should recieve $-A_t$ by time t, and 0 thereafter, in order to be fairly compensated. But if at time t the oracle has not spoken, and knowing this the stock owner decides to continue until time $t + s$, then he should be paid by that time an additional amount $-(A(t+s) - A(t))$ not to use the oracle.

For another interpretation, suppose a gambling house introduces the game "watch $B(t)$ and receive S_c at time Q_c." This can be implemented since the house may know $B(t)$, $t \le c$, in advance. Then $-A(t)$ gives the fair charge for playing the game until time t. We note that $A(t)$ can be calculated from $B(t)$ without using any future information (except the fact that $t \le Q_c$, at least until time Q_c when the game is over).

ADDED REMARKS: After completing an initial draft of this paper, it was brought to our attention by Chris Rogers that this example is a special case of those treated abstractly by M. Barlow in [**1**], by T Jeulin and M. Yor in [4]. The formula for the compensator from [**1**, Prop. 3.7] (to which the one from [4] is equivalent) is

$$A_{t\wedge Q_c} = -\int_0^{t\wedge Q_c} (1 - A^o_{u-})^{-1} d\langle B, A^o - \hat{A}\rangle_u$$

where A^o_u is the optional projection of $I_{[Q_c,\infty)}(u)$ and $\hat{A}_u$ is it dual optional projection. From this it is clear that $A_{t\wedge Q_c}$ is Lebesgue–absolutely–continuous, which provided a check on our calculations. More importantly, it is not very difficult to calculate A^0, and then to derive $\hat{A}$ from A^0 by using Ito's Formula, thus obtaining a shortened proof of Proposition 2.1 (as Professor Rogers has shown me). Indeed, we have

$$\begin{aligned} A_t^0 &= E(I_{[Q_c,\infty)}(t)|\mathcal{F}_t) \\ &= P(Q_c < t|\mathcal{F}_t) \\ &= \sqrt{\frac{2}{\pi(c-t)}}\int_0^{S(t)-X(t)}\left(\exp -\frac{y^2}{2(c-t)}\right) dy, \end{aligned}$$

and it follows by Ito's Formula that

$$\begin{aligned} A_t^0 &- \sqrt{\frac{2}{\pi}}\int_0^t \frac{dS(u)}{\sqrt{c-u}} \\ &= -\sqrt{\frac{2}{\pi}}\int_0^t \frac{1}{\sqrt{c-u}}\exp -\frac{(S(u)-B(u))^2}{2(c-u)} dB(u). \end{aligned}$$

Then by optional stopping we have $\hat{A}_t = \sqrt{\frac{2}{\pi}}\int_0^t \frac{dS(u)}{\sqrt{c-u}}$, and Proposition 2.1 follows.

Finally, an expression somewhat resembling that of Proposition 2.1, but containing an additional term, is found in [**3**, p. 49]. The problem considered there, in which $\sigma(Q_c)$ is adjoined immediately at $t = 0$, is quite different from ours. The compensator of $B(t)$ for $t \geq Q_c$ is also given, which would be the same as for our problem.

In view of these facts, we might not want to publish our own calculations, except for the following considerations. First, our method is in no way limited to "honest" times, as is that of [**1**] and [4], and it does not depend on these results, or on Ito's Formula. Second, it may be of use to indicate the type of calculations which our method leads to, even though they become quite tedious in the present case. Third, since the result is now known by other methods, we can omit the final pages of checking that the three "o-terms" do not contribute to the answer.

PROOF: We continue to let $\mathcal{F}_t$ denote the usual augmentation of $\cap_{\epsilon>0}\sigma(X_s, s < t+\epsilon)$. To construct $\lambda R_\lambda(X_t)(= X_t$ for $t \geq Q_c)$, we need to calculate $E(X_{t+s}|\mathcal{F}_t)$ over $\{t < Q_c\}$. It is easy to see that the conditioning reduces to being given the pair (X_t, S_t), but to write S_t as given we need to introduce a further notation to distinguish it from the future maximum. We write $S_o(t)$ for S_t when given in a conditional probability. Then for $s \leq c-t$ we have $E^0(X_{t+s}|\mathcal{F}_t) = E^{B(t)}(X_s|S_{c-t} > S_o(t))$. Setting $B(t) = x$ for brevity, we will need the P^x joint density of (Q_{c-t}, S_{c-t}) from L. Shepp [11, (1.6)]. In the variables (θ, y) it is $\frac{1}{\pi}\frac{(y-x)\exp(-\frac{(y-x)^2}{2\theta})}{\theta^{\frac{3}{2}}(c-t-\theta)^{\frac{1}{2}}}$, $0<\theta<c-t$, $y>x$. Thus, for $s \le c-t$,

$$E^x(X_s|S_{c-t} > S_o(t)) \bullet P^x(S_{c-t} > S_o(t))$$
$$= \frac{1}{\pi}\int_0^s \left(\int_{S_o(t)}^\infty \left(y(y-x)\exp\frac{-(y-x)^2}{2\theta}\right) dy\right) \theta^{-\frac{3}{2}}(c-t-\theta)^{-\frac{1}{2}} d\theta$$
$$(2.1) \qquad + \frac{1}{\pi}\int_s^{c-t} \left(\int_{S_o(t)}^\infty E^x(B(x)|Q_{c-t} = \theta, S_{c-t} = y)\right.$$
$$\left. \bullet (y-x)\exp\frac{-(y-x)^2}{2\theta} dy\right) \theta^{-\frac{3}{2}}(c-t-\theta)^{-\frac{1}{2}} d\theta.$$

Let us denote these two double integral terms by T_1 and T_2, respectively. We integrate by parts in T_1 to obtain

$$T_1 = \frac{1}{\pi}\int_0^s \left(S_o(t)\exp - \frac{(S_o(t)-x)^2}{2\theta} + \int_{S_o(t)-x}^\infty \exp(-\frac{y^2}{2\theta})dy\right) \bullet (\theta(c-t-\theta))^{-\frac{1}{2}} d\theta.$$

In order to find the contribution of T_1 to $\lambda^2 R_\lambda(X_u)$ in (1.2), note that for $s \ge c-t$ the contribution of T_2 is 0, and that of T_1 is the same as for $s = c-t$ (since $X_{c+t} = X_c$). Integrating by parts twice, we obtain

$$\lambda^2 \int_0^\infty e^{-\lambda s} T_1(s) ds = \lambda^2 \int_0^{c-t} e^{-\lambda s} T_1(s) ds + \lambda e^{-\lambda(c-t)} T_1(c-t)$$
$$(2.2) \qquad = \frac{\lambda}{\pi}\int_0^{c-t} e^{-\lambda s}\left[S_o(t)\exp\frac{-(S_o(t)-x)^2}{2s} ds\right.$$
$$\left. + \int_{S_o(t)-x}^\infty \exp -\frac{y^2}{2s} dy\right] (s(c-t-s))^{\frac{-1}{2}} ds.$$

Continuing with this term, but reintroducing the variable u from (1.2) in place of t (so that T_1 depends on s, u, and x, where $x = X(u) = B(u)$ for $u \le Q_c$) we now take $t \le Q(c)$ and calculate pathwise

$$(2.3) \qquad \lim_{\lambda\to\infty} \int_0^t \lambda^2 \int_0^\infty e^{-\lambda s} T_1(s; u, B(u)) ds du,$$

in lieu of $\lim_{\lambda\to\infty} \int_0^t \lambda^2 R_\lambda X_u$ in (1.2). Actually, from (2.1) there is also a denominator $P^{X(u)}(S_{c-u} > S_o(u))$ to be included in the integrand, but this term is awkward when we need L^1–limits, and it does not involve λ. Therefore, we set $T_K = Q_c \wedge \inf(t : P^{B(t)}\{S_{c-t} > S_o(t)\} \le K^{-1})$, and (for fixed K) we replace t by $t^* = t \wedge T_K$ (note that T_K is an $\mathcal{F}_t$–stopping time and $T_K = Q_c$ as $K \to \infty$), so that the denominator is bounded away from 0 by K^{-1} for $0 < u \le t^*$. Then it does not affect the convergence as $\lambda \to \infty$ in (2.3). It is easy[1] to see that this will also be unaffected if we restrict the ds–integral to $0 < s \le \epsilon$, which in turn allows us to replace the term $(c-t-s)^{-\frac{1}{2}}$ by $(c-t)^{-\frac{1}{2}}$ in (2.2) as $\lambda \to \infty$, and then again allow $s \to \infty$ in (2.2). This leaves

$$(2.4)\quad \lim_{\lambda\to\infty} \frac{\lambda}{\pi} \int_0^{t^*} (c-u)^{-\frac{1}{2}} \int_0^\infty e^{-\lambda s} \left[S_o(u) \exp -\frac{(S_o(u)-X_u)^2}{2s} + \int_{S_o(u)-X_u}^\infty \exp -\frac{y^2}{2s} dy \right] s^{-\frac{1}{2}} ds du$$

Now the ds integration leads to the usual resolvent kernel $(2\lambda)^{-\frac{1}{2}} \exp -\sqrt{2\lambda}x$ of Brownian motion, and (2.4) becomes

$$(2.5)\quad \lim_{\lambda\to\infty} \sqrt{\frac{\lambda}{\pi}} \int_0^{t^*} (c-u)^{-\frac{1}{2}} \left[S_o(u) \exp -\sqrt{2\lambda}(S_o(u)-X_u) + (2\lambda)^{-\frac{1}{2}} \exp -\sqrt{2\lambda}(S_o(u)-X_u) \right] du$$

For $u < t^*$, we have $S_o(u) - X_u = S_o(u) - B_u$ which is equivalent to $|B_u|$ in law, and hence has a continuous local time $\ell^+(u,x)$; $x \ge 0$. Using this, and approximating (2.5) by Riemann sums, it becomes

$$(2.6)\quad \begin{aligned} &\lim_{\lambda\to\infty} \sqrt{\frac{\lambda}{\pi}} \lim_{n\to\infty} \sum_{k=0}^{n-1} (c - \frac{k}{n}t^*)^{-\frac{1}{2}} \int_{\frac{k}{n}t^*}^{\frac{k+1}{n}t^*} [S_o(\frac{k}{m}t^*). \\ &\exp -\sqrt{2\lambda}(S_o(u)-X_u) + (2\lambda)^{-\frac{1}{2}} \exp -\sqrt{2\lambda}(S_o(u)-X_u)]du \\ &= \lim_{\lambda\to\infty} \sqrt{\frac{\lambda}{\pi}} \lim_{n\to\infty} \sum_{k=0}^{n-1} (c - \frac{k}{n}t^*)^{-\frac{1}{2}} \int_0^\infty [S_o(\frac{k}{n}t^*) \exp -\sqrt{2\lambda}x \\ &+ (2\lambda)^{-\frac{1}{2}} \exp -\sqrt{2\lambda}x (\ell^+(\frac{k+1}{n}t^*, x) - \ell^+(\frac{kt^*}{n}, x))dx. \end{aligned}$$

For each λ, this is dominated by

$$\frac{1}{\sqrt{2\pi}} (c-t^*)^{-\frac{1}{2}} (S_o(t^*)\sqrt{2\lambda}+1) \int_0^\infty \exp(-\sqrt{2\lambda}x) \ell^+(t^*,x) dx$$

in such a way that as $\lambda \to \infty$, using continuity of $S_o(u)$,we have convergence both pathwise and in L_1 to

[1] We will use several times the observation that, if $\int_0^\infty e^{-\lambda s} g(s) ds < \infty$ for a $g \ge 0$, then $\lim_{\lambda\to\infty} \lambda^k \int_0^\epsilon e^{-\lambda s} g(s) ds$ exists if and only if $\lim_{\lambda\to\infty} \lambda^k \int_0^\infty e^{-\lambda s} g(s) ds$ exists, and then the two limits are equal for every $\epsilon > 0$.

$$(2.7) \qquad \frac{1}{\sqrt{2\pi}} \int_0^{t^*} (c-u)^{-\frac{1}{2}} S_o(u) d\ell^+(u,0).$$

In more detail, to interchange the limits we observe that as $n \to \infty$ we have a Cauchy sequence in L^1, uniformly in $\lambda > \frac{1}{2}$, by reason of the uniform bound

$$\frac{1}{\sqrt{2\pi}}(c-t^*)^{-\frac{1}{2}}(S_o(t^*)+1)(\sup_x \ell^+(t^*,x)) \in L^1.$$

Finally, to include the conditioning in (2.1) into the contribution to $\lim_{\lambda\to\infty} \int_0^{t^*} \lambda^2 R_\lambda X_u du$ we also need to incorporate a denominator $P^{X(u)}(S_{c-u} > S_o(u))$ into (2.4). But since this is bounded away from 0 for $u < t^*$, and in (2.7) $\ell^+(u,0)$ increases only when $X(u) = S_o(u)$, this factor just becomes 1 in the limit, and may be ignored for $u \le t^*(= t \wedge T_K)$. Thus (2.7) is the limiting contribution of T_1 to the compensator A_{t^*} of (1.2), except for a change of sign. (The integrand term X_u in (1.2) must thus be added to minus the contribution of $\lambda \int_0^{c-u} e^{-\lambda s} T_2(s) ds$, with $T_2(s)$ from (2.1)).

The first task in evaluating T_2 is to estimate $E^x(B(s)|Q_c = \theta, S_c = y)$; $s < \theta < c$. There is no difficulty to write the exact expression, but it is a little complicated. We note that when $x = B(0)$, $z = B(c)$, $\theta = Q_c$ and $y = S_c$ are all given, the path $B(s)$, $0 \le s \le c$, breaks into independent parts $0 \le s \le \theta$ and $\theta \le s \le c$. For the second part $y - B(\theta + s)$ is just the excursion of the reflected Brownian motion $S_s - B_s$ straddling c. It is well-known from excursion theory (see for example [5, Theorem 5.2.7 and Lemma 5.2.8]) that, conditional on the value of $y - B(c)$, this process is equivalent to a Bessel bridge $Besbr_3(s)$ from 0 to $y - B(c)$, $0 \le s \le c - \theta$. The process needed here, however, is $y - B(\theta - s)$, $0 \le s \le \theta$. But if $z = B(c)$ is given, while Q_c and S_c are unknown, $B(t)$ in $0 \le t \le c$ becomes a Brownian bridge from x to z. It is equivalent in law to $B(c-t) + (2\frac{t}{c} - 1)(z - x)$, $0 \le t \le c$, and it follows that if $\theta = Q_c$ and $y = S_c$ are also given, then $y - B(\theta - s)$ is also a $Besbr_3(s)$, from 0 to $y - x$. This does not depend on z, so we can compute, using the Bes_3 transition density $p_3(t,x,y) = \frac{y}{x}(2\pi t)^{-\frac{1}{2}}(exp\frac{-(x-y)^2}{2t} - exp\frac{-(x+y)^2}{2t})$, $t > 0$,

$$\begin{aligned} E^x(B(s)|Q_c = \theta, S_c = y) &= E(y - Besbr_3(\theta - s)) \\ &= y - p_3^{-1}(\theta, 0, y-x) \int_0^\infty z p_3(\theta - s, 0, z) p_3(s, z, y-x) dz. \end{aligned}$$

Denoting $y - x = w$, this gives for $s > \theta$

(2.8) $E^x(B(s)|Q_c = \theta, S_c = y)$

$$= y - \frac{\theta^{\frac{3}{2}} \exp((2\theta)^{-1}w^2)}{w\sqrt{2\pi}(\theta - s)^{\frac{3}{2}}\sqrt{s}} \int_0^\infty z^2 exp\frac{-z^2}{2(\theta - s)}(exp\frac{(w-z)^2}{2s} - exp\frac{-(w+z)^2}{2s})dz$$

$$= y - (\frac{\theta}{\theta - s})^{\frac{3}{2}} \frac{\exp((2s\theta)^{-1}(s-\theta)w^2)}{w\sqrt{2\pi s}} \int_0^\infty z^2 exp\frac{-\theta z^2}{2s(\theta - s)}(exp\frac{wz}{s} - exp\frac{-wz}{s})dz$$

$$= y - \frac{\exp((2s\theta)^{-1}(s-\theta)w^2)}{w\sqrt{2\pi s}}$$

$$[\int_0^\infty u^2 \left(exp\frac{-(u^2 - 2\sqrt{(1-\theta^{-1}s)}wu)}{2s} - exp\frac{-(u^2 + 2\sqrt{1-\theta^{-1}s}wu)}{2s} \right) du]$$

$$= y - \frac{1}{w\sqrt{2\pi s}} \int_0^\infty u^2 \left(exp\frac{-(u - \sqrt{1-\theta^{-1}s}w)^2}{2s} - exp\frac{-(u + \sqrt{1-\theta^{-1}s}w)^2}{2s} \right) du$$

We are concerned with the behavior of this last as $s \to 0+$ (corresponding to $\lambda \to \infty$ in (1.2)). It is trivial to see that the limit of the 2nd term concentrates to w so that the expression has limit $y - w = x$, as expected, but this is the x that is subtracted in the integrand of (1.2), and (as $\lambda \to \infty$) we need the other terms that are left over. Considering the 2nd term as the obvious difference, the *second* integral is equal to

(2.9)

$$\frac{-1}{w\sqrt{2\pi s}} \left[\int_0^\infty u \left(u + \sqrt{\frac{\theta - s}{\theta}} w \right) exp\frac{-(u + \sqrt{1-\theta^{-1}s}w)^2}{2s} du \right.$$

$$\left. -w\sqrt{\frac{\theta - s}{\theta}} \int_0^\infty \left(u + w\sqrt{\frac{\theta - s}{\theta}} - w\sqrt{\frac{\theta - s}{\theta}} \right) exp\frac{-(u + \sqrt{1-\theta^{-1}s}w)^2}{2s} du \right]$$

$$= -\left(\frac{s}{w\sqrt{2\pi}} \int_L^\infty e^{-\frac{1}{2}v^2} dv \right)$$

$$+ \left(\sqrt{\frac{(\theta - s)s}{2\pi\theta}} exp\frac{-(\theta - s)w^2}{\theta s} \right) - \left(\frac{\theta - s}{\theta} w \int_L^\infty \frac{e^{-\frac{1}{2}v^2}}{\sqrt{2\pi}} dv \right),$$

where $L = w\sqrt{\frac{\theta - s}{\theta s}}$. Introducing a notation for these 3 terms, we write (2.9) as $-R_1 + R_2 - R_3$. On the other hand, the *first* integral becomes

$$\frac{1}{w\sqrt{2\pi s}} \int_0^\infty \left[\left(u - \sqrt{\frac{\theta - s}{\theta}} w \right)^2 + \left(2\sqrt{\frac{\theta - s}{\theta}} wu - \frac{\theta - s}{\theta} w^2 \right) \right]$$

$$exp\frac{-(u - \sqrt{1-\theta^{-1}s}w)^2}{2s} du.$$

Breaking this as a sum of two integrals as indicated the first one is a variance if the lower limit is extended to $-\infty$, which entails the error

$$\frac{-1}{w\sqrt{2\pi s}}\int_{L\sqrt{s}}^{\infty} x^2 e^{-\frac{1}{2s}x^2}dx = -(R_1+R_2).$$

So this equals $\frac{s}{w} - (R_1+R_2)$. The second is

$$\frac{1}{\sqrt{2\pi s}}\sqrt{\frac{\theta-s}{\theta}}\int_0^{\infty}\left[2\left(u-w\sqrt{\frac{\theta-s}{\theta}}\right)+w\sqrt{\frac{\theta-s}{\theta}}\right] exp\frac{-(u-\sqrt{1-\theta^{-1}s}w)^2}{2s}du,$$

and writing this in turn as a sum of two integrals, the first equals

$$2\sqrt{\frac{(\theta-s)s}{\theta}}\int_{-L}^{\infty} v\frac{e^{-\frac{1}{2}v^2}}{\sqrt{2\pi}}ds = 2R_2,$$

and the second is

$$w\left(\frac{\theta-s}{\theta}\right)\int_{-L}^{\infty}\frac{e^{-\frac{1}{2}v^2}}{\sqrt{2\pi}}dv = w\left(\frac{\theta-s}{\theta}\right) - R_3.$$

Adding all of these terms, we obtain

$$\begin{aligned} &2R_2 - 2R_3 + 2R_1 + \frac{s}{w} + w\left(\frac{\theta-s}{\theta}\right) \\ &= 2\sqrt{\frac{(\theta-s)s}{2\pi\theta}}exp\frac{-(\theta-s)w^2}{2\theta s} - 2w\left(\frac{\theta-s}{\theta}\right)\int_L^{\infty}\frac{e^{-\frac{v^2}{2}}}{\sqrt{2\pi}}dv \\ &\quad - \frac{2s}{w\sqrt{2\pi}}\int_L^{\infty} e^{-\frac{1}{2}v^2}dv + \frac{s}{w} + w\left(\frac{\theta-s}{\theta}\right). \end{aligned} \tag{2.10}$$

The first three terms are $o(s^k)$ as $s\to 0$ for any $k>0$ if $w>0$, and (2.8) reduces to

$$\begin{aligned} E^x(B(s)|Q_c=\theta, S_c=y) &= y - \left(\frac{s}{y-x}\right) \\ &\quad - (y-x)\left(1-\frac{s}{\theta}\right) - 2R_2 + 2R_3 + 2R_1 \\ &= x - \frac{s}{y-x} + (y-x)\frac{s}{\theta} - 2R_2 + 2R_3 + 2R_1. \end{aligned} \tag{2.11}$$

Let us return afterwards to the three o terms, and first complete the contribution to the compensator based only on x and the $O(s)$ terms of (2.11). Now the total contribution to T_2 up to time t^* (using (2.1)) is

$$\begin{aligned} \lim_{\lambda\to\infty}\lambda\int_0^{t^*}\Bigg[X_u - \frac{\lambda}{\pi}\int_0^{c-u} e^{-\lambda s}\int_s^{c-u}\int_{S_o(u)}^{\infty} \\ E^{X(u)}(B(s)|Q_{c-u}=\theta, S_{c-u}=y)P^{X(u)}(S_{c-u}>S_o(u))^{-1} \\ (y-X_u)\exp-\frac{(y-X_u)^2}{2\theta}dy)\theta^{-\frac{3}{2}}(c-u-\theta)^{-\frac{1}{2}}d\theta ds\Bigg] du. \end{aligned}$$

The term $X(u)$ in $E^{X(u)}(B(s)|Q_{c-u}=\theta, S_{c-u}=y)$ from (2.11) is combined with the former X_u to contribute

$$\lim_{\lambda\to\infty}\lambda\int_0^{t^*}X_u\left(1-\lambda\int_0^{c-u}e^{-\lambda s}\frac{P^{X(u)}(Q_{c-u}>s \text{ and } S_{c-u}>S_o(u))}{P^{X(u)}(S_{c-u}>S_o(u))}ds\right)du$$
$$=\lim_{\lambda\to\infty}\lambda\int_0^{t^*}X_u\int_0^{c-u}e^{-\lambda s}P^{X(u)}(Q_{c-u}<s|S_{c-u}>S_o(u))dsdu.$$

We now proceed much as in (2.6) to write this as

$$\pi^{-1}\lim_{\lambda,n\to\infty}\lambda\sum_{k=0}^{n-1}(c-\frac{k}{n}t^*)^{-\frac{1}{2}}\int_{\frac{k}{n}t^*}^{\frac{k+1}{n}t^*}$$
$$\psi(u)X(u)\int_0^\infty\lambda e^{-\lambda s}\int_0^s\theta^{-\frac{1}{2}}\exp-\frac{(S_o(u)-X(u))^2}{2\theta}d\theta dsdu$$
$$=\pi^{-1}\lim_{\lambda,n\to\infty}\lambda\sum_{k=0}^{n-1}(c-\frac{k}{n}t^*)^{-\frac{1}{2}}\int_{\frac{k}{n}t^*}^{\frac{k+1}{n}t^*}$$
$$\psi(u)X(u)\int_0^\infty s^{-\frac{1}{2}}e^{-\lambda s}\exp-\frac{(S_o(u)-X(u))^2}{2s}dsdu,$$

where $\psi(u)=(P^0\{S_{c-u}>S_o(u)-X(u)\})^{-1}$. Continuing, this becomes in terms of the local time $\ell^+(t,x)$,

$$=\sqrt{\frac{2}{\pi}}\lim_{\lambda,n\to\infty}\sum_{k=0}^{n-1}(c-\frac{k}{n}t^*)^{-\frac{1}{2}}X_{\frac{k}{n}t^*}\frac{1}{\sqrt{2\lambda}}\int_0^\infty dx\bullet$$
$$e^{-\sqrt{2\lambda}x}\left(\ell^+\left(\frac{k+1}{n}t^*,x\right)-\ell^+\left(\frac{k}{n}t^*,x\right)\right)P^0\{S_{c-u}>x\}^{-1}$$
$$=\frac{1}{\sqrt{2\pi}}\lim_{n\to\infty}\sum_{k=0}^{n-1}(c-\frac{k}{n}t^*)^{-\frac{1}{2}}X_{\frac{k}{n}t^*}\left(\ell^+\left(\frac{k+1}{n}t^*,0\right)-\ell^+\left(\frac{k}{n}t^*,0\right)\right)$$
$$=\frac{1}{\sqrt{2\pi}}\int_0^{t^*}(c-u)^{-\frac{1}{2}}X_u d\ell^+(u,0).$$

The justification of the interchange of limits using also $P^{X(u)}(S_{c-u}>S_o(u))>K^{-1}$ for $u<t^*$, is the same as for (2.7), and the limit is both pathwise and in L^1. Since $d\ell^+(u,0)$ increases only when $X_u=S_o(u)$, this term *cancels* the contribution $(-(2.7))$ of T_1.

The remaining two terms $-\frac{s}{y-x}+(y-x)\frac{s}{\theta}$ from (2.11) contribute (with $\psi(u)$

as before)

$$\lim_{\lambda\to\infty}\frac{\lambda^2}{\pi}\int_0^{t^*}du\psi(u)\int_0^{c-u}dsse^{-\lambda s}\int_s^{c-u}d\theta\theta^{-\frac{3}{2}}(c-u-\theta)^{-\frac{1}{2}}\bullet$$

$$\left(\int_{S_o(u)}^{\infty}\left(\frac{1}{y-X_u}-\frac{y-X_u}{\theta}\right)(y-X_u)\exp-\frac{(y-X_u)^2}{2\theta}dy\right)$$

$$=\lim_{\lambda\to\infty}\frac{1}{\pi}\int_0^{t^*}du\psi(u)\int_0^{\infty}dvve^{-v}\int_{\frac{v}{\lambda}}^{c-u}$$

$$d\theta\theta^{-\frac{3}{2}}(c-u-\theta)^{-\frac{1}{2}}\left(\int_{S_o(u)}^{\infty}\left(1-\frac{(y-X_u)^2}{\theta}\right)\exp-\frac{(y-X_u)^2}{2\theta}dy\right)$$

$$=\frac{1}{\pi}\int_0^{t^*}du\psi(u)\int_0^{c-u}d\theta\theta^{-\frac{3}{2}}(c-u-\theta)^{-\frac{1}{2}}$$

$$\left(\int_{S_o(u)}^{\infty}\left(1-\frac{(y-X_u)^2}{\theta}\right)\exp-\frac{(y-X_u)^2}{2\theta}dy\right)$$

$$=-\frac{1}{\pi}\int_0^{t^*}du\psi(u)(S_o(u)-X_u)\int_0^{c-u}d\theta\theta^{-\frac{3}{2}}(c-u-\theta)^{-\frac{1}{2}}\exp-\frac{(S_o(u)-X_u)^2}{2\theta}$$

$$=-\sqrt{\frac{2}{\pi}}\int_0^{t^*}(c-u)^{-\frac{1}{2}}\exp-\frac{(S_o(u)-X_u)^2}{2(c-u)}\psi(u)du,$$

where we integrated out θ from the joint density in (θ, y) at $y = S_o(u) - X_u$ for the last step. Setting $H(u,v) = \ell n\left(\int_v^{\infty}\exp-\frac{y^2}{2(c-u)}dy\right)$, this term reduces to $\int_0^{t^*}\frac{\partial}{\partial v}H(u, S_o(u)-X(u))du$, but a local time representation is foiled by the inhomogeniety in u. We note the intuitive meaning of the integrand as the P^{X_u} conditional density of S_{c-u} at $S_o(u)$ given that it exceeds $S_o(u)$.

It remains to show that the three o–terms in (2.11) do not contribute to A_t. This required a further lengthy calculation, involving the same methods used already plus some rather intricate analysis. In view of the Further Remarks following Proposition 2.1, we have decided to spare the reader the details.

References

[1]. M. T. Barlow, *Study of a filtration expanded to include an honest time*, Z. Wahrscheinlichkeitstheorie verw. Geb. **44** (1978), 307–323.

[2]. C. Dellacherie and P.–A. Meyer, *Probabilités et Potentiel, Chap. V–VIII*, Hermann, Paris.

[3]. Thierry Jeulin, *Semi–Martingales et Grossissement d'une Filtration*, Lect. Notes in Math.. Springer–Verlag, Berlin.

[4]. T. Jeulin and M. Yor, *Grossissement d'une filtration et semimartingales: Formules explicites*, Seminaire de Probabilités XII. Springer–Verlag, Berlin.

[5]. F. B. Knight, *Essentials of Brownian Motion and Diffusion*, Math. Surveys **18** (1981). Amer Math. Society, Providence.

[6]. F. B. Knight, *A post-predictive view of Gaussian processes*, Ann. Scient. Ec. Norm. Sup. 4e series **t16** (1983), 541-566.
[7]. F. B. Knight, *Essays on the Prediction Process*, Lecture Notes and Monograph Series, S. Gupta, Ed., Inst. Math. Statistics **1** (1981). Hayward, Cal.
[8]. P.–A. Meyer, *Probability and Potentials*, Blaisdell Pub. Co.. 1966.
[9]. P.–A. Meyer, *A remark on F. Knight's paper*, Ann. Scient. Éc. Norm. Sup. 4e series **t16** (1983), 567–569.
[10]. K. M. Rao, *On decomposition theorems of Meyer*, Math. Scand. **24** (1969), 66–78.
[11]. L. A. Shepp, *The joint density of the maximum and its location for a Wiener process with drift*, J. Appl. Prob. **16** (1979), 423–327.

Professor Frank B. Knight
Department of Mathematics
University of Illinois
1409 West Green Street
Urbana, Illinois 61801
U.S.A.

Rate of Growth of Local Times of Strongly Symmetric Markov Processes

MICHAEL B. MARCUS

Let S be a locally compact metric space with a countable base and let $X = (\Omega, \mathcal{F}_t, X_t, P^x)$, $t \in R^+$, be a strongly symmetric standard Markov process with state space S. Let m be a σ–finite measure on S. What is actually meant by "strongly symmetric" is explained in [**MR**] but for our purposes it is enough to note that it is equivalent to X being a standard Markov process for which there exists a symmetric transition density function $p_t(x, y)$, (with respect to m). This implies that X has a symmetric 1–potential density

$$u^1(x,y) = \int_0^\infty e^{-t} p_t(x,y)\,dt \tag{1}$$

We assume that

$$u^1(x,y) < \infty \qquad \forall x,y \in S \tag{2}$$

which implies that there exists a local time $L = \{L_t^y, (t,y) \in R^+ \times S\}$ for X which we normalize by setting

$$E^x\left(\int_0^\infty e^{-t}\,dL_t^y\right) = u^1(x,y) \tag{3}$$

It is easy to see, as is shown in [**MR**], that $u^1(x,y)$ is positive definite on $S \times S$. Therefore, we can define a mean zero Gaussian process $G = \{G(y), y \in S\}$ with covarience

$$E(G(x)G(y)) = u^1(x,y) \qquad \forall x,y \in S$$

The processes X and G, which we take to be independent, are related through the 1–potential density $u^1(x,y)$ and are referred to as associated processes.

There is a natural metric for G

$$d(x,y) = (E(G(x) - G(y))^2)^{1/2} = (u^1(x,x) + u^1(y,y) - 2u^1(x,y))^{1/2} \tag{4}$$

which, obviously, is a function of the 1–potential density of the Markov process associated with G. We make the following assumptions about G. Let $Y \subset S$ be countable and let $y_0 \in Y$. Assume that

(5)
$$\lim_{\substack{y \to y_0 \\ y \in Y}} d(y, y_0) = 0$$

(6)
$$\sup_{\substack{d(y,y_0) \geq \delta \\ y \in Y}} G(y) < \infty \qquad \text{a.s.} \quad \forall \delta > 0$$

(7)
$$\lim_{\delta \to 0} \sup_{\substack{d(y,y_0) \geq \delta \\ y \in Y}} G(y) = \infty \qquad \text{a.s.}$$

and let

(8)
$$a(\delta) = E\left(\sup_{\substack{d(y,y_0) \geq \delta \\ y \in Y}} G(y) \right)$$

Note that by (7) $\lim_{\delta \to 0} a(\delta) = \infty$. In Theorem 1 we present some estimates on the rate at which L_y^t goes to infinity as y goes to y_0.

THEOREM 1. *Let X and G be associated processes as described above, so that, in particular, (5) (6) and (7) are satisfied on a countable subset Y of S. Let $L = \{L_t^y, (t, y) \in R^+ \times S\}$ be the local time of X. Then*

(9)
$$\overline{\lim_{\delta \to 0}} \sup_{\substack{d(y,y_0) \geq \delta \\ y \in Y}} \frac{L_t^y}{a(\delta)} \geq 2(L_t^{y_0})^{1/2} \qquad \forall t \in R^+ \quad \text{a.s.}$$

and

(10)
$$\overline{\lim_{\delta \to 0}} \sup_{\substack{d(y,y_0) \geq \delta \\ y \in Y}} \frac{L_t^y}{a^2(\delta)} \leq 1 \qquad \forall t \in R^+ \quad \text{a.s.}$$

where $a(\delta)$ is given in (8).

Theorem 1 shows that (6) holds with $G(y)$ replaced by L_t^y whatever the value of t and that (7) holds with $G(y)$ replaced by L_t^y as long as $L_t^{y_0} > 0$. But we know from [**MR**] Theorem IV that these statements are equivalent. So we could just as well have given the hypotheses in terms of the local time. However, since there is such an intimate relationship between the local time of X and the Gaussian process associated with X and since the critical function $a(\delta)$ is given in terms of the Gaussian process, there is no reason not to give conditions on the associated Gaussian process as hypotheses for properties of the local time.

Obviously there is a big gap between (9) and (10). On the other hand these estimates which are a consequence of a great deal of work developed in [MR] are the best that we can obtain. We present them because we think that they are new results and hope that they will stimulate further investigation of this problem.

Equivalent upper and lower bounds for $a(\delta)$ have been obtained by Fernique and Talagrand. See [T] (7) and Theorem 1. (We say that functions $f(\delta)$ and $g(\delta)$ are equivalent, and write $f(\delta) \approx g(\delta)$, as $\delta \to 0$, (resp. as $\delta \to \infty$) if there exist constants $0 < c_1 \le c_2 < \infty$ such that $c_1 f(\delta) \le g(\delta) \le c_2 f(\delta)$ for all $\delta \in [0, \delta']$, for some $\delta' > 0$, (resp. for all $\delta \in [\lambda', \infty)$, for some $\lambda' < \infty$)). We will use a part of this result in the examples below.

Before we go on to the proof of Theorem 1 let us discuss some applications. What we are examining here is a local time which is unbounded at a point but bounded away from the point. One source of examples comes from symmetric Markov chains with a single instantaneous state in the neighborhood of which the local time blows up. Processes of this sort were considered in [MR] Section 10. In fact (9) is a general statement of a result which was obtained for special cases in [MR] Theorem 10.1, (10.11). An abundant source of Markov processes with unbounded local times are certain real valued Lévy processes. See [B] and also [MR]. However the local times of these processes are unbounded on all intervals. Still we can apply the Theorem to these processes by looking at them on a nowhere dense sequence in their state space which has a single limit point. By choosing sequences converging to the limit point at different rates we can get an idea of how quickly the local time blows up. The following Corollary of Theorem 1 gives some examples.

COROLLARY 2. *Let X be a symmetric real valued Lévy process such that*

$$Ee^{i\lambda X(t)} = e^{-t\psi(\lambda)} \tag{11}$$

where

$$\psi(\lambda) \approx \lambda(\log \lambda)^{1+\alpha} \tag{12}$$

at infinity. Let $y_k = \exp(-(\log k)^\beta)$, $k = 1, \dots, \infty$, where $0 < \beta < \infty$, $0 < \alpha < 1$ and $\beta\alpha < 1$. Let $\bar\beta = \beta \vee 1$ and let $Y = \{\{y_k\}_{k=2}^{\infty}, 0\}$. Then we have

$$\overline{\lim_{k\to\infty}} \frac{L_t^{y_k}}{(\log k)^{(1-\alpha\bar\beta)/2}} \ge C(L_t^0)^{1/2} \qquad \forall t \in R^+ \quad \text{a.s.} \tag{13}$$

for some constant $C > 0$ and

$$\overline{\lim_{k\to\infty}} \frac{L_t^{y_k}}{(\log k)^{1-\alpha\beta}} \le C' \qquad \forall t \in R^+ \quad \text{a.s.} \tag{14}$$

for some constant $C' < \infty$. Equivalently, we have

$$\overline{\lim_{\delta\to 0}} \sup_{\substack{d(y,0)\ge\delta \\ y\in Y}} \frac{L_t^y}{(\log 1/\delta)^{(1-\alpha\beta)/(2\beta)}} \ge C(L_t^0)^{1/2} \qquad \forall t \in R^+ \quad \text{a.s.} \tag{15}$$

and

$$\overline{\lim_{\delta\to 0}} \sup_{\substack{d(y,0)\ge\delta \\ y\in Y}} \frac{L_t^y}{(\log 1/\delta)^{(1-\alpha\beta)/\beta}} \le C' \qquad \forall t \in R^+ \quad \text{a.s.} \tag{16}$$

Writing the limits as in (15) and (16) gives a clearer idea of how these sequences blow up in the neighborhood of zero.

We will first give the proof of Theorem 1 and next, in Lemma 3, derive some results on the suprema of Gaussian sequences. These results, along with Theorem 1, will immediately give Corollary 2.

PROOF OF THEOREM 1: The statement in (9) follows from Theorem 6.4 [**MR**] as modified in the proof of Theorem 10.1 of the same paper. In fact (9) is what is actually proved in Theorems 6.4 and Theorem 10.1 even though it is stated in (10.11) of [**MR**] in a special case. The main motivation for this note is to give a clearer statement of what is actually proved in Theorems 6.4 and 10.1 of [**MR**]. The only point that might be confusing is that $Y \cap \{d(y, y_0) \ge \delta\}$ is taken to be finite in the Theorems in [**MR**]. This is to insure that (6) of this paper is satisfied. In this note (6) is imposed as an hypothesis. (This enables us to apply Theorem 1 to a process that is defined on a countable dense set and has an isolated point at which it goes to infinity).

The statement in (10) is not given in [**MR**] but follows easily along the same line as many of the results contained in that paper. By [**F**] Theorem 3.2.1 we see that

$$\lim_{\delta\to 0} \sup_{\substack{d(y,y_0)\ge\delta \\ y\in Y}} \frac{G^2(y)}{m^2(\delta)} = 1 \qquad \text{a.s.} \tag{17}$$

where

$$m(\delta) = \text{median of } \left(\sup_{\substack{d(y,y_0)\ge\delta \\ y\in Y}} |G(y)| \right) \tag{18}$$

Therefore, by [MR] Lemma 4.3 for almost all ω with respect to the probability space supporting G

$$\lim_{\delta\to 0}\sup_{\substack{d(y,y_0)\geq\delta\\ y\in Y}}\frac{L_t^y+\frac{G^2(y,\omega)}{2}}{m^2(\delta)}=\frac{1}{2}\qquad\text{for almost all } t\in R^+ \text{ a.s.}$$

where the almost sure is with respect to the probability space of X, i.e. P^x almost sure, for all $x\in S$. Recall that L and G are independent. Now if we take an ω for which (17) holds we see that

$$\text{(19)}\qquad \overline{\lim_{\delta\to 0}}\sup_{\substack{d(y,y_0)\geq\delta\\ y\in Y}}\frac{L_t^y}{m^2(\delta)}\leq 1\qquad\text{for almost all } t\in R^+ \text{ a.s.}$$

Since (2), (5) and (6) imply that $\sup_{y\in Y}EG^2(y)<\infty$ it follows from [F] Corollary 2.2.2 that we can replace $m(\delta)$ by $a(\delta)$ in (19). Finally we can replace "almost all t" by "all t" since L_t^y is monotone in t and thus we get (10). (By much more elementary considerations we note that

$$\text{(20)}\qquad m(\delta)\leq 2E\sup_{\substack{d(y,y_0)\geq\delta\\ y\in Y}}|G(y)|\leq 4E\sup_{\substack{d(y,y_0)\geq\delta\\ y\in Y}}G(y)=4a(\delta)$$

which also gives (10) but with the constant 16).

We will now give some specialized results on the rate of growth of the expected value of some Gaussian sequences that are suited to the problem at hand.

LEMMA 3. *Let $\{y_k\}_{k=1}^\infty$ be a sequence of real numbers such that $\lim_{k\to\infty}y_k=0$. Let $Y=\{\{y_k\}_{k=1}^\infty,0\}$ and let $\{G(y),y\in Y\}$ be a mean zero Gaussian process such that for some $h'>0$*

$$\text{(22)}\qquad \rho_1^2(h)\leq E(G(h)-G(0))^2\leq\rho_2^2(h)\qquad \forall h\in[0,h']$$

where $\rho(h)$ is non–decreasing on $[0,h']$. Assume that

$$\text{(23)}\qquad y_{k-1}-y_k\downarrow\quad\text{as}\quad k\uparrow\infty$$

and

$$\text{(24)}\qquad \rho_2(y_k)(\log k)^{1/2}\uparrow\quad\text{as}\quad k\uparrow\infty$$

Then

$$\text{(25)}\qquad C_1\rho_1(y_{k-1}-y_k)(\log k)^{1/2}\leq E\left(\sup_{1\leq j\leq k}G(y_j)\right)\leq C_2\rho_2(y_k)(\log k)^{1/2}$$

for all k sufficiently large and constants $0 < C_1 \le C_2 < \infty$ independent of $\{y_k\}_{k=1}^{\infty}$. (C_1 is an absolute constant). Furthermore, assume that $\rho_2^2(h) = E(G'(h) - G'(0))^2$ for some mean zero Gaussian process G' and that $y_1 \le 1$. Then we also have that

$$E\left(\sup_{1\le j\le k} G(y_j)\right) \le C_3\left(\int_{y_{k-1}-y_k}^{1} \frac{\rho_2(v)}{v(\log 1/v)^{1/2}}\,dv + 1\right) \tag{26}$$

for some absolute constant C_3.

PROOF: To obtain the left–hand–side of (25) we use the Sudakov bound for the expected value of the supremum of Gaussian processes, stated for the problem under consideration. Let $Y_\delta = \{y \in Y : d(y, y_0) \ge \delta\}$ and let $N(Y_\delta, \epsilon)$ be the minimum number of closed balls in the pseudo-metric d (see (4)), that covers Y_δ. Then there exists a universal constant K such that

$$K \sup_{\epsilon>0} \epsilon(\log N(Y_\delta, \epsilon))^{1/2} \le E \sup_{y\in Y_\delta} G(y) \tag{27}$$

(See [LT] for this and other results on Gaussian processes that are not given a specific reference). Now, let $Y_\delta = \{y \in Y : d(y, 0) \ge y_k\}$, i.e. $\delta = y_k$. It is obvious that $N(Y_\delta, \rho_1(y_{k-1} - y_k)/2) = k$ and hence we get the left–hand–side of (25).

To obtain the right–hand–side of (25) we use the Borel–Cantelli Lemma applied to the sequence $\{(G(y_j) - G(0))/(d(0, y_j)(3\log j)^{1/2})\}_{j=2}^{\infty}$ to see that

$$\sup_{2\le j\le\infty} \frac{G(y_j) - G(0)}{d(0, y_j)(3\log j)^{1/2}} < \infty \qquad \text{a.s.}$$

Thus by [JM] II Corollary 4.7

$$E\left(\sup_{2\le j\le\infty} \frac{G(y_j) - G(0)}{d(0, y_j)(3\log j)^{1/2}}\right) = C$$

for some finite constant C. Therefore

$$E\left(\sup_{2\le j\le k} G(y_j) - G(0)\right) \le C \sup_{2\le j\le k} d(0, y_j)(3\log j)^{1/2} \tag{28}$$

Using (24) and (28) we obtain the right–hand–side of (25). (Note, the existence of $G(0)$ implies that $EG^2(0) < \infty$).

To obtain (26) we use the following result of Fernique, stated for the problem at hand. (See [T] or [LT]). Let $Y_k = \{y_j\}_{j=1}^{k}$ and let m be a probability measure on Y_k. Then, for $\{H(y), y \in Y_k\}$ a Gaussian process

$$E\left(\sup_{y\in Y_k} H(y)\right) \le K' \sup_{x\in Y_k} \int_0^{\infty} \left(\log \frac{1}{m(B(x,\epsilon))}\right)^{1/2} d\epsilon \tag{29}$$

for some absolute constant K', where $B(x,\epsilon) = \{y \in Y_k : d(y,x) \le \epsilon\}$ and d is defined for H as in (4). It follows by [JM] II Lemma 4.4 that

$$E\left(\sup_{1\le j\le k} G(y_j)\right) \le E\left(\sup_{1\le j\le k} G'(y_j)\right) \tag{30}$$

We will show that (26) holds for G'. (We do this to use the regularity hypotheses imposed on ρ_2). Let

$$m(\{y_j\}) = \frac{y_{j-1}-y_j}{y_1-y_k} \qquad 3\le j\le k$$
$$m(\{y_j\}) = \frac{y_1-y_2}{2(y_1-y_k)} \qquad j=1,2$$

Define

(31)

$$\begin{aligned}
I_k &= \int_0^\infty \left(\log \frac{1}{m(B(y_k,\epsilon))}\right)^{1/2} d\epsilon \\
&\le \rho_2(y_{k-1}-y_k)\left(\log\frac{2}{y_{k-1}-y_k}\right)^{1/2} \\
&\quad + \sum_{j=1}^{k-2}\left(\log\frac{2}{y_j-y_k}\right)^{1/2}(\rho_2(y_j-y_k)-\rho_2(y_{j+1}-y_k)) \\
&\le \rho_2(y_{k-1}-y_k)\left(\log\frac{2}{y_{k-1}-y_k}\right)^{1/2} + \int_{y_{k-1}}^1 \left(\log\frac{2}{u-y_k}\right)^{1/2} d\rho_2(u-y_k) \\
&\le \rho_2(y_{k-1}-y_k)\left(\log\frac{2}{y_{k-1}-y_k}\right)^{1/2} + \int_{y_{k-1}-y_k}^1 \left(\log\frac{2}{v}\right)^{1/2} d\rho_2(v) \\
&= \int_{y_{k-1}-y_k}^1 \frac{\rho_2(v)}{v(\log 2/v)^{1/2}}\,dv + \rho_2(1)(\log 2)^{1/2}
\end{aligned}$$

Using the same methods one can check that $I_j \le 2I_k$ for $1\le j<k$. Thus we get (26) from (29), (30) and (31).

Proof of Corollary 2: Since X is a symmetric Lévy process and $(1+\psi(\lambda))^{-1} \in L^1(d\lambda)$, it's 1–potential density is finite and satisfies

$$u^1(x,y) = u^1(0,x-y) = u^1(0,y-x)$$

Moreover

$$d^2(0,h) = 2(1-u^1(0,h)) = \frac{2}{\pi}\int_0^\infty \frac{1-\cos\lambda h}{1+\psi(\lambda)}\,d\lambda$$

(See [MR] Section 9 and the references therein). It is easy to see that

$$d^2(0,h) \approx (\log 1/h)^{-\alpha} \qquad \text{as } h\to 0$$

Therefore in order to calculate $a(\delta)$ we can use Lemma 3, applied to the Gaussian process associated with X, with $\rho_1^2(h) = c_1(\log 1/h)^{-\alpha}$ and $\rho_2^2(h) = c_2(\log 1/h)^{-\alpha}$ for some constants $0 < c_1 \le c_2 < \infty$. Note that under the hypotheses of Lemma 2

$$a(y_k) = E(\sup_{1\le j\le k} G(y_j))$$

and

$$\sup_{\substack{d(y,0)\ge y_k \\ y\in Y}} L_t^y = \sup_{j\le k} L_t^{y_k}$$

Note also that

$$\log \frac{1}{y_k} \approx (\log k)^{\beta} \qquad \text{as } k \to \infty$$

and

$$\log \frac{1}{y_{k-1} - y_k} \approx (\log k)^{\bar{\beta}} \qquad \text{as } k \to \infty$$

When $\beta \ge 1$ we use (25) and Theorem 1 to obtain (13) and (14) and when $\beta \le 1$ we use the left–hand–side of (25), (26) and Theorem 1 to obtain (13) and (14). Since $\rho^2(h)$ is concave in for $h \in [0, h']$ for some $h' > 0$ it satisfies the requirements for (26). Also note that it doesn't matter in (13) and (14) if we write $\overline{\lim}_{k\to\infty} \sup_{j\le k}$ or simply $\overline{\lim}_{k\to\infty}$. We get (15) and (16) from (13) and (14), (writen in the form $\overline{\lim}_{k\to\infty} \sup_{j\le k}$) simply by taking $\delta = y_k$ and observing that $\log k = (\log 1/\delta)^{1/\beta}$.

It is interesting that we need both upper bounds in (25) and (26) to make these simple estimates. The result in (25) is completely elementary and is well known. The bound in (26) is more subtle. But it is not as good as the one in (25) if $\{y_k\}$ goes to zero quickly enough.

REFERENCES

B Barlow, M.T. Necessary and sufficient conditions for the continuity of local time of Levy processes, Ann. Probability 16, 1988, 1389–1427.

F Fernique, X. Gaussian random vectors and their reproducing kernal Hilbert spaces, Technical Report Series No. 34, University of Ottawa, 1985.

JM Jain, N.C. and Marcus, M.B. Continuity of subgaussian processes, In: Probability on Banach spaces, Advances in Probability Vol. 4, 1978, 81–196, Marcel Dekker, NY.

LT Ledoux, M. and Talagrand, M. Probability in Banach spaces, preprint; to appear as a book published by Springer Verlag, New York.

MR Marcus, M.B. and Rosen, J. Sample path properties of the local times of strongly symmetric Markov processes via Gaussain processes, preprint

T Talagrand, M. Regularity of Gaussian processes, Acta Math. 159, 1987, 99–149.

Michael B. Marcus, Department of Mathematics, Texas A&M University, College Station, TX 77843

On the Continuity of Measure-Valued Processes

EDWIN PERKINS

Let $Y = (\Omega, \underline{F}, \underline{F}_t, Y_t, P^x)$ be a Hunt process with Borel semigroup P_t taking values in a topological Lusin space $(E,\underline{E})$ (E is homeomorphic to a Borel subset of a compact metric space and $\underline{E}$ is its Borel σ-field). $M_F(E)$ and $M_S(E)$ denote the spaces of finite measures and finite signed measures, respectively, on $(E,\underline{E})$ with the weak (i.e. weak*) topology. The $(Y,-\lambda^2/2)$ - superprocess is a continuous $M_F(E)$-valued Borel strong Markov process X. If $m_0 \in M_F(E)$ the law Q_{m_0} on $C([0,\infty),M_F(E))$ of X starting at $X_0 = m_0$ is uniquely determined by (write $X_t(\phi)$ for $\int\phi(x)X_t(dx)$)

$$Q_{m_0}(\exp(-X_t(\phi))) = \exp(-m_0(V_t\phi)) \qquad \phi\in bp\underline{E}$$

($bp\underline{E}$ is the set of bounded non-negative measurable functions on E) where $V_t\phi$ is the unique solution of

$$V_t\phi(x) = P_t\phi(x) - \int_0^t P_s((V_{t-s}\phi)^2/2)(x)\,ds \qquad t \ge 0,\ x \in E$$

(see Fitzsimmons (1988, (4.7), (2.3)).

Although the weak continuity of X only implies $X_t(\phi)$ is continuous for each bounded continuous real-valued ϕ on E (ϕ in $C_b(E)$), $X_t(\phi)$ is continuous on $(0, \infty)$ a.s. for all $\phi \in b\underline{E}$ (the space of bounded measurable functions from E to $\mathbb{R}$) for a large class of Y's. This fact seems to have first been noticed by Reimers (1989a, Thm 7.3) who proved it when Y is a Brownian motion by means of a nonstandard construction of X. He used it to show

$$X_t(A)=0 \text{ for all } t>0 \quad Q_{m_0}\text{-a.s. if and only if A is Lebesgue null.}$$

In this article we give an elementary standard proof of this stronger continuity for a broad class of Hunt processes, Y.

The only observation in this article which is perhaps non-trivial is Theorem 1 which is stated in the more general setting of an arbitrary $M_S(E)$-valued process $\hat{X}$ where $(E,\underline{E})$ is any measurable space. Under suitable hypotheses, this result applies the Garsia-Rodemich-Rumsey (G.R.R.) theorem to establish the a.s.-continuity of $\hat{X}_t(\phi)$ for $\phi \in b\underline{E}$. The catch is that a direct application of the G.R.R. theorem only gives a continuous version of $\hat{X}_t(\phi)$. The solution is that the G.R.R. theorem gives an explicit modulus of continuity which is preserved under bounded pointwise convergence and allows us to bootstrap up from the continuous functions. This result is applied to the superprocess X by considering $\hat{X} = X_t - X_0(P_t\cdot)$.

The existence of an elementary standard proof of this continuity result should not deter anyone from studying Reimers' nonstandard construction of X. The argument given here was motivated by the nonstandard proof and the nonstandard perspective gives many other insights into the nature of this process and solutions of other stochastic p.d.e.'s. For example Reimers (1989a) gives the only direct and rigorous connection between X and the stochastic p.d.e.

$$\frac{\partial u}{\partial t} = \frac{\Delta u}{2} + \sqrt{u}\,\dot{W}$$

in dimensions other than 1 (see Reimers (1989b) or Konno-Shiga (1988) for the one-dimensional case).

Notation. If $\{\phi_n\} \subset b\underline{E}$ we write $\phi_n \overset{bp}{\to} \phi$ when ϕ_n converges to ϕ in the bounded pointwise sense. If $\underline{C} \subset b\underline{E}$, $\overline{\underline{C}}^{bp}$ denotes the bounded pointwise closure of $\underline{C}$. Let $b_1\underline{E} = \{\phi \in b\underline{E}: \sup_{x\in E}|\phi(x)| = \|\phi\| \le 1\}$.

Theorem 1. Let $(E,\underline{E})$ be a measurable space, $\{\hat{X}_t: t\in[0,N]\}$ be an $M_S(E)$-valued process and $\underline{C} \subset b_1\underline{E}$. Assume $\Psi:[0,\infty) \to [0,\infty)$ is an increasing convex function increasing to ∞ at ∞ and $p:[0,\infty)\to[0,\infty)$ is an increasing function such that $\lim_{u\to 0+} p(u) = 0$. We suppose

(1) If $\Gamma_{\phi,q}(\omega) = \int_0^N\int_0^N \Psi(|\hat{X}_u(\phi) - \hat{X}_t(\phi)|/p(|u-t|))^q du\,dt$ then there are $c_0 > 0$ and $q > 1$ such that $P(\Gamma_{\phi,q}) \le c_0$ for all $\phi\in\underline{C}$.

(2) $\hat{X}_t(\phi)$ is continuous on $[0,N]$ a.s. for all $\phi\in\underline{C}$.

(3) $\int_0^N \Psi^{-1}(\gamma r^{-2})dp(r) < \infty$ for all $\gamma > 0$.

Then for any $\phi \in \underline{\bar{C}}^{bp}$, $\hat{X}_t(\phi)$ is continuous on $[0,N]$ a.s. and

(4) $|\hat{X}_u(\phi) - \hat{X}_t(\phi)| \leq 8 \int_0^{|u-t|} \Psi^{-1}(\Gamma_{\phi,1}\, r^{-2})dp(r)$ for all $u,t \in [0,N]$,

(5) $P(\Gamma_{\phi,q}) \leq c_0$ and so $\Gamma_{\phi,1} < \infty$ a.s.

Proof. Let $\underline{H}$ be the set of functions ϕ in $b\underline{E}$ for which (4) and (5) hold. $\underline{C} \subset \underline{H}$ by (2) and the theorem of Garsia-Rodemich-Rumsey (Garsia (1970)). Assume $\{\phi_n\} \subset \underline{H}$ and $\phi_n \xrightarrow{bp} \phi$. Then $\hat{X}_u(\phi_n) \to \hat{X}_u(\phi)$ for all $u \in [0,N]$ and a double application of Fatou's Lemma gives

$$P(\Gamma_{\phi,q}) \leq \underline{\lim}_{n\to\infty} P(\Gamma_{\phi_n,q}) \leq c_0 \quad \text{(by (5))}.$$

$P(\Gamma_{\phi_n,q}) \leq c_0$ implies $\{\Psi(|\hat{X}_u(\phi_n)-\hat{X}_t(\phi_n)|/p(|u-t|)): n\in\mathbb{N}\}$ is uniformly integrable with respect to du dt dP on $[0,N]^2 \times \Omega$ and hence

$$\lim_{n\to\infty} P\Big(\int_0^N\int_0^N |\Psi(|\hat{X}_u(\phi_n)-\hat{X}_t(\phi_n)|/p(|u-t|))-\Psi(|\hat{X}_u(\phi)-\hat{X}_t(\phi)|/p(|u-t|))|\,du\,dt\Big) = 0 .$$

Therefore there is a subsequence $\{n_k\}$ such that

(6) $$\lim_{k\to\infty} \Gamma_{\phi_{n_k},1} = \Gamma_{\phi,1} \quad \text{a.s.}$$

Let $k \to \infty$ in

$$|\hat{X}_u(\phi_{n_k}) - \hat{X}_t(\phi_{n_k})| \leq 8 \int_0^{|u-t|} \Psi^{-1}(\Gamma_{\phi_{n_k},1}r^{-2})dp(r)$$

(recall $\phi_{n_k} \in \underline{H}$) to conclude

$$|\hat{X}_u(\phi) - \hat{X}_t(\phi)| \leq 8 \underline{\lim}_{k\to\infty} \int_0^{|u-t|} \Psi^{-1}(\Gamma_{\phi_{n_k},1}r^{-2})dp(r)$$

$$= 8 \int_0^{|u-t|} \Psi^{-1}(\Gamma_{\phi,1}r^{-2})dp(r)$$

(the last by (6)). We have proved $\phi\in\underline{H}$ and hence $\underline{H}$ is closed under bounded pointwise convergence. This proves (4) and (5) for all ϕ in $\underline{\bar{C}}^{bp}$ and the a.s. continuity of $\hat{X}_t(\phi)$ follows from this and (3).∎

We next state a simple special case of Theorem 1 which may be easier to apply in practice. It's what one would have obtained by applying the usual proof of Kolmogorov's continuity criterion rather than the G.R.R. theorem.

Corollary 2. Let $(E,\underline{E})$ be a measurable space, $\{\hat{X}_t: t\in[0,N]\}$ be an $M_S(E)$-valued process and $\underline{C} \subset b_1\underline{E}$. Assume $p>1$, $\delta, c_0 > 0$ satisfy

$$P(|\hat{X}_u(\phi) - \hat{X}_t(\phi)|^p) \leq c_0|u-t|^{1+\delta} \text{ for all } u,t\in[0,N],\ \phi\in\underline{C},$$

and $\hat{X}_t(\phi)$ is continuous on $[0,N]$ a.s. for all $\phi\in\underline{C}$. Then for any $\phi\in\underline{\bar{C}}^{bp}$

and any $0<\eta<\delta/p$ there is a $\rho(\phi,\eta,\omega) > 0$ a.s. such that $|\hat{X}_u(\phi) - \hat{X}_t(\phi)| \le |u-t|^{\eta}$ for all $u,t\in[0,N]$ satisfying $|u-t| < \rho(\phi,\eta,\omega)$.

<u>Proof</u>. This is a simple application of Theorem 1. Take $\Psi(r) = r^{p'}$, $p(u) = u^{\epsilon}$ and $q = p/p'$ where $p'\in(1,p)$ is sufficiently close to p, and $\epsilon < (2+\delta)/p$ is sufficiently close to $(2+\delta)/p$.∎

Consider now the $(Y,-\lambda^2/2)$-superprocess X, and the associated $M_S(E)$-valued process

$$\hat{X}_t(\phi) = X_t(\phi) - X_0(P_t\phi) \qquad \phi \in b\underline{E} .$$

Then there is an orthogonal martingale measure (see Walsh(1986, Ch.2)) Z on $E \times [0,\infty)$ such that

$$\langle Z(\phi)\rangle_t = \int_0^t X_s(\phi^2)ds \qquad \phi \in b\underline{E}$$

and

$$\text{(7)} \qquad \hat{X}_t(\phi) = \int_0^t \int P_{t-s}\phi(x)dZ(s,x) \text{ a.s. for all } t\ge 0,\ \phi \in b\underline{E}$$

(see Fitzsimmons (1989, (2.13)). Note that $Q_{m_0}(X_t(\phi))=m_0(P_t\phi) = X_0(P_t\phi)$ Q_{m_0}-a.s. (e.g. by (7)) and so the weak continuity of X_t implies that of $\hat{X}$.

It is a now a routine exercise to use (7) to verify the hypotheses of Theorem 1 for suitable Ψ and p. Let

$$h(v,\delta) = \sup_{||\phi||\le 1}||P_{v+\delta}\phi - P_v\phi|| = \sup_{x\in E}|P^x(Y_{v+\delta}\in\cdot) - P^x(Y_v\in\cdot)|$$

where $|\nu|$ is the total variation of $\nu\in M_S(E)$. Note that $h(\cdot,\delta)$ is decreasing by the semigroup property and $h\le 2$. If $N>0$, let

$$p_N(r) = \sup_{r'\le r} \left(r' + \int_0^N h(v,r')^2 dv\right)^{1/2}$$

and if $\phi \in b_1\underline{E}$ let

$$\Gamma_{N,\phi,q} = \int_0^N \int_0^N \exp\{q|\hat{X}_u(\phi)-\hat{X}_t(\phi)|/6N^{1/2}p_N(|u-t|)\}dudt,\ q > 0.$$

As $X_t(1)$ is the diffusion on $[0,\infty)$ with generator $(x/2)\, d^2/dx^2$ absorbed at 0, the following lemma is a simple application of the maximal inequality for the submartingale $\exp(\theta X_t(1))$ (see Knight (1981,p.100) for its transition density). Let $X_N^*(1)=\sup_{s\le N} X_s(1)$.

<u>Lemma 3</u> $Q_{m_0}(X_N^*(1)>\lambda) \le \exp\{-\lambda/2N\}$ for all $\lambda \ge 4m_0(E)$.

Theorem 4. Assume

(8) $$\int_0^1 p_1(r)r^{-1}dr < \infty .$$

(a) For any $\phi \in b_1\underline{E}$, $Q_{m_o}(\Gamma_{N,\phi,4/3}) \le 49\, N^2 \exp\{2m_o(E)/N\}$ and in particular $\Gamma_{N,\phi,1} < \infty$ for all $N \in \mathbf{N}$ Q_{m_o} - a.s.

(b)

(9) $$|\hat{X}_u(\phi) - \hat{X}_t(\phi)| \le \|\phi\|\, 96N^{1/2}[\log \frac{1}{|u-t|}\, p_N(|u-t|) + \int_0^{|u-t|} p_N(r)r^{-1}dr + (\log \Gamma_{N,\phi/\|\phi\|,1})p_N(|u-t|)]$$

for all $u,t \in [0,N]$, $N \in \mathbf{N}$ Q_{m_o} - a.s. for each $\phi \in b\underline{E}$.

The right-hand of (9) approaches 0 as $|u-t|\downarrow 0$ and therefore for each $\phi \in b\underline{E}$, $\hat{X}_t(\phi)$ is a.s. continuous on $[0,\infty)$, $X_t(\phi)$ is a.s. continuous on $(0,\infty)$, and $X_t(\phi)$ is a.s continuous at 0 if and only if $m_o(P_t\phi)$ is.

Proof (a) If $0 \le u \le t < N$ and $\phi \in b_1\underline{E}$, then using (7) we have for any $K > 0$

$$Q_{m_o}(|\hat{X}_u(\phi) - \hat{X}_t(\phi)|/p_N(u-t) \ge x)$$

$$\le Q_{m_o}(|\int_0^t\int P_{u-s}\phi(x) - P_{t-s}\phi(x)dZ(s,x)| \ge xp_N(u-t)/2)$$

$$+ Q_{m_o}(|\int_t^u\int P_{u-s}\phi(x)dZ(s,x)| \ge xp_N(u-t)/2)$$

$$\le 4\exp\{-x^2p_N(u-t)^2/8K\}$$

$$+ Q_{m_o}(\int_0^t X_s((P_{u-s}\phi - P_{t-s}\phi)^2)ds > K)$$

$$+ Q_{m_o}(\int_t^u X_s((P_{u-s}\phi)^2)ds > K)$$

(e.g. by (37.12) of Rogers-Williams (1987))

$$\le 4\exp\{-x^2p_N(u-t)^2/8K\} + Q_{m_o}(X_N^*(1)\int_0^t h(t-s,u-t)^2ds > K)$$

$$+ Q_{m_o}(X_N^*(1)(u-t) > K)$$

$$\le 4\exp\{-x^2p_N(u-t)^2/8K\} + \exp\{-K(2N\, p_N(u-t)^2)^{-1}\}$$

$$+ \exp\{-K(2N(u-t))^{-1}\}$$

by Lemma 3 providing that

(10) $$K \ge 4m_o(E)p_N(u-t)^2.$$

Let $K = x\, N^{1/2}p_N(u-t)^2/2$. If $x \ge 8m_o(E)N^{-1/2}$, then (10) holds and the above estimate implies

$$Q_{m_o}(|\hat{X}_u(\phi) - \hat{X}_t(\phi)|/p_N(u-t) \ge x) \le 6\exp\{-x/(4N^{1/2})\}.$$

(a) now follows by a trivial calculation.

(b) Let $\underline{C}$ denote the class of continuous functions in $b_1\underline{E}$ and let $N \in \mathbf{N}$. We first check the hypotheses of Theorem 1 on $[0,N]$ with $\Psi(r) = \exp\{r/6N^{1/2}\}$, $q = 4/3$ and $p(r) = p_N(r)$. (a) implies (1). (2) follows from the weak continuity of $\hat{X}$. The monotonicity of $h(\cdot,r)$ implies $p_N(r)^2 \le N\, p_1(r)^2$ and so (8) shows

(11) $$\int_0^N p_N(r) r^{-1} dr < \infty.$$

An integration by parts shows that if $\gamma > 0$ then for $0<\delta<N$.

$$\int_\delta^N \ln(\gamma/r^2) dp_N(r) + p_N(\delta)\ln(\gamma/\delta^2) = \ln(\gamma/N^2) p_N(N) + 2\int_\delta^N p_N(r) r^{-1} dr.$$

The right-hand side approaches a finite limit as $\delta \downarrow 0$ hence so does the left side and (3) holds. This together with (11) and

$$p_N(\delta)\ln(\gamma/\delta^2) \le \int_0^\delta \ln(\gamma/r^2) dp_N(r)$$

shows that the right-hand side of (9) approaches 0 as $|u-t| \downarrow 0$. We may now apply Theorem 1 to conclude that for all $\phi \in b_1\underline{E} = \bar{\underline{C}}^{bp}$ and all $u,t \in [0,N]$,

$$\begin{aligned} |\hat{X}_u(\phi) - \hat{X}_t(\phi)| &\le 48\, N^{1/2} \int_0^{|u-t|} \ln(\Gamma_{N,\phi,1} r^{-2}) dp_N(r) \\ &= 48\, N^{1/2}[(\ln(\Gamma_{N,\phi,1}) + 2\ln(|u-t|^{-1}))p_N(|u-t|) \\ &\quad + 2\int_0^{|u-t|} r^{-1} dp_N(r)]. \end{aligned}$$

This implies (9) for $\phi \in b_1\underline{E}$ and hence for all ϕ in $b\underline{E}$ (consider $\phi/\|\phi\|$). Since $p_N(r)$ approaches 0 as $r \downarrow 0$ and $h(\cdot,r)$ is decreasing it follows that $P_t\phi$ is $\|\ \|$-continuous in $t \in (0,\infty)$ for any $\phi \in b\underline{E}$. The remaining assertions in (b) are therefore obvious.∎

<u>Corollary 5</u>. Assume $P_t\phi(x) = \int p_t(x,y)\phi(y)\nu(dy)$ for all $t>0$, $\phi \in b\underline{E}$ where ν is a measure on E, and suppose

(12) $$\sup_x \int |p_u(x,y) - p_t(x,y)|\nu(dy) \le c_1 (u-t) t^{-1} \quad \text{for all } 0<t<u.$$

(a) For any $\phi \in b\underline{E}$ Q_{m_o}-a.s. $X_t(\phi)$ is continuous on $(0,\infty)$ and is continuous at 0 if and only if $m_o(P_t\phi)$ is continuous at 0.

(b) For all ϕ in $b\underline{E}$ and N in $\mathbf{N}$ there is a random variable $K(N,\phi)$ (finite Q_{m_o}-a.s.) such that

$$\begin{aligned} |X_u(\phi) - X_t(\phi)| \le \|\phi\|\, 96\, N^{1/2}(4c_1+1)^{1/2}[(\log(1/|u-t|))|u-t|^{1/2} \\ + K(N,\phi)|u-t|^{1/2}] + |m_o(P_u\phi) - m_o(P_t\phi)| \end{aligned}$$

for all u,t in $[0,N]$ Q_{m_o}-a.s.

(c) If in addition $\int p_{t_o}(x,y)dm_o(x) > 0$ ν-a.a.y for some $t_o>0$, then for any $A\in\underline{E}$

$X_t(A)=0$ for all $t>0$ Q_{m_o}-a.s. if and only if $\nu(A)=0$.

Proof. (12) implies that $h(v,r) \le \min(c_1 rv^{-1},2)$ and hence $p_N(r) \le (4c_1+1)^{1/2} r^{1/2}$. (a) and (b) are now immediate from Theorem 4 with $K(N,\phi) = \log(\Gamma_{N,\phi/\|\phi\|,1}) + 2$.

(c) Suppose $\nu(A) = 0$. Then for each $t>0$ $Q_{m_o}(X_t(A)) = m_o(P_t 1_A) = 0$. Therefore $X_t(A)=0$ for all $t\in\mathbb{Q}^{>0}$ Q_{m_o}-a.s. and hence $X_t(A)=0$ for all $t>0$ by the a.s.-continuity of $X_t(A)$ on $(0,\infty)$.

Conversely if t_o is as above and $X_{t_o}(A)=0$ Q_{m_o} a.s., then

$$0 = Q_{m_o}(X_{t_o}(A)) = \int\int_A p_{t_o}(x,y)\nu(dy)m_o(dx)$$

and hence $\nu(A) = 0$ by the hypothesis on p_{t_o}. ∎

Corollary 6. If P_t is the semigroup of the symmetric α-stable process in $\mathbb{R}^d$ $(\alpha\in(0,2])$ and hence X_t is the super-α-stable process, then the hypotheses (and hence conclusions) of Corollary 5 hold for any $m_o \ne 0$ with $d\nu = dx$ and

$$c_1 = 2d\alpha^{-1} 2^{(d/\alpha-1)^+} + 1.$$

Proof. If $p_t(y)$ is the density of the symmetric α-stable process in $\mathbb{R}^d$ at time t then $p_t(y)$ is a strictly positive decreasing function of $|y|$ satisfying $p_t(y) = p_{t/c}(yc^{-1/\alpha})c^{-d/\alpha}$ for all $c > 0$. Using these facts it is easy to see that for $t, r > 0$

$$\begin{aligned}|p_t(y) - p_{t+r}(y)| &= |p_t(y) - p_t(y(1+r/t)^{-1/\alpha})(1+r/t)^{-d/\alpha}| \\ &\le p_t(y)[1 - (1+r/t)^{-d/\alpha}] + p_t(y(1+r/t)^{-1/\alpha}) - p_t(y)\end{aligned}$$

(consider $p_t(y) > p_{t+r}(y)$ or $p_t(y) < p_{t+r}(y)$ separately). Integrate out y to find

$$\begin{aligned}\int|p_t(y) - p_{t+r}(y)|dy &\le 1 - (1+r/t)^{-d/\alpha} + (1+r/t)^{d/\alpha}-1 \\ &\le 2d\alpha^{-1}\, 2^{(d/\alpha-1)^+}(r/t) \qquad 0<r\le t.\end{aligned}$$

If $r>t$ use the trivial upper bound 2 to complete the derivation of (12). ∎

If $X_t(\phi)$ is Q_{m_o} - a.s. continuous on $(0,\infty)$ for each $m_o \in M_F(E)$ and $\phi\in b\underline{E}$ then, taking mean values, we see that $t \to P_t\phi(x)$ is continuous on

$(0,\infty)$ for each $x \in E$ and $\phi \in b\underline{E}$ (the necessary uniform integrability is given by Lemma 3).

Open Problem. Is the converse true?

The hypothesis (8) of Theorem 4 implies (use the fact that $h(\cdot,r)$ is non-decreasing)

$$\int_0^1 \|P_{t+r}\phi - P_t\phi\| \, r^{-1}dr < \infty \text{ for all } \phi \in b\underline{E},\ t>0,$$

and, in particular, $t \to P_t\phi$ is a norm-continuous on $(0,\infty)$.

References

P. Fitzsimmons (1988). Construction and regularity of measure-valued Markov branching processes. Israel J. Math 64, 337-361.

P. Fitzsimmons (1989). Correction and addendum to: Construction and regularity of measure-valued Markov branching processes, Israel J. Math, to appear.

A. Garsia (1970). Gaussian processes with multidimensional time parameter, 6th Berkeley Symposium on Math., Statistical probability vol. 2, 366-374, University of California Press, Berkeley.

F. Knight (1981). Essentials of Brownian motion and diffusion, Amer. Math. Soc., Providence.

N. Konno and T. Shiga (1988). Stochastic partial differential equations for some measure-valued diffusions, Probab. Theory Rel. Fields 79, 201-226.

M. Reimers (1989a). Hyperfinite methods applied to the critical branching diffusion, Probab. Theory Rel. Fields 81, 11 - 28.

M. Reimers (1989b). One dimensional stochastic partial differential equations and the branching measure diffusion, Probab. Theory Rel. Fields 81, 319-340.

L.C.G. Rogers and D.Williams (1987). Diffusions, Markov Processes and martingales volume 2: Itô Calculus, Wiley, Chichester.

J.B. Walsh (1986). An introduction to stochastic partial differential equations. Lecture Notes in Mathematics 1180, Springer-Verlag, New York.

Edwin Perkins

Department of Mathematics

University of British Columbia

Vancouver, B.C. V6T 1Y4

Canada

A Remark on Regularity of Excessive Functions for Certain Diffusions

Z. R. POP-STOJANOVIC

In an earlier paper [4], the first author has shown that a diffusion process whose potential kernel satisfies certain analytic conditions, has all of its excessive harmonic functions, which are not identically infinite, continuous. In a subsequent paper [5], the same author has shown that under these conditions the excessiveness of its nonnegative harmonic functions is automatic. In this paper we are showing that a regularity condition for the excessive functions introduced here, will imply that the Riesz measure does not charge the semi-polar sets of the process.

Let $X = (\Omega, F, F_t, X_t, \theta_t, P^x)$ denote a transient diffusion, i.e., a strong Markov process with continuous sample paths on a locally compact Hausdorff state space $(E, \mathcal{E})$ with a countable base. Following [2,3], we are assuming the existence of a potential kernel with following properties.

Let

$$U(x,dy) = u(x,y)\xi(dy)$$

denote this kernel where ξ is a Radon measure, and the potential density function u is such that:

(a) For every x, and for every y, function $(x,y) \to u^{-1}(x,y)$ is finite and continuous; in particular, this implies $u(x,y) > 0$ for all (x,y).

(b) $u(x,y) = \infty$ if and only if $x = y$.

(Other notations and concepts throughout this paper are generally as in Blumenthal and Getoor [1]).

Remark. It can be shown that our assumptions imply the existence of a strong Markov dual. Hence, one can refer to the Chapter 6 of [1] and to apply techniques developed there in order to obtain our results presented here. However, it is not clear that all the assumptions of that Chapter are consequences of the assumptions made here. Thus, we choose to follow the direct path.

The next two Propositions are rather expected and we are presenting them without the proofs.

Proposition 1. Let s be an excessive function (for X), and (T_n) a sequence of terminal times tending to infinity as $n \to \infty$. We can write

$$s = p + h$$

where p and h are excessive, $P_{T_n} h = h$ for all n, and $P_{T_n} p \downarrow 0$ almost everywhere, as $n \to \infty$.

Proposition 2. Let (T_n) be a sequence of terminal times increasing to infinity as $n \to \infty$. One has

$$u(x,y) = v(x,y) + w(x,y)$$

where v and w are excessive functions for each y, $P_{T_n} v(.,y) \downarrow 0$ almost everywhere for each y as $n \to \infty$, and

$$P_{T_n} w(x,y) = w(x,y)$$

for all x,y. Moreover, the set of y's such that w(.,y) is not identically zero is a polar set.

The proof of this Proposition follows directly from Proposition 1.

Now, toward our main goal we have the following definition.

Definition. We say that an excessive function s is of class

(D) if for each sequence (T_n) of stopping times increasing to infinity as $n \to \infty$,

$$P_{T_n} s \downarrow 0$$

almost everywhere as $n \to \infty$. We say that an excessive function s is <u>regular</u> if for every sequence of stopping times (T_n) increasing to a stopping time T, as $n \to \infty$,

$$P_{T_n} s \downarrow P_T s ,$$

almost everywhere as $n \to \infty$.

Using the two given Propositions and the representation of excessive functions we have the following result.

<u>Theorem.</u> A potential is of class (D) if and only if its Riesz measure does not charge polar sets. A potential is <u>regular</u> if and only if its Riesz measure does not charge semi-polar sets.

<u>Proof.</u> We shall prove here the second statement of this theorem. The proof is presented in two steps.

<u>Step 1.</u> Here, we are showing the following: if g is a function such that

$$\int g(x)u(x,y)dx \le 1,$$

then for every Borel set A, the set

(1) $$B = \{y;\ y \in A,\ \int g(x)P_{\{A\}}u(x,y)dx < \int g(x)u(x,y)dx\}$$

is a semi-polar set.

To see this, it is sufficient to show that every compact subset of B is a semi-polar set. Let L be a compact subset of B. We can write $L = S + Q$ where S is a semi-polar set and Q is finely closed with all its points being regular points for Q.

Let

$$s = P_Q 1.$$

The Riesz measure μ of s is concentrated on $\overline{Q}$ (see Corollary 3, p. 178 in [2]). In particular, it is concentrated on L. Clearly, $P_Q s = s$

which implies that for every x,

(2) $$P_Q u(x,y) = u(x,y)$$

for μ - almost all y. By Fubini theorem (2) holds for almost all x, and μ - almost all y. In particular, since $P_A u(x,y) \geq P_Q u(x,y)$,

$$\int g(x) P_A u(x,y) dx = \int g(x) u(x,y) dx$$

for μ - almost all y. But this contradicts (1) because μ is concentrated on $L \subset B$. □

Step 2. Here, we are showing that if μ does not charge semi-polar sets, then $s = U\mu$ is regular.

Indeed, since the sum of regular potentials is a regular potential and, since μ is a Radon measure, we may assume μ to be a finite measure. Let g be a function such that $\int g(x)dx = 1$ and

$$\int g(x)s(x)dx < \infty .$$

Given $\varepsilon > 0$ and a sequence (T_n) of stopping times which increases to a stopping time T as $n \to \infty$. Let $Uf_k \uparrow s$ as $k \to \infty$. By Egorov theorem there are a compact set K and a positive integer m such that if $s_K = U\mu_K$, one has

(3) $$s \leq Uf_m + \varepsilon \text{ on } K \text{ and } \int (s - s_K) g(x) dx \leq \varepsilon .$$

Since μ does not charge semi-polar sets it follows from the first observation that $P_K s_K = s_K$. This fact together with (3) implies

(4) $$s_K \leq Uf_m + \varepsilon \text{ everywhere, and } \int (s - s_K) g(x) dx \leq \varepsilon.$$

Now, by using the first inequality in (4), one gets:

(5) $$\int \lim_n P_{T_n} s_K(x) g(x) dx \leq \varepsilon + \int \lim_n P_{T_n} Uf_m(x) g(x) dx = \varepsilon + \int g(x) P_T Uf_m(x) dx \leq \varepsilon + \int g(x) P_T s(x) dx,$$

where the regularity of Uf_m and the fact that $Uf_m \leq s$ have been used.

On the other hand, since $s - s_K$ is an excessive function,

$$P_{T_n}(s - s_K) \leq s - s_K$$

for all n, so by Fatou lemma and the second inequality in (4), one gets:

$$\int g(x)\lim_n P_{T_n} s(x)dx \le \varepsilon + \int g(x)\lim_n P_{T_n} s_K(x)dx.$$

This inequality and (5) imply that

$$\int g(x)\lim_n P_{T_n} s(x)dx \le 2\varepsilon + \int g(x)P_T s(x)dx,$$

which in turn implies, with ε arbitrary and $\lim P_{T_n} s(x) \ge P_T s(x)$, that for almost all x, $\lim_n P_{T_n} s(x) = P_T s(x)$ as desired.

The proof of the second statement of the Theorem follows now from these two observations just proved. □

REFERENCES

[1] R. M. Blumenthal and R. K. Getoor, Markov Processes and Potential Theory, New York, Academic Press, 1968.

[2] K. L. Chung and M. Rao, A new setting for Potential Theory, Ann. Inst. Fourier 30, 1980, 167 - 198.

[3] K. L. Chung, Probabilistic approach in Potential Theory to the equilibrium problem, Ann. Inst. Fourier 23, 1973.

[4] Z. R. Pop-Stojanovic, Continuity of Excessive Harmonic Functions for certain Diffusions, Proc. the AMS, Vol. 103, num. 2, 1988, 607-611.

[5] ________________ Excessiveness of Harmonic Functions for Certain Diffusions, Journal of Theoretical Probability, Vol. 2, No. 4, 1989, 503 - 508.

Z.R. Pop-Stojanovic
Department of Mathematics
University of Florida
Gainesville, Florida 32611

$A(t,B_t)$ is not a Semimartingale

L.C.G. ROGERS and J.B. WALSH

1. Introduction. Let $(B_t)_{t\geq 0}$ be Brownian motion on $\mathbb{R}$, $B_0 = 0$, and for each real x define

$$A(t,x) \equiv \int_0^t I_{(-\infty,x]}(B_s)ds = \int_{-\infty}^x L(t,y)dy,$$

where $\{L(t,y) : t \geq 0, y \in \mathbb{R}\}$ is the local time process of B. The process $A(t,x)$ enters naturally into the study of the Brownian excursion filtration (see Rogers & Walsh [1],[2], and Walsh [4]). In [2], it was necessary to consider the occupation density of the process $Y_t \equiv A(t,B_t)$, which would have been easy if Y were a semimartingale; it is not, and the aim of this paper is to prove this.

To state the result, we need to set up some notation. Let $(X_t)_{0\leq t\leq 1}$ be the process $A(t,B_t) - \int_0^t L(u,B_u)dB_u$, and define for $j,n \in \mathbb{N}$

$$\Delta_j^n X \equiv X(j2^{-n}) - X((j-1)2^{-n}), \qquad j \leq 2^n,$$

and

$$V_p^n \equiv \sum_{j=1}^{2^n} |\Delta_j^n X|^p.$$

THEOREM 1. *For any $p > 4/3$,*

$$V_p^n \xrightarrow[\text{a.s.}]{L^1} 0 \qquad (n \to \infty). \tag{1}$$

For any $p < 4/3$,

$$\limsup_{n\to\infty} V_p^n = +\infty \quad \text{a.s.} \tag{2}$$

This proves conclusively that X (and hence Y) cannot be a semimartingale, because if it were, it could be written as $X = M + A$, where M is a local martingale, A is a finite-variation process (both continuous since X is; see Rogers & Williams [4], VI.40). Now since $V_2^n \xrightarrow{\text{a.s.}} 0$, M must be zero, and $X = A$; but $\overline{\lim} V_1^n = +\infty$ rules out the possibility that X is finite-variation, as we shall see.

In outline, the proof runs as follows. Firstly, we estimate $E|\Delta_j^n X|^p$ above and deduce from this that $EV_p^n \to 0$ for any $p > 4/3$; in fact, the L^1 convergence is sufficiently rapid that $V_p^n \xrightarrow{\text{a.s.}} 0$. Next we estimate $E|\Delta_j^n X|^p$ below, and combine the estimates to prove that $EV_{4/3}^n$ is bounded away from 0 and from infinity. The upper bound allows us to prove that $\{V_{4/3}^n : n \geq 1\}$ is uniformly integrable, and hence that $P(\limsup V_{4/3}^n > 0) > 0$. From this, by Hölder's inequality, we prove that for any $p < 4/3$, $P[\limsup V_p^n = +\infty] > 0$. Finally, an application of the Blumenthal $0-1$ law allows us to conclude.

In the forthcoming paper, we analyse the exact 4/3-variation of X completely, and prove that it is $\gamma \int_0^t L(s, B_s)^{2/3} ds$, from which the present conclusions (and more) follow. (Here, γ is $4\pi^{-\frac{1}{2}}\Gamma(7/6)E(\int L(1,x)^2 dx)^{2/3}$.) The proof of this is a great deal more intricate, however, and this paper shows how to achieve the lesser result with less effort.

2. Upper bounds. To lighten the notation, we are going to perform a scaling so that there is only one parameter involved. It is elementary to prove that for any $c > 0$, the following identities in law hold:

$$(L(t,x))_{t\geq 0, x\in\mathbb{R}} \stackrel{\mathcal{D}}{=} \left(cL\left(\frac{t}{c^2}, \frac{x}{c}\right)\right)_{t\geq 0, x\in\mathbb{R}}; \tag{3}$$

$$(A(t,x))_{t\geq 0, x\in\mathbb{R}} \stackrel{\mathcal{D}}{=} \left(c^2 A\left(\frac{t}{c^2}, \frac{x}{c}\right)\right)_{t\geq 0, x\in\mathbb{R}}; \tag{4}$$

$$(X_t)_{t\geq 0} \stackrel{\mathcal{D}}{=} \left(c^2 X_{t/c^2}\right)_{t\geq 0}. \tag{5}$$

Hence $V_p^n \stackrel{\mathcal{D}}{=} N^{-p} \sum_{j=1}^{N} |X_j - X_{j-1}|^p$, where $N \equiv 2^n$. We can write the increment $X_{j+1} - X_j$ in the form

$$(6)\, X_{j+1} - X_j = \int_j^{j+1} I_{\{B_u \le B_{j+1}\}} du + \int_{B_j}^{B_{j+1}} L(j,x)dx - \int_j^{j+1} L(s,B_s)dB_s$$

$$= \int_j^{j+1} I_{\{B_u \le B_{j+1}\}} du + \int_{B_j}^{B_{j+1}} \{L(j,x)$$

$$- L(j,B_j)\}dx - \int_j^{j+1} \{L(s,B_s) - L(j,B_j)\}dB_s .$$

Let us write

$$\int_j^{j+1} I_{\{B_u \le B_{j+1}\}} du \equiv Z_{j,1},$$

$$\int_{B_j}^{B_{j+1}} \{L(j,x) - L(j,B_j)\}dx \equiv Z_{j,2},$$

$$\int_j^{j+1} \{L(s,B_s) - L(j,B_s)\}dB_s \equiv Z_{j,3},$$

$$\int_j^{j+1} \{L(j,B_s) - L(j,B_j)\}dB_s \equiv Z_{j,4},$$

so that

$$(7) \qquad X_{j+1} - X_j = Z_{j,1} + Z_{j,2} - Z_{j,3} - Z_{j,4}.$$

We now estimate various terms. For $p \ge 2$, with c denoting a variable constant

$$\text{(i)} \qquad |Z_{j,1}| \equiv |\int_j^{j+1} I_{\{B_u \le B_{j+1}\}} du| \le 1;$$

$$\text{(ii)} \qquad E|Z_{j,3}|^p \equiv E|\int_j^{j+1} (L(j,B_s) - L(s,B_s))dB_s|^p$$

$$\le cE(\int_j^{j+1} (L(j,B_s) - L(s,B_s))^2 ds)^{p/2}$$

$$\le cE \int_j^{j+1} |L(j,B_s) - L(s,B_s)|^p ds$$

$$= c \int_0^1 EL(u,0)^p du,$$

by reversing the Brownian motion from (s, B_s);

$$\le c.$$

(iii) By Tanaka's formula,

$$L(t,x)-L(t,0)=|B_t-x|-|x|-|B_t|-\int_0^t(\mathrm{sgn}(B_s-x)-\mathrm{sgn}(B_s))dB_s,$$

and

$$||B_t-x|-|x|-|B_t||\le 2(|B_t|\wedge|x|),$$

so we have the estimation

$$E|L(t,x)-L(t,0)|^p\le c\{|x|^p\wedge t^{p/2}+E|\int_0^t I_{\{0<B_s<|x|\}}ds|^{p/2}\};$$

but

$$\begin{aligned}E|\int_0^t I_{(0<B_s<|x|)}ds|^{p/2}&=E|\int_0^{|x|}L(t,y)dy|^{p/2}\\&=t^{p/2}E(\int_0^{|x|/\sqrt{t}}L(1,y)dy)^{p/2},\\&\quad\text{using the scaling relationship (3);}\\&\le t^{p/2}\left(\frac{|x|}{\sqrt{t}}\right)^{p/2-1}E\int_0^{|x|/\sqrt{t}}L(1,y)^{p/2}dy\\&\le ct^{p/2}\left(\frac{|x|}{\sqrt{t}}\right)^{p/2-1}\frac{|x|}{\sqrt{t}}\\&=c|x|^{p/2}t^{p/4}.\end{aligned}$$

Hence for $p\ge 2$

$$E|L(t,x)-L(t,0)|^p\le c\{|x|^p\wedge t^{p/2}+|x|^{p/2}t^{p/4}\}. \tag{8}$$

(iv)
$$\begin{aligned}E|Z_{j,2}|^p&\equiv|\int_{B_j}^{B_{j+1}}\{L(j,x)-L(j,B_j)\}dx|^p\\&=E|\int_0^{W_1}\{L(j,x)-L(j,0)\}dx|^p,\end{aligned}$$

where W is a Brownian motion independent of $(B_s)_{0\le s\le j}$;

$$\begin{aligned}&=E|\int_0^{|W_1|}\{L(j,x)-L(j,0)\}dx|^p\\&\le E(\int_0^\infty I_{(x\le|W_1|)}|L(j,x)-L(j,0)|^p dx|W_1|^{p-1})\\&=\int_0^\infty dxE|L(j,x)-L(j,0)|^pE(|W_1|^{p-1};|W_1|>x),\end{aligned}$$

and the function $\Phi_p(x) \equiv E(|W_1|^{p-1}; |W_1| > x)$ decreases rapidly, so

$$\leq c \int_0^\infty ((|x| \wedge \sqrt{j})^p + |x|^{p/2} j^{p/4}) \Phi_p(x) dx, \qquad \text{by (iii)}$$
$$\leq c(1 + j^{p/4}).$$

(v) $$E|Z_{j,4}|^p \equiv E| \int_j^{j+1} (L(j, B_s) - L(j, B_j)) dB_s|^p$$
$$\leq cE(\int_0^1 (L(j, W_s) - L(j, 0))^2 ds)^{p/2},$$

where W is a Brownian motion independent of $(B_s)_{0 \leq s \leq j}$;

$$\leq cE \int_0^1 |L(j, W_s) - L(j, 0)|^p ds$$
$$= c \int g_1(y) E|L(j, y) - L(j, 0)|^p dy,$$

where g_1 is the Green function of Brownian motion on $[0, 1]$;

$$\leq c \int g_1(y) \{(|y| \wedge \sqrt{j})^p + |y|^{p/2} j^{p/4}\} dy, \qquad \text{by (iii);}$$
$$\leq c(1 + j^{p/4}).$$

Thus of the four terms in (7) making up $X_{j+1} - X_j$, the pth moments of $Z_{j,1}$ and $Z_{j,3}$ are bounded, and the pth moments of $Z_{j,2}$ and $Z_{j,4}$ grow at most like $1 + j^{p/4}$. (Notice that the bounds for the pth moments, proved only for $p \geq 2$, extend to all $p > 0$ by Hölder's inequality.) We shall soon show that this is the true growth rate. Firstly, though, we complete the upper bound estimation by replacing $X_{j+1} - X_j$ by something more tractable, namely

(9) $$\xi_j \equiv \int_{B_j}^{B_{j+1}} L(j, x) dx - \int_j^{j+1} L(j, B_s) dB_s$$
$$\equiv \int_{B_j}^{B_{j+1}} \{L(j, x) - L(j, B_j)\} dx - \int_j^{j+1} \{L(j, B_s) - L(j, B_j)\} dB_s.$$

To see that this is negligibly different from $X_{j+1} - X_j$, observe the elementary inequality valid for all $p \geq 1$, and $a, b \in \mathbb{R}$:

(10) $$||b|^p - |a|^p| \leq |b - a| p(|a|^{p-1} \vee |b|^{p-1}).$$

Now since $\xi_j = Z_{j,2} - Z_{j,4} = X_{j+1} - X_j - Z_{j,1} + Z_{j,3}$, we conclude from (10) that

$$\begin{aligned}
&E||\xi_j|^p - |X_{j+1} - X_j|^p| \\
&\qquad \leq pE\{|Z_{j,1} - Z_{j,3}|(|\xi_j|^{p-1} \vee |X_{j+1} - X_j|^{p-1})\} \\
&\qquad \leq p(E|Z_{j,1} - Z_{j,3}|^a)^{1/a}(E\{|\xi_j|^{b(p-1)} + |X_{j+1} - X_j|^{b(p-1)}|)^{1/b} \\
&\qquad\quad \text{for any } a, b > 1 \text{ such that } a^{-1} + b^{-1} = 1; \\
&\qquad \leq c(1 + j^{(p-1)/4}),
\end{aligned}$$

using the estimates (i), (ii), (iv) and (v). Thus since $V_p^n \stackrel{\mathcal{D}}{=} N^{-p} \sum_{j=1}^N |X_j - X_{j-1}|^p$, we have for $p > 1$

$$\begin{aligned}
E|N^{-p} \sum_{j=0}^{N-1} (|\xi_j|^p - |X_{j+1} - X_j|^p)| \\
&\leq cN^{-p} \sum_{j=0}^{N-1} (1 + j^{(p-1)/4}) \\
&\leq c(1 + N^{-3(p-1)/4}) \\
&\to 0 \qquad \text{as } N \to \infty,
\end{aligned}$$

so for each $p > 1$, $V_p^n - \tilde{V}_p^n \to 0$ in L^1, where

$$\begin{aligned}
\tilde{V}_p^n &\equiv \sum_{j=1}^N | \int_{B((j-1)2^{-n})}^{B(j2^{-n})} L((j-1)2^{-n}, x)dx - \int_{(j-1)2^{-n}}^{j2^{-n}} L((j-1)2^{-n}, B_s)dB_s|^p \\
&\stackrel{\mathcal{D}}{=} N^{-p} \sum_{j=1}^N |\xi_{j-1}|^p.
\end{aligned}$$

Henceforth, we shall concentrate on $\tilde{V}_p^n$, that is, on the ξ_j. Notice that we can say immediately that for $p > 4/3$

$$\begin{aligned}
EV_p^n &= N^{-p} E \sum_{j=1}^N |X_j - X_{j-1}|^p \\
&\leq cN^{-p} \sum_{j=1}^N (1 + j^{p/4}) \\
&\leq CN^{-p}(1 + N^{1+p/4}) \\
&\leq cN^{-3p/4+1}
\end{aligned}$$

so that not only does $V_p^n \to 0$ in L^1, but also the convergence is geometrically fast in n, so there is even almost sure convergence. This proves the statement (1) of Theorem 1.

3. Lower bounds. We can compute

$$
\begin{aligned}
E(\xi_j|\mathcal{F}_j) &= E[\int_{B_j}^{B_j+1} L(j,x)dx|\mathcal{F}_j] \\
&= \int_0^\infty \{L(j,B_j+x) - L(j,B_j-x)\}\overline{\Phi}(x)dx,
\end{aligned}
$$

where $\overline{\Phi}(x) \equiv P(B_1 > x)$ is the tail of the standard normal distribution;

$$
\begin{aligned}
&\stackrel{\mathcal{D}}{=} \int_0^\infty \{L(j,x) - L(j,-x)\}\overline{\Phi}(x)dx \\
&= \int_0^\infty (|B_j - x| - |B_j + x|)\overline{\Phi}(x)dx \\
&\qquad + 2\int_0^\infty (\int_0^j I_{[-x,x]}(B_s)dB_s)\overline{\Phi}(x)dx
\end{aligned}
$$

by Tanaka's formula.

We estimate the pth moment of each piece in turn, the first being negligible in comparison with the second. Indeed, since $||B_j - x| - |B_j + x|| \leq 2|x|$, the first term is actually bounded, and for the second we compute

$$
\int_0^\infty (\int_0^j I_{[-x,x]}(B_s)dB_s)\overline{\Phi}(x)dx = \int_0^j f(B_s)dB_s,
$$

where $f(x) \equiv \int_{|x|}^\infty \overline{\Phi}(y)dy$, so that by the Burkholder-Davis-Gundy inequalities, the pth moment of the second term is equivalent to

$$
\begin{aligned}
E(\int_0^j f(B_s)^2 ds)^{p/2} &= E(\int f(x)^2 L(j,x)dx)^{p/2} \\
&= j^{p/4} E(\int f(x)^2 L(1, x/\sqrt{j})dx)^{p/2} \\
&\sim j^{p/4} E(\int f(x)^2 L(1,0)dx)^{p/2}
\end{aligned}
$$

as $j \to \infty$. Thus we have for each $p \geq 1$ that

$$
E|\xi_j|^p \geq E|E(\xi_j|\mathcal{F}_j)|^p \geq c_p j^{p/4}, \tag{11}
$$

which, combined with the bounds of §2, implies that for each $p \geq 1$ there are constants $0 < c_p < C_p < \infty$ such that for all $j \geq 0$

(12) $$c_p \leq \frac{E|\xi_j|^p}{1 + j^{p/4}} \leq C_p.$$

Hence in particular

(13) $$0 < \liminf_{n\to\infty} EV_{4/3}^n \leq \limsup_{n\to\infty} EV_{4/3}^n < \infty,$$

and for each $p < 4/3$

(14) $$\lim_{n\to\infty} EV_p^n = +\infty,$$

making the conclusion of the Theorem look very likely.

4. The final steps. We shall begin by proving that $\{V_{4/3}^n : n \geq 0\}$ is uniformly integrable. Indeed, for each $p \geq 1$

$$\begin{aligned}
\|V_p^n\|_2 &= \|N^{-p} \sum_{j=1}^{N} |\xi_{j-1}|^p\|_2 \\
&\leq N^{-p} \sum_{j=1}^{N} \||\xi_{j-1}|^p\|_2 \\
&\leq cN^{-p} \sum_{j=1}^{N} (1 + j^{p/4})
\end{aligned}$$

by (12). Hence for $p = 4/3$, the sequence (V_p^n) is bounded in L^2, therefore uniformly integrable. Hence

(15) $$P(\limsup_n V_{4/3}^n > 0) > 0,$$

because otherwise $V_{4/3}^n \to 0$ a.s., and hence in L^1 (by uniform integrability), contradicting (13). Now define

$$V_p^n(t) \equiv \sum_{j=1}^{[2^n t]} |\Delta_j^n X|^p,$$

and let

$$F_k \equiv \{\limsup_{n\to\infty} \sum_{j=1}^{2^{n-k}} |\Delta_j^n X|^{4/3} > 0\},$$

an event which is $\mathcal{F}(2^{-k})$-measurable. Notice that $F_{k+1} \subseteq F_k$; and by Brownian scaling, all the F_k have the same probability, which is positive by (15). By the Blumenthal $0-1$ law, $P(F_k) = 1$ for every k, and hence for each $t > 0$

$$P\left[\limsup_{n\to\infty} V_{4/3}^n(t) > 0\right] = 1. \tag{16}$$

Now suppose that X were of finite variation, so that there exist stopping times $T_k \uparrow 1$ such that $V_1(T_k) \equiv\uparrow \lim_{n\to\infty} V_1^n(T_k) \le k$. Choose $a > 1 > \alpha > 0$ such that $4a\alpha/3 = 1$, and let b be the conjugate index to a ($b^{-1} + a^{-1} = 1$). By Hölder's inequality,

$$V_{4/3}^n(T_k) \le (V_1^n(T_k))^{1/a}(V_{4b(1-\alpha)/3}^n(T_k))^{1/b}$$

and since $4b(1-\alpha)/3 > 4/3$, the second factor on the right-hand side goes to zero a.s. as $n \to \infty$. The first factor remains bounded as $n \to \infty$, by definition of T_k. Hence $V_{4/3}^n(T_k) \xrightarrow{\text{a.s.}} 0$ as $n \to \infty$, which is only consistent with (16) if each T_k is zero a.s., which is impossible since $T_k \uparrow 1$.

References

[1] L.C.G. ROGERS and J.B. WALSH. Local time and stochastic area integrals. To appear in *Ann. Probab.*

[2] L.C.G. ROGERS and J.B. WALSH. The intrinsic local time sheet of Brownian motion. Submitted to *Probab. Th. Rel. Fields.*

[3] L.C.G. ROGERS and D. WILLIAMS. *Diffusions, Markov Processes and Martingales, Vol.2*. Wiley, Chichester, 1987.

[4] J.B. WALSH. Stochastic integration with respect to local time. *Seminar on Stochastic Processes 1982*, pp. 237–302. Birkhäuser, Boston, 1983.

L.C.G. Rogers
Statistical Laboratory
16 Mill Lane
Cambridge CB2 1SB
GREAT BRITAIN

J.B. Walsh
Department of Mathematics
University of British Columbia
Vancouver, B.C.
V6T 1Y4 CANADA

Self-Intersections of Stable Processes in the Plane: Local Times and Limit Theorems

JAY S. ROSEN

1. Introduction

X_t will denote the symmetric stable process of index $\beta>1$ in $\mathbb{R}^2$, with transition density $p_t(x)$ and λ-potential

$$G_\lambda(x) = \int_0^\infty e^{-\lambda t} p_t(x)\, dt.$$

We recall that

$$(1.1) \qquad G_0(x) = \frac{\Gamma(\frac{2-\beta}{2})}{\Gamma(\beta/2)} \cdot \frac{1}{2^\beta \pi} \cdot \frac{1}{|x|^{2-\beta}}$$

To study the k-fold self-intersections of X we will attempt to give meaning to the formal expression

$$(1.2) \qquad \int\cdots\int_{0\le t_1\le\ldots\le t_k\le t} \delta(X_{t_2} - X_{t_1})\cdots\delta(X_{t_k} - X_{t_{k-1}})$$

Let $f\ge 0$ be a continuous function supported in the unit disc, and set

$$f_\epsilon(x) = \frac{1}{\epsilon^2} f(x/\epsilon)$$

If we think of f_ϵ as an approximate δ function, we are led to consider

---

*This research supported in part by NSF DMS- 8802288

$$(1.3)\quad \alpha_{k,\epsilon}(t) = \int_{0\le t_1\le\dots\le t_k\le t}\cdots\int dt_1 \prod_{i=2}^{k} f_\epsilon(X_{t_i} - X_{t_{i-1}})dt_i$$

as an approximation to (1.2).

As $\epsilon \to 0$, $\alpha_{k,\epsilon}(t)$ will diverge (due to the contributions near the 'diagonals' $\{t_i=t_j\}$). To get a non-trivial limit we must 'renormalize', which in our case means subtracting from $\alpha_{k,\epsilon}(t)$ terms involving lower order intersections. Thus, we define the approximate renormalized self-intersection local time,

$$(1.4)\qquad \gamma_{k,\epsilon}(t) = \sum_{j=1}^{k}(-h_\epsilon)^{k-j}\binom{k-1}{j-1}\alpha_{j,\epsilon}(t)$$

$$= \int_{0\le t_1\le\dots\le t_k t}\cdots\int dt_1 \prod_{i=2}^{k}\left[f_\epsilon(X_{t_i} - X_{t_{i-1}})dt_i - h_\epsilon\, \delta_{t_{i-1}}(dt_i)\right]$$

where

$$(1.5)\qquad h_\epsilon = \int f_\epsilon(x)G_0(x)d^2x$$
$$= \frac{1}{\epsilon^{2-\beta}}\int G_0(x)\ f(x)\ d^2x.$$

Note that $\gamma_{1,\epsilon}(t) = t$.

Following Dynkin [1988B], to reduce our anlaysis to managable proportions, rather than study $\gamma_{k,\epsilon}(t)$ for fixed t, we study $\gamma_{k,\epsilon}(\zeta)$ where ζ is an independent exponential random variable

$$(1.6)\qquad P(\zeta>t) = e^{-\lambda t}$$

We will find that $\gamma_{k,\epsilon}(\zeta)$ converges, as $\epsilon \to 0$, if and only if β is sufficiently large. We recall that X has k-fold self-intersections if and only if $k(2-\beta)<2$.

Theorem 1: If $(2k-1)(2-\beta)<2$, then $\gamma_{k,\epsilon}(\zeta)$ converges in L^2 to a non-trivial random variable denoted by $\gamma_k(\zeta)$. Moreover, we have

$$\|\gamma_{k,\epsilon}(\zeta) - \gamma_k(\zeta)\|_2 \leq c\ \epsilon^{\alpha/2}$$

where $$\alpha = 2-(2k-1)(2-\beta)>0$$

Aside from the intrinsic interest of $\gamma_k(\zeta)$ as a measure of k-fold intersections, we hope to show in future work that $\gamma_k(\zeta)$ arises naturally in the asymptotic expansion for the area of the 'stable sausage'

$$S_\epsilon = \left\{ x \in \mathbb{R}^2 \,\middle|\, \inf_{0\leq s<\zeta} \|X_s - x\| \leq \epsilon \right\}$$

generalizing the work of LeGall [1988] for Brownian motion. We also note our previous work involving a different form of renormalization, Rosen [1986]. The simplifications arising from the present form of renormalization will be most helpful in what follows.

When the condition of Theorm 1 is not satisfied, $\gamma_{k,\epsilon}(\zeta)$ will not converge in L^2. Instead, appropriately normalized, we get a central limit type theorem involving L, a random variable with density $\frac{1}{2}e^{-|x|}$, [known as Laplace's first law].

Theorem 2: If $(2k-1)(2-\beta) = 2$ then

$$\frac{\gamma_{k,\epsilon}(\zeta)}{\sqrt{\lg(1/\epsilon)}} \xrightarrow{\text{(dist.)}} \left[\sqrt{\frac{c(\beta,k)}{\lambda}}\right] L$$

where

$$c(\beta,k) = 2\pi \left[\frac{\Gamma\left[\frac{2-\beta}{2}\right]}{\Gamma(\beta/2)} \cdot \frac{1}{2^\beta \pi}\right]^{2k-1}.$$

Remark: (i) compare (1.1).

(ii). If B_t denotes a real Brownian motion then B_ζ and $\frac{1}{\sqrt{2\lambda}} L$ have the same law. This provides a conceptual link between Theorem 2 and Rosen [1988], Yor [1985].

Theorem 3: If $(2k-1)(2-\beta)>2$ but $(2(k-1)-1)(2-\beta)<2$ then

$$\epsilon^{\alpha/2}\gamma_{k,\epsilon}(\zeta) \xrightarrow{(dist.)} \left[\sqrt{\frac{c(\beta,k)}{\lambda}}\,\right] L$$

where $\alpha = (2k-1)(2-\beta)-2>0$ and $c(\beta,k)$ is an explicit constant.

Remark: In the proof of Theorem 3, we will find that

$$c(\beta,k) = \lim_{\epsilon\to 0} \frac{\lambda\epsilon^{\alpha}}{2} E(\gamma^2_{k,\epsilon}(\zeta)),$$

and we will give an explicit formula for $c(\beta,k)$.

For more information on self-intersection local times see the survey of Dynkin [1988A] and the references therein.

2. Preliminaries

We have formulated our theorems in terms of $\gamma_{k,\epsilon}(t)$, an expression which does not involve λ, the parameter of the exponential time ζ. In our proofs, it will be more convenient to work with

$$(2.1)\qquad \Gamma_{k,\epsilon}(t) = \sum_{j=1}^{k}(-H_\epsilon)^{k-j}\begin{bmatrix}k-1\\ j-1\end{bmatrix}\alpha_{k,\epsilon}(t)$$

$$= \int_{0\le t_1\le\cdots\le t_k\le t}\cdots\int dt_1 \prod_{j=1}^{k}\left[f_\epsilon(X_{t_i}-X_{t_{i-1}})\, dt_i - H_\epsilon\delta_{t_{i-1}}(dt_i)\right]$$

which differs from $\gamma_{k,\epsilon}(t)$, (1.4) in that $h_\epsilon=\int f_\epsilon(x)G_0(x)dx$ is replaced by

(2.2) $$H_\epsilon = \int f_\epsilon(x) G_\lambda(x)\,dx$$

It is easily checked that

(2.3) $$\gamma_{k,\epsilon}(t) = \sum_{j=1}^{k} \left[-(h_\epsilon - H_\epsilon)\right]^{k-j} \binom{k-1}{j-1} \Gamma_{j,\epsilon}(t).$$

This expression will allow us to derive results about the γ's from results on the Γ's.

The main point of this section is to derive a useful expression for

(2.4) $$E\Big[\prod_{j=1}^{n} \Gamma_{k,\epsilon_j}(\zeta)\Big]$$

$$=E\Big[\int\cdots\int \prod_{j=1}^{n} dt_1^j \prod_{i=2}^{k}\Big[f_{\epsilon_j}\Big[X_{t_i^j} - X_{t_{i-1}^j}\Big]dt_i^j - H_{\epsilon_j}\delta_{t_{i-1}^j}(dt_i^j)\Big]\Big]$$

$$=\sum_D I(D)$$

where

(2.5) $$I(D)$$

$$=E\Big[\int_D\cdots\int \prod_{j=1}^{n} dt_1^j \prod_{i=2}^{k}\Big[f_{\epsilon_j}\Big[X_{t_i^j}-X_{t_{i-1}^j}\Big]dt_i^j - H_\epsilon \delta_{t_{i-1}^j}(dt_i^j)\Big]\Big]$$

and D runs over the set of orderings of the nk+1 points $0, t_i^j$; $1\le i\le k$; $1\le j\le n$; such that $0\le t_1^j \le t_2^j \le \cdots \le t_k^j$ for all j.

Fix D. We call a set S of t's **elementary**, relative to D, if

(2.6) $$S=\left\{t_i^j,\ t_{i+1}^j, \cdots,\ t_{i+\ell}^j,\ t_{\tilde{i}}^{\tilde{j}}\right\}$$

and S satisfies

a) $t_{i+\ell}^j \le t_{\tilde{i}}^{\tilde{j}}$

b) no other t's come between t_i^j and $t_{\tilde{i}}^{\tilde{j}}$ in D, (except t_{i+m}^j, $2\le m\le \ell$)

c) S is maximal in the sense that the t preceeding t_i^j in D is not t_{i-1}^j.

With such an elementary sequence S, (2.6), we associate a function $H_s(Y)$ of nk variables

$$Y = \left\{Y_i^j;\ 1\leq i\leq k;\ 1\leq j\leq n\right\}$$

by the formula

$$(2.7)\quad H_s(Y) = G_\lambda(y_{i+1}^j)\cdots G_\lambda(y_{i+\ell}^j)\Delta^\ell_{y_{i+1}^j,\ldots,y_{i+\ell}^j} G_\lambda(Y_i^j - Y_{\bar{i}}^{\bar{j}})$$

Here $y_{i+1}^j \doteq Y_{i+1}^j - Y_i^j$, etc.

$$\Delta^\ell_{a_1,\ldots,a_\ell} F = \Delta_{a_1}\Delta_{a_2}\cdots\Delta_{a_\ell} F$$

and

$$\Delta_a F(x) = F(x+a) - F(x)$$

In particular, if $S = \left\{t_i^j,\ t_{\bar{i}}^{\bar{j}}\right\}$ has only two elements, then the above reduces to

$$(2.8)\qquad H_s(Y) = G_\lambda(Y_{\bar{i}}^{\bar{j}} - Y_i^j)$$

Let $\epsilon(D)$ denote the elementary sequences in D. Our formula for I(D) is

$$(2.9)\qquad I(D) = \int\cdots\int \left[\prod_{\substack{i=2\\ \forall j}}^{k} f_{\epsilon_j}(y_i^j)\right] \prod_{S\in\mathcal{E}(D)} H_S(Y)\, dY$$

as is easily checked, using (2.2).

The following lemma, proven in Section 7 is basic.

Lemma 1: Let $\beta > 1$, then

$$(2.10)\qquad 0 \leq G_\lambda(z) \leq c\left[G_0(z) \wedge \frac{1}{|z|^3}\right].$$

If $|z| \geq 2\ell\epsilon$, then

$$(2.11)\qquad \sup_{|a_i|\leq\epsilon}\left|\Delta^\ell_{a_1,\ldots,a_\ell} G_\lambda(z)\right| \leq c\left[\frac{\epsilon}{|z|}\right]^\ell G_0(z)\, R(z)$$

$$(2.12) \qquad \sup_{|a_i| \le \epsilon} \left[\prod_{i=1}^{\ell} G_\lambda(a_i)\right] \left|\Delta^\ell_{a_1,\dots,a_\ell} G_\lambda(z)\right|$$
$$\le c\; G_0^{1+\ell}(z) \left[\frac{\epsilon}{|z|}\right]^{(\beta-1)\ell} R(z)$$

where $R(z)$ is a bounded monotone decreasing integrable function. (In fact we can take $R(z) = \frac{1}{1+z^{1+\beta}}$).

If $S \in \epsilon(D)$ has the form (2.6), we say that S has length ℓ, and write $\ell(S) = \ell$. For this S, (2.7) and lemma 1 mean that

$$(2.13) \qquad \left|H_S(Y)\right| \le c\; G_0^{\ell(S)+1}(Z) \left[\frac{\epsilon_j}{|Z|}\right]^{(\beta-1)\ell(S)} R(z)$$

whenever $Z = Y^{\bar{j}}_{\bar{i}} - Y^j_i$ satisfies $|Z| \ge 2\ell\epsilon_j$.

3. Proof of Theorem 1

From now on, λ is fixed and $G(x)$ without a subscript will refer to $G_\lambda(x)$. Similarly, we write $\gamma_{k,\epsilon}$ for $\gamma_{k,\epsilon}(\zeta)$, etc.

We first show that to prove theorem 1, it suffices to prove the following analogue for Γ.

Proposition 1: If $(2-\beta)(2k-1)<2$, then $\Gamma_{k,\epsilon}$ converges in L^2 to a non-trivial random variable denoted by Γ_k. Moreover, we have

$$(3.1) \qquad \|\Gamma_{k,\epsilon} - \Gamma_k\|_2 \le c\; \epsilon^{\alpha/2}$$

where $\alpha = 2-(2k-1)(2-\beta) > 0$.

To see that proposition 1 implies theorem 1, define

$$(3.2) \qquad H(x) = G_0(x) - G_\lambda(x) = \int_0^\infty (1-e^{-\lambda t}) p_t(x)\,dt$$
$$= \frac{1}{(2\pi)^2} \int e^{ipx} \frac{\lambda}{p^\beta(\lambda+p^\beta)}\, d^2p.$$

Since $\beta > 1$, $H(x)$ is continuous, bounded and

$$(3.3)\qquad |h_\epsilon - H_\epsilon - H(o)| = |\int f_\epsilon(x)\ \left[H(x) - H(o)\right]dx$$

$$= |\ \frac{1}{(2\pi)^2}\int f_\epsilon(x)\left[\int (e^{ipx}-1)\ \frac{\lambda}{p^\beta(\lambda+p^\beta)}\ d^2p\right]dx$$

$$\leq c\left[\int f_\epsilon(x)\ |x|^\delta\ dx\right]\left[\int \frac{p^\delta}{p^\beta(\lambda+p^\beta)}\ d^2p\right]$$

$$\leq c\ \epsilon^\delta \int \frac{p^\delta}{p^\beta(\lambda+p^\beta)}\ d^2p$$

for any $0 \leq \delta \leq 1$.

Thus,

$$(3.4)\qquad |h_\epsilon - H_\epsilon - H(o)| \leq \begin{cases} c\epsilon, & \text{if } \beta > 3/2 \\ c\epsilon^{2\beta-2-\bar{\delta}}, & \text{if } \beta \leq 3/2 \end{cases}$$

for any $\bar{\delta} > 0$.

We write

$$2\beta - 2 - \bar{\delta}$$
$$= 2-2(2-\beta) - \bar{\delta}$$
$$= \frac{1}{2}(2-(2k-1)(2-\beta)) + 1 - \bar{\delta} + (k - \frac{5}{2})(2-\beta)$$
$$> \frac{1}{2}(2-(2k-1)(2-\beta))$$

since $k \geq 2$, and $\bar{\delta} > 0$ can be chosen small.

Since, obviously

$1 > \frac{1}{2}(2-(2k-1)(2-\beta))$, (3.4) gives

$$(3.5)\qquad |h_\epsilon - H_\epsilon - H(o)| \leq c\ \epsilon^{(2-(2k-1)(2-\beta))/2}$$

so that (2.3) and proposition 1 now imply Theorem 1, with

$$(3.6)\qquad \gamma_k \doteq \sum_{j=1}^{k} (-H(o))^{k-j}\begin{bmatrix}k-1\\ j-1\end{bmatrix}\Gamma_j$$

Proposition 1 will follow from

Proposition 2: If $(2-\beta)(2k-1) < 2$, then for

$$0 < \epsilon \le \bar{\epsilon} \le 2\epsilon < 1$$

we have

$$\|\Gamma_{k,\epsilon} - \Gamma_{k,\bar{\epsilon}}\|_2 \le c\ \bar{\epsilon}^{\alpha/2} \tag{3.7}$$

where $\alpha = 2 - (2k-1)(2-\beta)$

For, assume proposition 2. Given any $0 < \epsilon < \bar{\epsilon} < 1$. choose $n \ge 0$ such that

$$\frac{\bar{\epsilon}}{2^{n+1}} < \epsilon \le \frac{\bar{\epsilon}}{2^n}.$$ Then by (3.7),

$$\begin{aligned} \|\Gamma_{k,\epsilon} - \Gamma_{k,\bar{\epsilon}}\|_2 &\le \sum_{i=0}^{n-1} \|\Gamma_{k,\bar{\epsilon}/2^i} - \Gamma_{k,\bar{\epsilon}/2^{i+1}}\|_2 \\ &\quad + \|\Gamma_{k,\bar{\epsilon}/2^n} - \Gamma_{k,\epsilon}\|_2 \\ &\le c\ \bar{\epsilon}^{\alpha/2} \sum_{i=0}^{n} \frac{1}{(2^{\alpha/2})^i} \\ &\le c\ \bar{\epsilon}^{\alpha/2}. \end{aligned} \tag{3.8}$$

This shows the L^2 convergence of $\Gamma_{k,\epsilon}$, and also establishes (3.1)

Proof of Proposition 2: According to section 2

$$\|\Gamma_{k,\epsilon} - \Gamma_{k,\bar{\epsilon}}\|_2^2 \tag{3.9}$$

$$= E((\Gamma_{k,\epsilon} - \Gamma_{k,\bar{\epsilon}})^2) = \sum_D I(D)$$

$$= \sum_D \int \cdots \int F_{\epsilon,\bar{\epsilon}}(y^1_\cdot) F_{\epsilon,\bar{\epsilon}}(y^2_\cdot) \prod_{S \in \mathcal{E}(D)} H_S(Y)\,dY$$

where

$$F_{\epsilon,\bar{\epsilon}}(y_\cdot) = \prod_{i=2}^{k} f_\epsilon(y_i) - \prod_{i=2}^{k} f_{\bar{\epsilon}}(y_i) \tag{3.10}$$

Fix D.

The ordering D, in a natural way, induces an ordering on $Y^1_\cdot$, $Y^2_\cdot$. Thus, if $t^m_i \le t^{\bar{m}}_{\bar{i}}$, we will say that Y^m_i comes before $Y^{\bar{m}}_{\bar{i}}$. This induces an order on $\in(D)$. We may assume that the first element in D is t^1_1, hence our first element of $\in(D)$ is $\{o, t^1_1\}$ giving rise to the factor $G(Y^1_1)$. Let $S = \left\{t^1_1,\dots,t^1_l,t^2_1\right\}$ be the next element in $\in(D)$. Let $Z = Y^2_1 - Y^1_1$.

We first show that the contribution to I(D) from the region $\left\{|Z| \le 4k\bar{\epsilon}\right\}$ is $O(\epsilon^\alpha)$.

To see this, we first integrate the Y's in reverse order; we start with the last Y and integrate successively until we reach Y^2_1 using the bound

$$\int f_\epsilon(x-a)G(x)\ dx \tag{3.11}$$

$$= c\int e^{ipa}\ \hat{f}(\epsilon\ p)\ \frac{1}{\lambda+p^\beta}\ d^2p$$

$$\le c\int |\hat{f}(\epsilon\ p)|\ \frac{1}{p^\beta}\ d^2p$$

$$\le \frac{c}{\epsilon^{(2-\beta)}}.$$

For the Y^2_1 integral we use

$$\int_{|Z|\le 4k\bar{\epsilon}} G(Y^2_1 - Y^1_m)\ dY^2_1 \tag{3.12}$$

$$\le \int_{|Z|\le 6k\bar{\epsilon}} G(Z)\ d^2Z$$

$$\le \int_{|Z|\le 6k\bar{\epsilon}} G_o(Z)\ d^2Z$$

$$\leq c \frac{\bar{\epsilon}^2}{\bar{\epsilon}^{(2-\beta)}}$$

The remaining $Y^1_\ell, Y^1_{\ell-1}, \ldots, Y^1_2$ integrals are handled using (3.11), and finally $\int G(Y^1_1)dY^1_1 = \frac{1}{\lambda}$.

Since there were $\leq 2k$ G factors in (3.9), we find that the contribution from the region $\left\{|Z| \leq 4k\bar{\epsilon}\right\}$ is

$$O\left[\frac{\epsilon^2}{\epsilon^{(2-\beta)(2k-1)}}\right] = O(\epsilon^\alpha)$$

Thus, for the remainder of our proof we can assume that $|Z| \geq 4K\bar{\epsilon}$. In view of (2.13), we can bound the integral $I(D)$ over $|Z| \geq 4k\bar{\epsilon}$ by

$$\int_{|Z|\geq 4k\bar{\epsilon}} G_o^{2K-1}(Z) \left[\frac{\bar{\epsilon}}{|Z|}\right]^{(\beta-1)\ell(D)} R(Z)dZ \tag{3.13}$$

where $\ell(D) = \sum_{S \in \mathcal{E}(D)} \ell(S)$.

If $\ell(D) \geq 2$, we can bound (3.13) by replacing $(\beta-1)\ell(D)$ with

$$2(\beta-1) = 2 - 2(2-\beta),$$

giving

$$\int_{|Z|\geq 4k\bar{\epsilon}} \bar{\epsilon}^{2-2(2-\beta)} \frac{1}{|Z|^{2+(2k-3)(2-\beta)}} dZ = O(\bar{\epsilon}^\alpha) \tag{3.14}$$

since $k \geq 2$.

We can thus assume that $\ell(D) \leq 1$. If $\ell(D) = o$, D must be the ordering

$$D_* = t^1_1 \leq t^2_1 \leq t^1_2 \leq t^2_2 \leq t^1_3 \leq \ldots \leq t^2_k \tag{3.15}$$

and then

$$I(D_*) = \int F_{\epsilon,\bar{\epsilon}}(y^1) F_{\epsilon,\bar{\epsilon}}(y^2) \prod_{i=1}^k G(Y^1_i - Y^2_{i-1}) G(Y^2_i - Y^1_i) dY \tag{3.16}$$

We note that

$$G(Z+a+b) = G(Z)+\Delta_a G(Z) + \Delta_b G(Z) + \Delta^2_{a,b} G(Z)$$

and we use this to expand $G(Y^2_i - Y^1_\ell)$, with $Z = Y^2_1 - Y^1_1$ as before

and
$$a = Y^1_1 - Y^1_\ell = -\sum_{j=2}^{l} y^1_j$$

$$b = Y^2_i - Y^2_1 = \sum_{j=2}^{i} y^2_j$$

We can thus write the product in I(D) as a sum of monomials in $G(Z)$, $\Delta_a G(Z)$ and $\Delta^2_{a,b} G(Z)$. If any monomial contains either a $\Delta^2 G$ factor, or 2 ΔG factors then we can use (2.11), in a manner similar to (3.13), (3.14) to show that the integral over $|Z|\geq 4k\bar{\epsilon}$ is $O(\bar{\epsilon}^\alpha)$.

But, because of the factor $F_{\epsilon,\bar{\epsilon}}(y^1_\cdot)\ F_{\epsilon,\bar{\epsilon}}(y^2_\cdot)$ in (3.16), it is clear that the integral will vanish if our monomial is of the form $G^{2k-1}(Z)$ or $G^{2k-1}(Z)\Delta_a G(Z)$.

A similar analysis applies to the case of $\ell(D) = 1$, completing the proof of proposition 2, hence of Theorem 1.

4. The second moment

In this section we calculate the asymptotics of $E(\Gamma^2_{k,\epsilon})$ as $\epsilon \to 0$. If $(2k-1)(2-\beta) < 2$, then the last section shows that

(4.1) $$E(\Gamma^2_{k,\epsilon}) \longrightarrow \frac{2}{\lambda} \int G^{2k-1}(z)\ d^2z.$$

Consider now the case $(2k-1)(2-\beta) = 2$, so that $\alpha = 0$. It is easily checked that all estimates of the previous section which were $O(\epsilon^\alpha)$, also hold in this case, i.e. are

$O(1)$, leading to

$$(4.2)\qquad E(\Gamma^2_{k,\epsilon}) = \frac{2}{\lambda} \int_{|Z|\geq 4k\epsilon} G^{2k-1}(z)d^2z + O(1)$$

$$= \frac{2}{\lambda} \int_{4k\epsilon\leq|z|\leq 1} G^{2k-1}(z)d^2z + O(1)$$

since $G(z)$ is bounded and integrable for $|z|\geq 1$.

As in (3.2), we write

$$(4.3)\qquad G(z) = G_0(z) - H(z)$$

with H bounded, and we find immediately that (using (1.1))

$$(4.4)\qquad E(\Gamma^2_{k,\epsilon}) = \frac{2}{\lambda} \int_{4k\epsilon\leq|Z|\leq 1} G_0^{2k-1}(z)d^2z + O(1)$$

$$= 2\,\frac{c(\beta,k)}{\lambda}\,\lg(1/\epsilon) + O(1)$$

where $c(\beta,k) = 2\pi\left[\frac{\Gamma\left[\frac{2-\beta}{2}\right]}{\Gamma(\beta/2)}\,\frac{1}{2^{\beta}\pi}\right]^{2k-1}$ as in Theorem 2.

We next consider the case where $(2k-1)(2-\beta) > 2$. Here we will see that all orderings D will contribute a term of order $\frac{1}{\epsilon^{\alpha}}$ (where now $\alpha = (2k-1)(2-\beta)-2>0$), plus terms of lower order.

Consider a fixed ordering D as before, and

$$(4.5)\qquad I(D) = \int\cdots\int F_\epsilon(y^1_\cdot)F_\epsilon(y^2_\cdot) \prod_{S\in\mathcal{E}(D)} H_S(Y)\ dY$$

with

$$(4.6)\qquad F_\epsilon(y_\cdot) = \prod_{i=2}^{k} f_\epsilon(y_i).$$

Assume for definiteness, as in section 3, that the first element in $\mathcal{E}(D)$ is $\left\{o,t^1_1\right\}$, so that we have a factor $G(Y^1_1)$ in (4.5). We change variables

$Y_{\cdot}^{1}, Y_{\cdot}^{2} \to X_i$, $i = 1,\dots,2r$ where X_i is the argument of the i'th G factor in I(D). More precisely, if the i'th interval in $D = \{0<t_1^1<\dots\}$ is $t_j^m < t_{\bar{j}}^{\bar{m}}$, then $X_i = Y_{\bar{j}}^{\bar{m}} - Y_j^m$. We integrate out $dX_1 = dY_1^1$ and write

$$(4.7)\quad I(D) = \frac{1}{\lambda}\int\cdots\int F_\epsilon(y_\cdot^1)F_\epsilon(y_\cdot^2)\prod_{S\in\bar{\mathcal{E}}(D)} H_S(Y)\, dX_2\dots dX_{2r}$$

where $\bar{\mathcal{E}}(D)$ is obtained from $\mathcal{E}(D)$ by removing the first sequence, $\{o,t_1^1\}$.

We write $G(z) = G_o(z) - H(z)$ as in (4.3), and use this to rewrite (4.7) as the sum of many terms. One term is

$$(4.8)\quad \frac{1}{\lambda}\int F_\epsilon(y_\cdot^1)F_\epsilon(y_\cdot^2)\prod_{S\in\bar{\mathcal{E}}(D)} H_S^o(Y)\, dX_2\cdots dX_{2k}$$

where H_s^o is defined by replacing each G in H_S with G_o. The other terms arising from (4.7) differ from (4.8) in that at least one G has been replaced by H. We first deal with (4.8), which will turn out to be the dominant term.

We scale in (4.8), and obtain

$$(4.9)\quad \frac{1}{\epsilon^\alpha}\frac{1}{\lambda}\int F(y_\cdot^1)F(y_\cdot^2)\prod_{S\in\bar{\mathcal{E}}(D)} H_S^o(Y)dX_2\cdots dX_{2k}$$

where now

$$F(y) = \prod_{i=2}^{k} f(y_i).$$

Let us show that the integral in (4.9) converges. If the first sequence in $\bar{\mathcal{E}}(D)$ is $\{t_1^1,t_2^1,\dots t_\ell^1,t_1^2\}$, set $Z = X_{\ell+1} = Y_1^2 - Y_\ell^1$. If $|Z| \geq 4k$, then by the H_S^o analogue of (2.13) we can bound our integral by

$$c\int_{|Z|\geq 4k} G_o^{2k-1}(z)\ \ dz < \infty.$$

If, on the other hand, $|Z| \leq 4k$. then all $|X_i| \leq 8k$, and using $\int_{|X|\le c} G_0(x)\,dx < \infty$ we can bound our integral by integrating in reverse order $dX_{2k},\ldots,d\ell$.

Next, consider a term arising from the expansion of (4.7), in which at least one of the G_0 factors of (4.8) has been replaced by $H(\cdot)$.

If $|Z| \le 4k\epsilon$, we first bound any $H(\cdot)$ factor by a constant, and then scale. We obtain an integral, which can be bounded as above (since now $|Z| \le 4k$) multiplied by $\frac{1}{\epsilon^{\bar{\alpha}}}$ with $\bar{\alpha} < \alpha$.

If $|Z| \ge 4k\epsilon$, then by (7.10) and (7.12) we find that for any ℓ, including $\ell = 0$, and $|a_i| \le \epsilon$,

(4.10) $$|\Delta^{\ell}_{a_1,\ldots,a_\ell} H(x)| \le c\left[\frac{|a_1|\cdot\ldots|a_\ell|}{x^\ell}\cdot\frac{1}{x^{2-\beta}}\right]\wedge 1$$

$$\le c\left[\frac{(|a_1|\cdot\ldots|a_\ell|)^{2-\beta}}{x^{(2-\beta)(\ell+1)}}\right]^{\delta} \qquad ,|X| \ge 2\ell\epsilon$$

for any $0 \le \delta \le 1$. Scaling with these bounds, gives a factor $\frac{1}{\epsilon^{\bar{\alpha}}}$ with $\bar{\alpha} < \alpha$ if $\delta < 1$, and an integral which can be bounded as long as δ is chosen close enough to 1 so that

$$(2k-1)(2-\beta)\ \delta > 2.$$

Thus we finally have

(4.11) $$E\left[\Gamma^2_{k,\epsilon}\right] = \frac{1}{\epsilon^\alpha}$$

$$\frac{1}{\lambda}\sum_D \int\cdots\int F(y^1_\cdot)F(y^2_\cdot)\prod_{S\in\bar{E}(D)} H^0_S(Y)\,dX_2\cdots dX_{2k}$$

$$+o\left[\frac{1}{\epsilon^\alpha}\right]$$

5. Proof of Theorem 2

We proceed by the method of moments. Since

$$\begin{cases} E(L^{2n}) &= (2n)! \\ E(L^{2n+1}) &= 0 \end{cases} \tag{5.1}$$

it suffices to show that

$$\begin{cases} E\left[\dfrac{\Gamma_{k,\epsilon}}{\sqrt{\lg(1/\epsilon)}}\right]^{2n} \longrightarrow (2n)! \left[\dfrac{c(\beta,k)}{\lambda}\right]^{n} \\ E\left[\dfrac{\Gamma_{k,\epsilon}}{\sqrt{\lg(1/\epsilon)}}\right]^{2n+1} \longrightarrow 0 \end{cases} \tag{5.2}$$

in order to get

$$\frac{\Gamma_{k,\epsilon}}{\sqrt{\lg(1/\epsilon)}} \xrightarrow{\text{(dist)}} \left[\sqrt{\frac{c(\beta,k)}{\lambda}}\right] L,$$

which then implies Theorem 2, by (2.3) and Theorem 1.

We recall from section 2 that

$$E(\Gamma^{m}_{k,\epsilon}) = \sum_{D} \int \cdots \int \left[\prod_{j=1}^{m} F_\epsilon(y^j_\cdot)\right] \prod_{S\in\mathcal{E}(D)} H_S(Y)\, dY \tag{5.3}$$

where D runs over all orderings of

$\left\{o, t^j_i,\ j=1,\dots,m; i=1\dots,k\right\}$

Let

$$U(D) = \bigcup_{j=1}^{m} \left[t^j_1,\ t^j_k\right] \tag{5.4}$$

U(D) naturally decomposes into the union of its components; U^1, $U^2,\dots,U^j$. If, say,

$$U^i = \bigcup_{\ell=1}^{p} \left[t_1^{j_\ell},\ t_k^{j_\ell}\right]$$

then we say that U^i has height p, and denote by D^i the ordering induced on

$$\left\{o,\ t_n^{j_\ell},\quad \ell = 1,\dots,p;\quad n = 1,\dots,k\right\}$$

by D. By translation invariance we find that

(5.5) $$I(D) = \prod_{i=1}^{j} I(D^i)$$

It is clear from this that if any component of U(D) has height 1, then I(D) = 0. Furthermore, from section 4 we know that if D^i has height 2, then

$$I(D^i) = \begin{cases} \frac{c(\beta,k)}{\lambda} \lg(1/\epsilon) + O(1), & \text{if } D = D_*, D_{**} \\ O(1) & \text{otherwise} \end{cases}$$

where D_* is given by (3.15), and D_{**} is obtained from D_* by permuting $t^1_.$ with $t^2_.$.

If m = 2n, and U(D) has n components of height 2, then the above allows us to compute I(D), and since there are (2n)! ways to permute the t^j's, we see that the contribution to (5.2) from orderings D with n components of height 2 is

$$(2n)! \left[\frac{c(\beta,k)}{\lambda}\right]^n (\lg 1/\epsilon)^n + O(\lg(1/\epsilon))^{n-1}$$

To complete the proof of (5.2) it suffices to show that if U(D) is connected and of height n > 2, then

(5.6) $$I(D) = o(\lg(1/\epsilon))^{n/2}$$

We will develop a three step procedure to prove (5.6).

We will refer to $Y^1_., Y^2_., \ldots, Y^n_.$ as n letters, and to Y^i_j as the j'th component of the letter $Y^i_.$. If $S \in \mathcal{E}(D)$ is of the form (2.6), i.e.,

(5.7) $$S = \left\{t^j_i, \ldots, t^j_{\ell+i}, t^{\bar{j}}_{\bar{i}}\right\}$$

and if $\ell > o$, then $H_S(Y)$, see (2.7), contains factors $G(y^j_{i+1}) \ldots G(y^j_{i+\ell})$, and we say that the letter $Y^j_.$ has ℓ isolated G factors. This terminology refers to the fact that in these factors $Y^j_.$ appears alone, without any other letter. Let

$$I = \left\{ i \,|\, Y^i_{\cdot} \text{ has isolated G factors} \right\}.$$

It is the presence of isolated G factors which complicates the proof of (5.6), and necessitates the three step procedure which we soon describe.

For each $S \in \mathcal{E}(D)$ of the form (5.7), (even if $\ell = o$) we write

$$(5.8)\quad H_S(Y) = H_S(Y) \left[1_{\left\{|y^j_1 - y^{\bar{j}}_1| \leq 4n\epsilon\right\}} + 1_{\left\{|Y^j_1 - Y^{\bar{j}}_1| > 4n\epsilon\right\}} \right]$$

and expand the product in (5.3) into a sum of many terms. We work with one fixed term. We then say that $Y^j_{\cdot}$ and $Y^{\bar{j}}_{\cdot}$ are G–close or G–separated depending on whether the first or second characteristic function in (5.8) appears in our integral. If $Y^j, Y^{\bar{j}}$ never appear together in any $H_S(Y)$, then they are neither G–close nor G–separated. (This determination of G–close, etc. is fixed at the onset, and is not amended during the proof.)

For ease of reference we spell out two simple lemmas.)

Lemma 2: Let $g_i(Z) \geq 0$ be monotone decreasing in $|Z|$. If

$$(5.9)\qquad \int_{|Z| \geq \epsilon} \prod_{i=1}^{p} g_i(Z) d^n Z \leq M(\epsilon).$$

then for any $a_1, \ldots, a_p$

$$(5.10)\qquad \int_{\{|Z-a_i| \geq \epsilon, \forall_i\}} \prod_{i=1}^{p} g_i(Z-a_i) d^n Z \leq pM(\epsilon).$$

Proof: The integral in (5.10) is bounded by

$$\sum_{j=1}^{p} \int_{\left\{\begin{array}{l}|Z-a_i|\geq\epsilon,\forall_i\\ |Z-a_i|\geq|Z-a_j|,\forall_i\end{array}\right\}} \prod_{i=1}^{p} g_i(Z-a_i)d^nZ$$

$$\leq \sum_{j=1}^{p} \int_{|Z-a_j|\geq\epsilon} \prod_{i=1}^{p} g_i(Z-a_j)d^nZ \leq p\, M(\epsilon)$$

by (5.9). []

<u>**Lemma 3**</u>:

$$(5.11) \qquad \int_{|Y_1|\leq\epsilon} F_\epsilon(y.) \prod_{i=1}^{\ell} G_0(Y_{j_i}-a_i)dY_1\ldots dY_k \leq c\epsilon^{2-\ell(2-\beta)}$$

Proof: See the discussion about (3.11), (3.12). []

If S is of the form (5.7), and if $Y^j_\cdot, Y^{\bar{j}}_\cdot$ are G-separated we recall the bound of (2.13):

$$(5.12) \qquad |H_S(Y)| \leq c\, G_0^{\ell(s)+1}(Z)\left[\frac{\epsilon}{|Z|}\right]^\delta R(Z)$$

where $Z = Y_1^j - Y_1^{\bar{j}}$, and $0 \leq \delta \leq (\beta-1)\ell(s)$ is at our disposal.

Let

$$(5.13) \quad I_0 = \left\{i\epsilon I|Y^i \text{ is not G-close to any } Y^j,\ j\epsilon I\right\}$$

$$(5.14) \quad I_1 = I-I_0$$

We briefly outline our three steps, and then return to spell out the details. We integrate out one letter at a time, in a manner which allows us to keep track of potential problems.

__Step 1:__ We integrate out Y^i, $i\epsilon I_o$ using (5.12) when applicable.

__Step 2:__ We integrate out the letters from I_1, using (5.11) whenever possible.

__Step 3:__ We integrate the letters from I^c, i.e. letters without isolated G–factors. This is the most straightforward case.

Before spelling out the details, we can immediately recognize a potential problem. After integrating several letters, we may, inadevertently, have integrated out all G–factors containing some other letter, not yet integrated. Its integral might then diverge. To remedy this, before integrating each letter we carry out the following.

__Preservation Step:__ Before integrating Y., we search for any two letters, say X.,Z. with components which are separated only by components of Y. Thus we may have factors of the form

(5.15)

$$G\left[X-Y_i\right]G\left[Y_{i+1}-Y_i\right]\cdots G\left[Y_{i+\ell}-Y_{i+\ell-1}\right]\Delta^{\ell}_{y_{i+1},\dots,y_{i+\ell}}G(Y-Z)$$

(if (5.12) is not applied) or (if (5.12) is applied) of the form

$$(5.16)\quad G\left[X-Y_i\right]\ G_o^{\ell(s)+1}\left[Y_1-Z_1\right]\left[\frac{\epsilon}{|Y_1-Z_1|}\right]^{\delta}R\left[Y_1-Z_i\right]$$

(We include the case X. = 0, i = 1).

In the case of (5.15), we write out $\Delta^{\ell}G$ as a sum of many terms, focus on one of them, say

$$G\left[Y_i+y_{j_1}+y_{j_2}+\dots+y_{j_p}-Z\right]$$

From (5.15) we select the factors

$$(5.17)\quad G\left[X-Y_i\right]G\left[y_{j_1}\right]\cdots G\left[y_{j_p}\right]G\left[Y_i + y_{j_1} + \cdots + y_{j_p} - Z\right]$$

Now

$$(5.18)\quad |X-Z| \le \left|\left[X-Y_i\right]-y_{j_1}-y_{j_2}\cdots-y_{j_p}+\left[Y_i+y_{j_1}+\cdots+y_{j_p}-Z\right]\right|$$

$$\le |X-Y_i|+|y_{j_1}|+\cdots+|y_{j_p}|+|Y_i+y_{j_1}+\cdots+y_p-Z|$$

Hence $|X-Z|$ is less than $(p+2)$ times the maximum of the terms on the right hand side of (5.18). Hence one of the factors in (5.17) can be bounded by a constant times $G(X-Z)$.

If we have the form (5.16), then necessarily $|Y_1-Z_1| \ge 4n\epsilon$. If $|X-Z_1| \le 4n\epsilon$, then we can bound

$$W(Y_1-Z_1) \doteq G_o(Y_1-Z_1)\left[\frac{\epsilon}{|Y_1-Z_1|}\right]^{\delta} R(Y_1-Z_1) \text{ by } W(X-Z_1).$$

Note that $W(\cdot)$ is integrable. If $|X - Z_1| \ge 4n\epsilon$, then we use

$$(5.19)\quad |X - Z_1| \le |X - Y_i| + |Y_i - Y_1| + |Y_1 - Z_1|$$

$$\le |X - Y_i| + k\epsilon + |Y_1 - Z|$$

so that

$$(5.20)\qquad |X - Z_1| \le 2(|X- Y_i| + |Y_1 - Z_1|)$$

so that as before we can replace either the first factor in (5.16) by $G(X - Z_1)$, or a factor $W(Y_1 - Z_1)$ by $W(X - Z_1)$.

Note that this step actually lowers the number of G-factors involving $Y.$ prior to integrating $Y..$ After integrating $Y.$, we find that we have not increased the number of G-factors involved with $X.$, (or Z).

One way to think of this preservation step, is to suppress all $Y.$'s, and 'link up' with G or W the remaining letters which are now adjacent. (The case $X. = 0$ is included). The upshot is that we never lose any letters prior to their integration.

We finally remark that in (5.15), (5.16) we took our first factor to be $G(X. - Y_i)$. If this factor is actually $W(X. - Y_i)$ the same analysis pertains.

We now give the details of our three steps.

<u>Step 1:</u> We apply the bound (5.12) whenever S is of the form (5.7), with $j \in I_0$ having isolated G–intervals (i.e. $\ell(S) \neq 0$) and $|Y_1^j - Y_1^{\bar{j}}| \geq 4\epsilon n$. This is the only place we will apply (5.12). Note that (5.12) does not increase the total number of G–factors in our integral (we count both G_λ and G_0), but may increase the number of G factors containing $Y^i_\cdot$. Let N_i denote this latter quantity. I claim that

(5.12) $$\sum_{i\epsilon I_0} N_i \leq 2k|I_0|.$$

To see this, let $\ell(i)$ denote the number of isolated G–factors containing Y^i in the original integral, i.e., prior to applying the bound (5.12). At that stage $Y^i_\cdot$ could not have appeared in more than $2k-\ell(i)$ G–factors. The effect of (5.12) is to replace certain of the $\ell(i)$ isolated G–factors each of which had contributed 1 to N_i and zero to any N_j, $j \neq i$, by G–factors which contribute 1 to N_i and, at most, 1 to one other N_j. This proves (5.12)

If some $N_i \leq 2k-1$ then as in section 4 the dY^i integral is bounded. For, since $i \in I_0$, Y^i has isolated G–factors – hence, either it is close to some other letter, in which case lemma 3 shows the integral to be O(1), or else we will have applied (5.12), in which case lemma 2, with $\delta > o$ small, will show our integral to be O(1) as seen in section 4. (But remember, we always apply the preservation step prior to integrating !).

We proceed in this manner integrating all Y^i with $N_i \leq 2k-1$, (after each integration we update the remaining N_j's).

If all remaining $N_i \geq 2k$, then since (5.21) still holds, showing that now all $N_i = 2k$. The analysis of (5.21), in fact, shows that in such a case isolated G–factors containing such Y^i must be contained in factors $H_S(Y)$ containing a remaining Y^j, $j \in I_o$ and to which (5.12) has been applied; in particular, $|Y_1^i - Y_1^j| \geq 4n$. In such a case we check that $Y^i_{\cdot}, Y^j$ cannot be contained together in all 2k factors, hence Y^i must be contained in at least one factor with another letter, say $Y^{\tilde{j}}$. If the preservation step does not directly reduce the number of G–factors containing Y^i, then, since $|Y_1^j - Y_1^i| \geq 4n\epsilon$, we can still bound one factor by $W(Y^j_{\cdot} - Y^{\tilde{j}}_{\cdot})$, by using the same approach as in the preservation step, arguing separately for $|Y_1^j - Y_1^{\tilde{j}}| \leq 4n\epsilon$ or $> 4n\epsilon$.

In this manner we integrate out all letters Y^i, $i \in I_o$.

<u>Step 2:</u> I_1 is naturally partitioned into equivalence classes $Q_1, \ldots, Q_q$, where $i \sim j$ if we can find a sequence

$$i = i_1, i_2, i_3, \ldots, i_\ell = j$$

with Y^{i_p} G–close to $Y^{i_{p+1}}$.

Consider Q_1. Choose a $j \in Q_1$ such that $\ell(j) \leq \ell(i)$, $\forall i \in Q_1$. All Y^i, $i \in Q_1$, are close to Y^j in the sense that $|Y^i_1 - Y^j_1| \leq 4n^2\epsilon$. We then use lemma 3 to integrate, in any order, all $Y^i_\cdot$,
$i \in Q_1$, $i \neq j$. Since $Q_1 \subsetneq I$, we have $\ell(i) \geq 1$ so that the contribution from the $dY^i_\cdot$ integral is at most

$$O\left[\epsilon^{2 - (2k-\ell(i))(2-\beta)}\right] = O\left[\epsilon^{(\ell(i)-1)(2-\beta)}\right] \tag{5.22}$$

The $dY^j_\cdot$ integral, which is done last, is at most

$$O\left[\epsilon^{-\ell(j)(2-\beta)}\right], \tag{5.23}$$

from the $\ell(j) \geq 1$ isolated G–factors.

Combining (5.22) and (5.23) with $\ell(j) \geq \ell(i) \geq 1$, we see that the total contribution from Q_1 is $O(1)$ unless either $\ell(i) = 1$, $\forall\, i \in Q_1$ or if some $\ell(i) > 1$, then necessarily $Q_1 = \{i,j\}$ and $\ell(i) = \ell(j)$. In the former case we can also integrate out all $i \neq j$ except for one – so in both cases we can reduce ourselves to $Q_1 = \{i, j\}$, $\ell(i) = \ell(j) \geq 1$. We call such a pair a twin. Y^i, Y^j are close to each other, and we can assume they are close to no remaining letter (otherwise (5.23) can be improved to (5.22)). We leave such twins to step three.

We handle $Q_2, \dots, Q_q$ similarly.

Step 3: We begin with the remaining letter, say Y^i, which appears at the extreme right. Because of this, Y^i appears in $\leq 2k-1$ G–factors. If Y^i were part of a twin, then it has at most $2k - \ell(i) - 1$ G–factors, as opposed to the $2k - \ell(i)$ assumed for (5.22). This controls the twin.

If Y^i is not part of a twin, then $i \in I^c$. If Y^i

appears in 2k-1 G-factors with Y^j, then the analysis of section 4, shows that the $dY^i\, dY^j$ integral is at most $O(\lg(1/\epsilon))$.

It Y^i appears with 2 letters, we already know how to reduce the number of G-factors, so that the dY^i integral is bounded. We proceed in this manner until all letters are integrated.

This analysis shows that (5.6) holds unless $I = \phi$, and the rightmost letter has all G-factors in common with one other letter - but then these two letters form a component, contradicting the assumption that U(D) is connected of height > 2. This completes the proof of theorem 2.

6. Proof of Theorem 3

Taking over the notation of section 5, it suffices to show that if U(D) is connected and of height $n > 2$, then

$$I(D) = o(\epsilon^{-\alpha})^{n/2} \tag{6.1}$$

where $\alpha = (2k-1)(2-\beta)-2$.

The situation here is more complicated than that of Theorem 2, since typically our integrals diverge and we must control the divergence. We make two major modifications. In (5.12) we now take $\delta = 0$, and in applying the preservation step, or any other time we bound a factor such as G or W with factors not involving X in order to reduce the number of factors involving X to $\leq 2k-2$, we only bound G^γ, W^γ where γ is close to, but not equal to, one. This will not significantly affect the order of our X. integral – but when we come to integrate the other letters, a situation which would have led to $O(\epsilon^{-\alpha})$ with $\gamma = 1$ will now lead to $o(\epsilon^{-\alpha})$. These modifications will be taken for granted in what follows.

As in the last section, we will find that we can associate a factor $O(\epsilon^{-\alpha/2})$ with each letter, while at least one letter will be associated with $o(\epsilon^{-\alpha/2})$. By the remarks in the previous paragraph, and as detailed in the sequel, this will occur if any factors associated with our letter were obtained through a preservation like step.

We will assume that $(2k-2)(2-\beta) > 2$. The other cases are similar, but simpler.

<u>Step 1:</u> As in (5.21), we have

(6.2)
$$\sum_{i \in I_0} N_i \leq 2k \; |I_0|$$

where N_i are the number of G–factors involving $Y^i_\cdot$, after application of (5.12).

If $N_i < 2k-1$ for any i, the dY^i integral is $o\left[\epsilon^{-[(2k-2)(2-\beta)-2]}\right] = o\left[\epsilon^{-\alpha/2}\right]$, since our assumption

$$(2k-3)(2-\beta) < 2 \text{ implies } (2k-2)(2-\beta)-2 < (2-\beta).$$

Now assume $N_i = 2k-1$. If Y^i is linked to at least two other letters, then as in section 5, we can reduce the number of factors involving Y^i, and now the dY^i integral is $o(\epsilon^{-\alpha/2})$. If Y^i is linked to only one other letter, say Y^j, then $N_i = 2k-1$ is possible only if all Y^i, Y^j's are contiguous. (We note for later that Y^j can be in I^c or I_0 but not in $I - I_0$). The dY^i integral is $O(\epsilon^{-\alpha})$, while the dY^j integral will be bounded.

We can assume that all remaining $N_i \geq 2k$, so that by (6.2), we actually have $N_i = 2k$. We recall that this can occur only if (5.12) is applied with pairs in I_0. We leave this for the next step.

<u>Step 2:</u> We begin integrating from the right. Let X denote the rightmost remaining letter.

If $X \epsilon I^c$, it has no isolated factors, and being rightmost can appear in at most 2k–1 G–factors (the extra factors arising from (5.12) have either been integrated away, or involve only letters from I_0). If there were actually < 2k–1 G–factors, then the dX integral would be $o(\epsilon^{-\alpha/2})$. If X is linked to two distinct letters, we can reduce the number of factors as before, while if all 2k–1 links are to the same letter, say Y, then Y is necessarily in I^c, and the dX integral is $O(\epsilon^{-\alpha})$, with the dY integral bounded.

If, as we integrate, we find the rightmost letter $X = Y^i \epsilon I_0$, we can check that $N_i = 2k$ is no longer possible, and we return to the analysis of step 1.

Let us now suppose that the remaining rightmost letter

$$X \epsilon I - I_0.$$

Then $X \epsilon Q_i$ for some i, say i = 1. Assume first that X is within $4k^2\epsilon$ of some letter in Q_1^c (we include o), then automatically an analogous statement holds for all letters in Q_1. Before applying this we consider all Q_1 as one letter and apply the preservation step to Q_1^c. This way, we do not attempt to preserve letters of Q_1 itself. By the definition of Q_1, each letter has at least one isolated

G–factor, hence $\leq$ 2k–1 G–factors, while X, being rightmost, must have $\leq$ 2k–2. We begin by integrating dX, giving $o(\epsilon^{-\alpha/2})$. Again, by the definition of Q_1, X had a G–factor in common with at least one other letter of Q_1, hence that letter now has $\leq$ 2k–2 G–factors and we can integrate it, again giving a contribution $o(\epsilon^{-\alpha/2})$. At any stage in our successive integration of the letters of Q_1, it must be that some remaining letter has had on G–factor removed – since Q_1 was defined by an equivalence relation. This gives a contribution $o(\epsilon^{-\alpha/2})$ for each letter of Q_1.

Assume now that $X \in Q_1$ is not within $4k^2\epsilon$ of any letter in Q^c, so that in fact no letter of Q_1 is within $4k\epsilon$ of any letters of Q_1^c. If $|Q_1| \geq 3$, we integrate dX. We can use lemma 3 since X is close to the remaining letters of Q_1. Being the rightmost letter, its contribution is $o(\epsilon^{-\alpha/2})$. Prior to the dX integration we preserve all other letters, including $Q_1 - X$. Because of this, it is now possible that the remaining letters in Q_1 no longer form an equivalence class, but it will always be true that they are within $4k\epsilon$ of each other and of no letters in Q_1^c.

We continue in this fashion and can assume that X is in (an updated) Q_1, with $Q_1 = \{X,Y\}$. If $\ell(Y) \leq \ell(X)$, we do the dX integral using lemma 3 for a contribution $0\left[\epsilon^{2-(2k-\ell(X)-1)(2-\beta)}\right]$. When we reach Y, we have $\ell(Y)$ isolated G–factors contributing $0\left[\epsilon^{-\ell(Y)(2-\beta)}\right]$, and $\leq 2k - 2\ell(Y) - 1$ G–factors which give a convergent integral by lemma 2. Thus, the total contribution is $0(\epsilon^{-\alpha})$ if $\ell(Y) = \ell(X)$, and $o(\epsilon^{-\alpha})$ if in fact $\ell(Y) < \ell(X)$.

If, on the other hand $\ell(X) < \ell(Y)$, we first do the dY integral using lemma 3. Y has at most $2k-\ell(Y)$ G-factors. If in fact this is $\leq 2k-\ell(Y)-1 \leq 2k-\ell(X)-2$ then the dY integral is

$$O\left[\epsilon^{2-[2k-\ell(X)-2](2-\beta)}\right] = o(\epsilon^{-\alpha/2})\ O(\epsilon^{\ell(X)(2-\beta)})$$

and the dX integral is $O\left[\epsilon^{-\ell(X)(2-\beta)}\right]$ as above.

Otherwise, we preserve Q_1^c, then if Y still has $2k-\ell(Y)$ G-factors, we first assume that at least one of these G-factors links Y with some $Z \neq X$. We bound $G(Y-Z) \leq c\, G(X-Z)$, and after the dY integral there remain $\ell(X)$ isolated G-factors for X and $\leq 2k - 2\ell(X) \leq 2k - 2$ G-factors linking X with other letters. Thus the dX integral is bounded by $O\left[\epsilon^{-\ell(X)(2-\beta)}\right]o(\epsilon^{-\alpha/2})$ and altogether the dX dY integral is $o(\epsilon^{-\alpha})$.

If none of the $2k-\ell(Y)$ G-factors involving Y, involve any letters $Z \neq X$, then all non-isolated G-factors must link X and Y, in particular those factors to the immediate right and left. Since X occurs on the immediate left of Y, we needn't bother preserving it from the Y integration; which is

$$O\left[\epsilon^{2-(2k-\ell(Y))(2-\beta)}\right] = O\left[\epsilon^{2-(2k-1)(2-\beta)}\right]\ O\left[\epsilon^{(\ell(X)(2-\beta)}\right]$$
$$= O(\epsilon^{-\alpha})\ O\left[\epsilon^{\ell(X)(2-\beta)}\right]$$

and the contribution from dXdY is $O(\epsilon^{-\alpha})$.

In this manner we see that $I(D) = O(\epsilon^{-\alpha/2})^n$.

<u>Step 3:</u> we must now show that in fact

(6.3) $$I(D) = o(\epsilon^{-\alpha/2})^n$$

Let us agree to call two letters X,Y totally paired if there are no other letters between them. From the above analysis, we know that (6.3) holds unless D is such that all letters X fall into one of the following three types.

1) $X \in I^c$, and X is totally paired.

2) $X \in I_0$, and X totally paired. We recall that it cannot be paired with a letter from $I - I_0$.

3a) $X \in I - I_0$, and $X \in Q_i$, $|Q_i| = 2$. If, say $Q_1 = \{X,Y\}$, then necessarily X,Y are G-close, hence have at least one common G-factor, and by the above we know that $\ell(X) = \ell(Y)$ and X,Y are far (i.e. not within $4k\epsilon$) from Q_1^c.

3b) $Q_i = \{X,Y\}$ with X,Y totally paired.

Consider now Xhe very first letter on the right, X. X cannot be totally paired, since that would mean we have a component of height 2, contrary to our assumption that U(D) is connected of height ≥ 3. Thus X is of type 3a, say $X \in Q_1 = \{X,Y\}$.

Once again, Q_1 cannot be totally paired, hence, proceeding from the right there is a first letter, call it Z interrupting X,Y. Following Z there may be other letters from Q_1^c – we let W be the last of these prior to the next X or Y. (Of course, we can have $Z \equiv W$).

We begin by trying to preserve this W from Q_1. If this step removes a G-factor involving X or Y we break up the analysis into three cases.

a) If the removed G-factor contained X, then X now has $\leq 2k-\ell(X)-2$ G-factors, leading to an $o(\epsilon^{-\alpha/2})$ contribution as in step 2.

b) If the removed G-factor linked Y, but Z links X, then

bound $G(X-Z) \leq c\, G(Y-Z)$. Now preserve Q_1^c from Q_1. Once again X has $\leq 2k-\ell(X)-2$ factors, and while apriori Y has gained an extra G-factor, this gain is compensated by the loss of the G-factors which X,Y have in common. Note: we didn't have to preserve Y from the dX integration, because we have the factor $G(Y-Z)$.

c) If both the removed G-factor and Z link to Y, then bound $G(Z-Y) \leq c\, G(Z-X)$. Preserve Q_1^c from Q_1, and do the dY integral first, since Y now has $\leq 2k-\ell(Y)-2$ factors. (In fact, the gain of $G(Z-X)$ is compensated by the loss of a factor in common with Y). In any event the X,Y integral is $o(\epsilon^{-\alpha})$.

We can thus assume that our attempt to preserve the above W didn't remove any G-factors from X or Y. This can only happen if there is another W linked to X or Y to the left. We use step 2 to bound the X,Y integral by $O(\epsilon^{-\alpha})$, and now show that our resultant removal of two G-factors involving W will yield a proof of (6.3).

If W is of type 1), 2) or 3b) this is obvious, since they require total pairing without any loss of G-factors. Thus, W is of type 3a, hence part of a pair $Q_2 = \{U,W\}$. If W is to the right of U, the analysis of step 2 gives the desired result. Even of W is to the left of U, W has at most $2k-\ell(W)-2$ G-factors so that the dW integral is

$$O\left[\epsilon^{2-[2k-\ell(W)-2](2-\beta)}\right] = o(\epsilon^{-\alpha/2})\; O\left[\epsilon^{\ell(W)(2-\beta)}\right]$$

The dU integral has $\ell(U) = \ell(w)$ isolated integrals, and $\leq 2k-2\ell(W) \leq 2k-2$ others - hence the total dU, dW integral is $o(\epsilon^{-\alpha})$. This completes the proof of Theorem 3.

7. Proof of Lemma 1

Proof of lemma 1: We have

$$(7.1)\qquad G(x) = \int_0^\infty e^{-\lambda t} p_t(x)\,dt \le \int_0^\infty p_t(x) = G_o(x)$$

which gives half of (a). We note that

$$(7.2)\qquad p_t(x) = \frac{1}{(2\pi)^2}\int e^{ip\cdot x} e^{-tp^\beta} d^2p,\quad t > o$$

is a positive, C^∞ function of x, and

$$(7.3)\qquad p_t(x) \le ct^{-2/\beta}$$

If $|x| \neq 0$, say $x_1 \neq 0$, then integrating by parts in (7.2) in the dp_1 direction gives

$$(7.4)\qquad p_t(x) = \frac{-1}{(2\pi)^2}\,\frac{i}{x_1}\int e^{ip\cdot x} t\beta p_1 p^{\beta-2} e^{-tp^\beta} d^2p$$

Substituting this into (7.1) we have

$$(7.5)\ G(x) = \frac{-1}{(2\pi)^2}\,\frac{i}{x_1}\int_0^\infty e^{-\lambda t} dt\left[\int e^{ip\cdot x} t\beta p_1 p^{\beta-2} e^{-tp^\beta} d^2p)\right]$$

$$= \frac{c}{x_1}\int e^{ip\cdot x} p_1 p^{\beta-2} dp\left[\int_0^\infty e^{-\lambda t} t e^{-tp^\beta} dt\right]$$

$$= \frac{c}{x_1}\int e^{ip\cdot x}\,\frac{p_1 p^{\beta-2}}{(\lambda+p^\beta)^2} d^2p$$

where interchanging the order of integration is easily justified by Fubini's theorem since $\beta>1$.

We write (7.5) as

$$(7.6)\qquad G(x) = \frac{c}{x_1}\int e^{ip\cdot x}\, r_{\beta-1,\beta+1}(p)\, dp$$

where the notation $r_{a,b}(p)$ will remind us that

$$r_{a,b}(p) \leq \begin{cases} cp^a & , |p|\leq 1 \\ c\,\frac{1}{p^b} & , |p|\geq 1 \end{cases}$$

We integrate by parts twice more to find

$$\text{(7.7)} \qquad G(x) = \frac{c}{x_1^3} \int e^{ip\cdot x}\, r_{\beta-3,\beta+3}(p)\, d^2p$$

which completes the proof of (a), since $r_{\beta-3,\beta+3}(p)$ is integrable.

Furthermore, by (7.7)

$$\text{(7.8)} \qquad \nabla G(x) = \frac{c}{x_1^3} \int e^{ip\cdot x}\, \vec{p}\; r_{\beta-3,\beta+3}(p) d^2p$$

$$= \frac{c}{x_1^3} \int e^{ip\cdot x}\, r_{\beta-2,\beta+2}(p) d^2p$$

and we can integrate by parts once more to find

$$\text{7.9)} \qquad \nabla G(x) = \frac{c}{x_1^4} \int e^{ip\cdot x} r_{\beta-3,\beta+3}(p) d^2p.$$

This procedure can be iterated, and shows that

$$\text{(7.10)} \qquad |\nabla^\ell G(x)| \leq \frac{c}{x^{\ell+3}}$$

This will provide a good bound for large x. For small x, we recall (3.2):

$$\text{(7.11)} \qquad G(x) = G_o(x) - H(x).$$

Of course, we have

$$\text{(7.12)} \qquad |\nabla^\ell G_o(x)| \leq \frac{c}{x^{2-\beta+\ell}}$$

and we intend to show that

$$\text{(7.13)} \qquad |\Delta^\ell_{a_1,\dots,a_\ell} H(x)| \leq |a_1||a_2|\cdots|a_\ell|\, \frac{c}{x^\ell}$$

$$\text{for } |a_i| \leq \epsilon, \quad |X| \geq 4\ell\epsilon$$

Altogether, this will give, for $|X| \geq 4\ell\epsilon$

$$\text{(7.14)} \qquad |\Delta^\ell_{a_1,\dots,a_\ell} G(x)| \leq |a_1|\cdots|a_\ell|\, \frac{c}{X^{2-\beta+\ell}}$$

Combined with (7.10) we have

$$(7.15)\quad |\Delta^{\ell}_{a_1,\ldots,a_\ell} G(x)\,| \le \frac{|a_1|\cdots|a_\ell|}{x^\ell}\, r_{\beta-2,3}(x)$$

which is (2.11).

We note that $r_{\beta-2,3}(x)$ is integrable.

From (7.15) we have, for $|x| \ge 4\ell\epsilon$

$$(7.16)\qquad \sup_{|a_i| \le \epsilon} \prod_{i=1}^{\ell} G(a_i)\,|\Delta^{\ell}_{a_1,\ldots,a_\ell} G(x)|$$

$$\le c \sup_{|a_i|\le\epsilon} \prod_{i=1}^{\ell} \left[\frac{1}{a_i^{(2-\beta)}}\; \frac{|a_i|}{|x|}\right] r_{\beta-2,3}(x)$$

$$\le c \sup_{|a_i|\le\epsilon} \prod_{i=1}^{\ell} \left[\frac{|a_i|}{|x|}\right]^{\beta-1} G_0^\ell(x)\; r_{\beta-2,3}(x)$$

$$\le c\, G_0^\ell(x) \left[\frac{\epsilon}{|x|}\right]^{(\beta-1)\ell} r_{\beta-2,3}(x)$$

which is (2.12).

We now prove (7.13), (but we first remark that if $\beta > 3/2$, then $H(x)$ is C^1 and the following analysis can be simplified considerably).

$$H(x) = \frac{1}{(2\pi)^2} \int e^{ip\cdot x}\, \frac{\lambda}{p^\beta(\lambda+p^\beta)}\, d^2p.$$

so that

$$(7.17)\qquad \Delta_a H(x) = c\int e^{ip\cdot x}\, \frac{(e^{ip\cdot a}-1)}{p^\beta(\lambda+p^\beta)}\, d^2p$$

We integrate by parts in the dp_1 direction to find

$$(7.18)\quad \Delta_a\, H(x) = \frac{c}{x_1}\int e^{ip\cdot x}\, \frac{d}{dp_1}\left[\frac{e^{ip\cdot a}-1}{p^\beta(\lambda+p^\beta)}\right] d^2p$$

$$= c\,\frac{a_1}{x_1}\, H(x+a)$$

$$+ \frac{c}{x_1} \int e^{ip\cdot x}(e^{ip\cdot a}-1)\; r_{-\beta-1,2\beta+1}(p)\,d^2p$$

Since $|e^{ip\cdot a}-1| \leq 2\,|p||a|$ we obtain (7.13) for $\ell = 1$.

Write $F(x;a)$ for the integral in (7.18) so that

$$\Delta_a H(x) = c\,\frac{a_1}{x_1}\,H(x+a) + \frac{c}{x_1}\,F(x;a) \tag{7.19}$$

Then,

$$\Delta_b\Delta_a H(x) = c\,a_1\left[\Delta_b\left[\frac{1}{x_1}\right]\right]H(x+a) + c\,\frac{a_1}{x_1+b}\,\Delta_b\,H(x+a) + c\left[\Delta_b\left[\frac{1}{x_1}\right]\right]F(x;a) + \frac{c}{x_1+b}\,\Delta_b\,F(x;a) \tag{7.20}$$

We study the last term

$$\Delta_b F(x;a) = \int e^{ip\cdot x}(e^{ip\cdot b}-1)(e^{ip\cdot a}-1)r_{-\beta-1,2\beta+1}(p)d^2p \tag{7.21}$$

Integrating by parts gives us

$$\Delta_b F(x;a) = c\,\frac{b_1}{x_1}\,F(x+b;a) + c\,\frac{a_1}{x_1}\,F(x+a;b) + \frac{c}{x_1}\int e^{ip\cdot x}(e^{ip\cdot a}-1)(e^{ip\cdot b}-1)r_{-\beta-2,2\beta+2}(p)d^2p \tag{7.22}$$

and as before this establishes (7.13) for $\ell=2$. Iterating this procedure proves (7.13) for all ℓ, completing the proof of lemma 2.

REFERENCES

[1] Dynkin, E. [1988A] Self-Intersection Gauge for Random Walks and For Brownian Motion. Annals of Probab., Vol. 16, No. 1, 1988.

[2] Dynkin, E.B. [1988B] Regularized Self-intersection Local Times of the Planar Browninan Motion, Ann. Probab., Vol. 16, No. 1, 1988.

[3] LeGall, J.-F. [1988] Wiener Sausage and Self-Intersection Local Times. Preprint.

[4] Rosen, J. [1986] A Renormalized Local Time for the Multiple Intersections of Planar Brownian Motion. Seminaire de Probabilities XX, Spring Lecture Notes in Mathematics, 1204.

[5] Rosen [1988] - Limit Laws for the Intersection Local Time of Stable Processes in $\mathbb{R}^2$. Stochastics, Vol. 23, 219-240.

[6] Yor, M. [1985] Renormalisation et convergence en loi pour les temps locaux d'intersection du mouvement Brownien dans $\mathbb{R}^3$, Seminaire de Probabilites XIX, 1983/4, J. Azema, M. Yor, Eds., Springer-Verlag, Berlin-Heidelberg-New York-Tokyo, 350-356 (1985).

Jay S. Rosen*
Department of Mathematics
College of Staten Island
City University of New York
Staten Island, New York 10301

On Piecing Together Locally Defined Markov Processes

C.T. SHIH

Let E be a noncompact, locally compact separable metric space and let E_n be relatively compact open sets increasing to E. Suppose that for each n we are given a right process X_t^n on E_n and assume these processes are consistent, in the sense that X_t^{n+1} killed at the exit from E_n is a process that is (equivalent to) a time change of X_t^n, (equivalently, has identical hitting distributions as X_t^n). We consider the problem of constructing a right process Y_t on E such that for each n the process Y_t killed at the exit from E_n is a time change of X_t^n . The problem was posed in Glover and Mitro [3].

The problem is solved here under a technical condition, that any path of X_t^n must have finite lifetime if in X_t^{n+1} the corresponding time-changed path continues, i.e. still lives, after exiting from E_n. Also, we require the paths of each X_t^n to have left limits up to, but not including, their lifetime.

Actually what will be proved is somewhat more general. It is not required that the state spaces E_n be increasing, but only that they form an open covering of E (in this case the exit distributions of X_t^n will also be given); of course, the X_t^n must be consistent in an obvious way. The precise result is stated as the Main Theorem in section 1.

The problem of piecing together Markov processes that are *equivalent* on the common parts of their state spaces is treated in Courrège and Priouret [1] and

Meyer [4]; see the remark following theorem 1.1 below.

We remark that, with the result in this article, the theorem in [5] on construction of right processes from hitting distributions extends to the nontransient case; that is, the transience condition needs only to hold locally.

It is our pleasure to acknowledge very valuable discussions with Joe Glover on this work.

1. Statement and Proof of the Main Result

Let $E_\Delta = E \cup \{\Delta\}$ be the one-point compactification of a locally compact separable metric space E, and $\mathcal{E}_\Delta$ be its Borel σ-algebra. All (right) processes X_t considered in this article have E_Δ as the state space, with Δ as the usual adjoined death point, and have (almost) all paths right continuous, and with left limits on $(0, T_\Delta)$, where $T_\Delta = \inf\{t \geq 0: X_t = \Delta\}$. X_t is said to have an open set $G \subset E$ as its proper state space, and we usually say that X_t is a process on G, if each $x \in E - G$ is absorbing, i.e. $X_0 = x$ implies $X_t \equiv x$ a.s.; the time $T_{E_\Delta - G} = \inf\{t \geq 0: X_t \notin G\}$ is called its lifetime ζ. (We remark, however, that a proper subset G' of G can also be a proper state space of X_t. But no confusion will arise.)

Let X_t be a process on G, and let H be an open subset of G. We denote by $X_t|_H$ the process X_t stopped at the exit from H, i.e. the process $X(t \wedge T_{E_\Delta - H})$. So $X_t|_H$ is the process obtained from X_t by changing every $x \in G - H$ into an absorbing point.

Let X_t^1, X_t^2 be two processes. We write $X_t^1 = X_t^2$ if they are equivalent (in the usual sense), and write $X_t^1 \sim X_t^2$ if they are time changes of each other.

Main Theorem. Let $\{E_n, n \geq 1\}$ be an open covering of E with (compact) closures $\overline{E}_n \subset E$. For each n let X_t^n be a (right) process with E_n as its proper state space. Assume that the X_t^n satisfy the following consistency condition: for all $m \neq n$

$$X_t^m|_{E_m \cap E_n} \sim X_t^n|_{E_m \cap E_n} \,.$$

Then there exists a (right) process Y_t on E such that for all n

$$Y_t|_{E_n} \sim X_t^n .$$

Remark 1. Note it is assumed that we are given, for sets E_n, the *stopped* (rather than *killed*) process X_t^n at the exit from E_n, up to a time change, of a certain process on E. The stopped process contains a bit more information than the killed process, namely the exit distributions. Note also the requirement that if a path of the stopped process X_t^n reaches a point in $E - E_n$ at the exit time from E_n, then of course this time is finite. In the case that $E_n \uparrow E$, we need only to be given the killed processes to know the stopped processes, because the exit distributions of the stopped process X_t^n are the weak limits of the corresponding exit distributions from E_n of the killed processes X_t^m as $m \to \infty$. However the above mentioned condition of the exit time of X_t^n being finite if a path is to continue beyond this time (in X_t^{n+1}) is nevertheless a restriction. [This restriction is not a real one if the following conjecture is true: every right process, which may be partly transient and partly recurrent, can be time-changed so that the lifetime of almost every path is finite except possibly when the path left limit does not exist there.]

Remark 2. Another case where we know the exit distributions from the killed processes is when the X_t^n are diffusions. In general, of course, we need to be given the stopped processes (again, up to a time change) in order to be able to construct the Y_t.

Remark 3. The theorem covers the case when E is compact (where Δ is an isolated point). This is the case, for example, for a Brownian motion or diffusion on a circle or sphere.

Remark 4. If E is noncompact, the process Y_t is not necessarily unique (unique up to a time change). The process Y_t we will obtain is minimal in the sense that, with $T_n = \inf\{t \geq 0 : Y_t \notin E_1 \cup \ldots \cup E_n\}$, $\lim_n T_n$ is its lifetime.

Remark 5. In the case where $E_n \uparrow E$, the proof is relatively short; see corollary 1.4. Actually it can be proved without theorems 1.1 and 1.2; see the remark after theorem 1.3.

Theorem 1.1. For $i = 1,2$ let Z_t^i be a (right) process on an open set G_i and assume $Z_t^1|_{G_1 \cap G_2} = Z_t^2|_{G_1 \cap G_2}$. Then there exists a (right) process $\tilde{Z}_t$ on $G_1 \cup G_2$ such that $\tilde{Z}_t|_{G_i} = Z_t^i$ for both i.

A proof of theorem 1.1 can be found in Meyer [4], which derives a certain general result and uses it to prove among other things (a variation of) the theorem of Courrège and Priouret [1] on piecing together Markov processes that are equivalent on the common parts of their state spaces. For completeness we include, in section 2, a proof, which is somewhat different from the one in [4]. The reference [4] was pointed out to us by Pat Fitzsimmons.

The process $\tilde{Z}_t$ in theorem 1.1 is not necessarily unique; however, we have the following uniqueness result, which is needed later.

Theorem 1.2. Let G_i, Z_t^i and $\tilde{Z}_t$ be as in theorem 1.1. Let G_3 be open with $\overline{G}_3 \subset G_2$. Then if F is open with $F \subset G_1 \cup G_3$ and $\hat{Z}_t$ is a (right) process on F such that $\hat{Z}_t|_{F \cap G_i} \sim Z_t^i|_{F \cap G_i}$ for $i = 1,2$, we have $\hat{Z}_t \sim \tilde{Z}_t|_F$.

This will be proved at the end of section 2.

Theorem 1.3. For $i = 1,2$ let W_t^i be a (right) process on an open set H_i with $H_1 \subset H_2$. Suppose that $W_t^2|_{H_1} \sim W_t^1$. Then for any open H with $\overline{H} \subset H_1$ there exists a (right) process Z_t on H_2 such that $Z_t \sim W_t^2$ and $Z_t|_H = W_t^1|_H$.

Proof. Let $W_t = W_t^2|_{H_1} = W^2(t \wedge T_{E_\Delta - H_1})$; then $W_t \sim W_t^1$. Let A_t be a (strictly increasing continuous) additive functional whose inverse time-changes W_t into W_t^1. Define

$$B_t = \int_0^t 1_H(W_s)dA_s + \int_0^t 1_{E_\Delta - H}(W_s)ds \ .$$

B_t is a well-defined strictly increasing continuous additive functional in W_t. Denote by Z_t^1 the time-changed process from W_t by the inverse of B_t. Clearly

$Z_t^1|_H = W_t^1|_H$. Let $G_1 = H_1$, $G_2 = H_2 - \overline{H}$ and $Z_t^2 = W_t^2|_{G_2}$. Then Z_t^1, Z_t^2 satisfy the conditions of theorem 1.1. Denote by Z_t the process $\tilde{Z}_t$ in theorem 1.1. Thus Z_t is a process on $G_1 \cup G_2 = H_2$, and $Z_t|_{G_1} = Z_t^1$, $Z_t|_{G_2} = Z_t^2 = W_t^2|_{G_2}$. The first of these equivalences implies $Z_t|_H = W_t^1|_H$. To show $Z_t \sim W_t^2$, let G_3 be open with $H_2 - H_1 \subset \overline{G}_3 \subset G_2$. Note $G_1 \cup G_3 = H_2$. Applying theorem 1.2 to $\hat{Z}_t = W_t^2$ and $F = H_2 = G_1 \cup G_3$ we have $W_t^2 \sim Z_t$. ■

Remark. Theorem 1.3 can be proved directly, i.e. without using theorems 1.2 and 1.3, as follows. Define stopping times T_n in W_t^2 by: $T_0 = 0$, and for $n \geq 1$

$$T_{2n-1} = \inf\{t \geq T_{2n-2}: W_t^2 \notin H_1\}\ ,\ T_{2n} = \inf\{t \geq T_{2n-1}: W_t^2 \in \overline{H}\}\ .$$

The fact that paths have left limits on $(0, T_\Delta)$ implies $T_n \uparrow T_\Delta$. Let B_t be as in the above proof, which is defined in W_t^2. Define in W_t^2

$$\begin{aligned} C_t &= \sum_{n=0}^{\infty} B_{t\wedge T_{2n+1} - t\wedge T_{2n}} \circ \theta(T_{2n}) + \sum_{n=1}^{\infty} \int_{t\wedge T_{2n-1}}^{t\wedge T_{2n}} 1_{E_\Delta - H}(W_s^2)ds \\ &= \sum_{n=0}^{\infty} \left(\int_0^{t\wedge T_{2n+1} - t\wedge T_{2n}} 1_H(W_s^2)dA_s \right) \circ \theta(T_{2n}) + \int_0^t 1_{E_\Delta - H}(W_s^2)ds \end{aligned}$$

where θ denotes the shift operator. It is not difficult to rigorously show that C_t is a strictly increasing continuous additive functional. The time-changed process Z_t from W_t^2 by the inverse of C_t then satisfies theorem 1.3. Note that based on this, corollary 1.4 (which establishes the special case of the main theorem where $E_n \uparrow E$) does not have to rely on theorems 1.1 and 1.2, as its proof uses only theorem 1.3.

Corollary 1.4. Let E_n be relatively compact open sets with $E_n \uparrow E$. For each n we are given a (right) process X_t^n on E_n such that $X_t^{n+1}|_{E_n} \sim X_t^n$. Then there exists a (right) process Y_t on E such that $Y_t|_{E_n} \sim X_t^n$ for all n.

Proof. Choose open sets E_n' with $\overline{E_n'} \subset E_n$ and $E_n' \uparrow E$. We will define processes Y_t^n on E_n such that $Y_t^{n+1}|_{E_n'} = Y_t^n|_{E_n'}$ and $Y_t^n \sim X_t^n$. The sequence of processes $Y_t^n|_{E_n'}$ then admits a projective limit process Y_t on E staisfying $Y_t|_{E_n'} = Y_t^n|_{E_n'}$ for all n. The property $Y_t|_{E_m} \sim X_t^m$ will follow because if $E_m \subset E_n'$, $Y_t|_{E_m} = Y_t^n|_{E_m} \sim X_t^n|_{E_m} \sim X_t^m$. To define the sequence Y_t^n, first let $Y_t^1 = X_t^1$, and apply theorem 1.3 with $H_1 = E_1$, $H_2 = E_2$, $H = E_1'$, $W_t^1 = Y_t^1$ and $W_t^2 = X_t^2$ to get a process Y_t^2 (which is the Z_t in the theorem) on $H_2 = E_2$ satisfying $Y_t^2|_{E_1'} = Y_t^1|_{E_1'}$ and $Y_t^2 \sim X_t^2$. In general, assuming that we have obtained a process Y_t^n on E_n satisfying $Y_t^n|_{E_{n-1}'} = Y_t^{n-1}|_{E_{n-1}'}$ and $Y_t^n \sim X_t^n$, apply theorem 1.3 with $H_1 = E_n$, $H_2 = E_{n+1}$, $H = E_n'$, $W_t^1 = Y_t^n$ and $W_t^2 = X_t^{n+1}$ to get a process Y_t^{n+1} on E_{n+1} satisfying $Y_t^{n+1}|_{E_n'} = Y_t^n|_{E_n'}$ and $Y_t^{n+1} \sim X_t^{n+1}$. The existence of the sequence Y_t^n thus follows from induction. ■

Theorem 1.5. Let J_1, J_2, J_3 be open sets with $\overline{J}_3 \subset J_2$. For $i = 1, 2$ let V_t^i be a (right) process on J_i such that $V_t^1|_{J_1 \cap J_2} \sim V_t^2|_{J_1 \cap J_2}$. Then i) there exists a (right) process V_t on $J_1 \cup J_3$ such that $V_t|_{J_1} = V_t^1$ and $V_t|_{J_3} \sim V_t^2|_{J_3}$; ii) if F is open with $F \subset J_1 \cup J_3$ and $\hat{V}_t$ is a (right) process on F such that $\hat{V}_t|_{F \cap J_i} \sim V_t^i|_{F \cap J_i}$ for $i = 1, 2$ then $\hat{V}_t \sim V_t|_F$.

Proof. Let J_4 be open with $\overline{J}_3 \subset J_4 \subset \overline{J}_4 \subset J_2$. Applying theorem 1.3 with $H_1 = J_1 \cap J_4$ and $H_2 = J_2$, $W_t^1 = V_t^1|_{J_1 \cap J_4}$, $W_t^2 = V_t^2$ we obtain a process Z_t on J_2 satisfying $Z_t \sim V_t^2$ and $Z_t|_{J_1 \cap J_4} = W_t^1 = V_t^1|_{J_1 \cap J_4}$. Next use theorem 1.1 with $G_1 = J_1, G_2 = J_4, Z_t^1 = V_t^1$ and $Z_t^2 = Z_t|_{J_4}$ to obtain a process $\tilde{Z}_t$ on $J_1 \cup J_4$ such that $\tilde{Z}_t|_{J_1} = V_t^1$ and $\tilde{Z}_t|_{J_4} = Z_t|_{J_4}$, the latter equivalence implying $\tilde{Z}_t|_{J_4} \sim V_t^2|_{J_4}$. Let $V_t = \tilde{Z}_t|_{J_1 \cup J_3}$. The V_t satisfies i). ii) follows from theorem 1.2 with G_i, Z_t^i as above, $G_3 = J_3$, and $\hat{Z}_t = \hat{V}_t$. ■

Proof of Main Theorem. Let $\{G_n, n \geq 1\}$ be an open covering of E with $G_1 = E_1$ and $\overline{G}_n \subset E_n$ for $n \geq 2$. We will define for each n a process Y_t^n on $F_n = G_1 \cup ... \cup G_n$ such that $Y_t^{n+1}|_{F_n} = Y_t^n$. The process Y_t will be the projective limit of the sequence Y_t^n, which satisfies $Y_t|_{F_n} = Y_t^n$ for all n and has lifetime $\lim_n T_{E_\Delta - F_n}$. Let $Y_t^1 = X_t^1$. Applying theorem 1.5 with $J_1 = F_1 = G_1 =$

$E_1, J_2 = E_2, J_3 = G_2$, and $V_t^1 = Y_t^1, V_t^2 = X_t^2$ we obtain a process Y_t^2 (which is the V_t in the theorem) on $J_1 \cup J_3 = F_1 \cup G_2 = F_2$ such that i) $Y_t^2|_{F_1} = Y_t^1$ and $Y_t^2|_{G_2} \sim X_t^2|_{G_2}$, and ii) if F is open with $F \subset F_1 \cup G_2 = F_2$ and $\hat{V}_t$ is a process on F with $\hat{V}_t|_{F\cap F_1} \sim Y_t^1|_{F\cap F_1}$ and $\hat{V}_t|_{F_1\cap G_2} \sim X_t^2|_{F\cap G_2}$, then $\hat{V}_t \sim Y_t^2|_F$. Using ii) with $F = E_3 \cap F_2$ and $\hat{V}_t = X_t^3|_F$, (note $X_t^3|_{E_3\cap F_1} \sim X_t^1|_{E_3\cap F_1} = Y_t^1|_{E_3\cap F_1}$ and $X_t^3|_{E_3\cap G_2} \sim X_t^2|_{E_3\cap G_2}$), we get $Y_t^2|_{E_3\cap F_2} \sim X_t^3|_{E_3\cap F_2}$. This permits us to apply theorem 1.5 to $J_1 = F_2, J_2 = E_3, J_3 = G_3$, and $V_t^1 = Y_t^2, V_t^2 = X_t^3$ to obtain Y_t^3. In general suppose Y_t^n is obtained as a process on $F_n = F_{n-1} \cup G_n = G_1 \cup ... \cup G_n$ such that i) $Y_t^n|_{F_{n-1}} = Y_t^{n-1}$ and $Y_t^n|_{G_n} \sim X_t^n|_{G_n}$, and ii) if F is open with $F \subset F_{n-1} \cup G_n = F_n$ and $\hat{V}_t$ is a process on F with $\hat{V}_t|_{F\cap F_{n-1}} \sim Y_t^{n-1}|_{F\cap F_{n-1}}$ and $\hat{V}_t|_{F\cap G_n} \sim X_t^n|_{F\cap G_n}$, then $\hat{V}_t \sim Y_t^n|_F$. Using ii) with $F = E_{n+1} \cap F_n$ and $\hat{V}_t = X_t^{n+1}|_F$ (and an appropriate induction) we have $X_t^{n+1}|_{E_{n+1}\cap F_n} \sim Y_t^n|_{E_{n+1}\cap F}$. Now applying theorem 1.5 with $J_1 = F_n, J_2 = E_{n+1}, J_3 = G_{n+1}$, and $V_t^1 = Y_t^n, V_t^2 = X_t^{n+1}$ we obtain Y_t^{n+1} on $F_{n+1} = F_n \cup G_{n+1}$ satisfying the corresponding i) and ii). Thus the existence of the sequence Y_t^n follows from induction. Finally we need to show that the projective limit process Y_t satisfies $Y_t|_{E_m} \sim X_t^m$. Choose n with $E_m \subset F_n$; then $Y_t|_{F_n} \sim Y_t^n$ implies $Y_t|_{E_m} \sim Y_t^n|_{E_m}$. But $Y_t^n|_{E_m} \sim X_t^m$, which follows by applying condition ii) of Y_t^n with $F = E_m, \hat{V}_t = X_t^m$, and using an appropriate induction on n. ■

2. Proofs of Theorems 1.1 and 1.2

To prove theorem 1.1, let Ω be the space of all right continuous functions from $[0, \infty)$ into E_Δ. Ω can serve as the sample space of both Z_t^i. Of course $Z_t^i(\omega) = \omega_t$. Let $P_i^x, x \in E_\Delta$, be the probability measure governing Z_t^i when it starts at x. Define

$$P^x = P_i^x \text{ if } x \in G_1 \ ; \ = P_2^x \text{ if } x \in G_2 - G_1 \ ;$$

$$= P_1^x = P_2^x = \text{ point mass at the } \omega \text{ with } \omega_t \equiv x \text{ if } x \in E_\Delta - G_1 \cup G_2 \ .$$

Let $Z_t(\omega) = \omega_t = Z_t^i(\omega)$. With $\zeta_i = T_{E_\Delta - G_i} = \inf\{t \geq 0 \colon Z_t^i \notin G_i\}$, the lifetime of Z_t^i, let

$$\zeta = \zeta_1 \text{ if } Z_0 \in G_1 \ ; \ = \zeta_2 \text{ if } Z_0 \in G_2 - G_1 \ ;$$
$$= \zeta_1 = \zeta_2 = 0 \text{ if } Z_0 \in E_\Delta - G_1 \cup G_2 \ .$$

Now set

$$Q(\omega, d\omega') = P^{Z(\zeta(\omega))}(d\omega')$$

(note $Z_\infty \equiv \Delta$ by convention). Q is a (transition) kernel in $(Q, \mathcal{F})$ where $\mathcal{F}$ is the usual completion of $\sigma(Z_t, t \geq 0)$ w.r.t. the measures $P^\mu = \int \mu(dx) P^x$. Next define

$$\tilde{\Omega} = \Omega \times \ldots \times \Omega \times \ldots, \quad \tilde{\mathcal{F}} = \mathcal{F} \times \ldots \times \mathcal{F} \times \ldots$$

and let $\tilde{P}^x, x \in E_\Delta$, be the probability measure on $(\tilde{\Omega}, \tilde{\mathcal{F}})$ satisfying

$$\begin{aligned}\tilde{P}^x\{(\omega_1, \ldots, \omega_n, \ldots)&: \omega_k \in \Lambda_k, 1 \leq k \leq n\} \\ &= \int P^x(d\omega_1) \int Q(\omega_1, d\omega_2) \int \ldots \\ &\ldots \int Q(\omega_{n-1}, d\omega_n) 1_{\Lambda_1 \times \ldots \times \Lambda_n}(\omega_1, \ldots, \omega_n) \ .\end{aligned}$$

With $\tilde{\omega} = (\omega_1, \ldots, \omega_n, \ldots)$ let

$$T_0(\tilde{\omega}) = 0 \ ; \ T_n(\tilde{\omega}) = \zeta(\omega_1) + \ldots + \zeta(\omega_n) \ , \ n \geq 1 \ ; \ \tilde{\zeta}(\omega) = \lim_n T_n(\tilde{\omega}) \ .$$

Finally define

$$\begin{aligned}\tilde{Z}_t(\tilde{\omega}) &= Z_{t - T_{n-1}(\tilde{\omega})}(\omega_n) \quad \text{if } T_{n-1}(\tilde{\omega}) \leq t < T_n(\tilde{\omega}) \\ &= \Delta \quad \text{if } t \geq \tilde{\zeta}(\tilde{\omega}) > T_n(\tilde{\omega}) \text{ for all } n \\ &= Z_\zeta(\omega_n) \quad \text{if } t \geq \tilde{\zeta}(\tilde{\omega}) = T_n(\tilde{\omega}) \text{ for some } n \geq 1 \ .\end{aligned}$$

By the construction we have an obvious Markov property of $\tilde{Z}_t$ at the times T_n, which reflects the Markov property of the discrete time process $\tilde{\omega} \to \omega_n$ on $(\Omega, \mathcal{F})$; this will be used below.

In order to show that $\tilde{Z}_t$ is a right process on $G_1 \cup G_2$, define for $\alpha > 0, f \in$

$b\mathcal{E}_\Delta$

$$U_i^\alpha f(x) = P_i^x \int_0^{\zeta_i} e^{-\alpha t} f(Z_t^i)dt \ ,$$

$$U^\alpha f(x) = P^x \int_0^{\zeta} e^{-\alpha t} f(Z_t)dt$$
$$= U_1^\alpha f(x) \text{ if } x \in G_1 \ ; \ = U_2^\alpha f(x) \text{ if } x \in G_2 - G_1 \ ; \ = 0 \text{ otherwise} \ ,$$

$$\tilde{U}^\alpha f(x) = \tilde{P}^x \int_0^{\tilde{\zeta}} e^{-\alpha t} f(\tilde{Z}_t)dt \ .$$

The Markov property of $\tilde{Z}_t$ at the time T_1 yields immediately

Lemma 2.1. For $x \in E_\Delta, \alpha > 0, f \in b\mathcal{E}_\Delta$

$$\tilde{U}^\alpha f(x) = \tilde{P}^x \int_0^{T_1} e^{-\alpha t} f(\tilde{Z}_t)dt + \tilde{P}^x e^{-\alpha T_1} \tilde{U}^\alpha f(\tilde{Z}_{T_1})$$
$$= U^\alpha f(x) + P^x e^{-\alpha\zeta} \tilde{U}^\alpha f(Z_\zeta)$$

Lemma 2.2. For $y \in G_1 \cap G_2, \alpha > 0, f \in b\mathcal{E}_\Delta$

$$\tilde{U}^\alpha f(y) = P_2^y \int_0^{\zeta_2} e^{-\alpha t} f(Z_t^2)dt + P_2^y e^{-\alpha\zeta_2} \tilde{U}^\alpha f(Z^2(\zeta_2)) \ .$$

Proof. Define $R = \inf\{t \geq 0 \colon Z_t \notin G_1 \cap G_2\}, \tilde{R} = \inf\{t \geq 0 \colon \tilde{Z}_t \notin G_1 \cap G_2\}$. Then

$$\tilde{U}^\alpha f(y) = \tilde{P}^y \int_0^{\tilde{R}} e^{-\alpha t} f(\tilde{Z}_t)dt + \tilde{P}^y [\int_{\tilde{R}}^{\tilde{\zeta}} e^{-\alpha t} f(\tilde{Z}_t)dt \ ; \ \tilde{Z}_{\tilde{R}} \in G_1 - G_2]$$
$$+ \tilde{P}^y [e^{-\alpha\tilde{R}} \tilde{U}^\alpha f(\tilde{Z}_{\tilde{R}}) \ ; \ \tilde{Z}_{\tilde{R}} \in G_2 - G_1]$$
$$= P_1^y \int_0^{R} e^{-\alpha t} f(Z_t^1)dt + \tilde{P}^y [e^{-\alpha\tilde{R}} \tilde{U}^\alpha f(\tilde{Z}_{\tilde{R}}) \ ; \ \tilde{Z}_{\tilde{R}} \in G_1 - G_2]$$
$$+ P_1^y [e^{-\alpha R} \tilde{U}^\alpha f(Z_R^1) \ ; \ Z_R^1 \in G_2 - G_1] \ ,$$

using the fact $P^y = P_1^y$ for $y \in G_1$ on the 1st and 3rd terms; and for the 2nd term, combining the Markov property of Z_t^1 at R with that of $\tilde{Z}_t$ at T_1. Since $Z_t^1|_{G_1 \cap G_2} = Z_t^2|_{G_1 \cap G_2}$ and since $P^z = P_2^z$ for $z \in G_2 - G_1$, the above

$$= P_2^y \int_0^{R} e^{-\alpha t} f(Z_t^2)dt + P_2^y [e^{-\alpha R} \tilde{U}^\alpha f(Z_R^2) \ ; \ Z_R^2 \in G_1 - G_2]$$
$$+ P_2^y [e^{-\alpha R} (U_2^\alpha f(Z_R^2) + P_2^{Z^2(R)} e^{-\alpha\zeta_2} \tilde{U}^\alpha f(Z_{\zeta_2}^2)) \ ; \ Z_R^2 \in G_2 - G_1]$$
$$= I + II + (III + IV) \ ,$$

where we have use Lemma 2.1 to obtain the third term. Now

$$I + III = P_2^y \int_0^{\zeta_2} e^{-\alpha t} f(Z_t^2)dt \ , \ II + IV = P_2^y e^{-\alpha\zeta_2} \tilde{U}^\alpha f(Z_{\zeta_2}^2)$$

completing the proof. ∎

Lemma 2.3. Let $x \in E_\Delta, s \geq 0, \tilde{\Lambda}$ be of the form $\tilde{\Lambda} = \{\tilde{Z}_{s_j} \in B_j, 1 \leq j \leq k;\ s < T_1\}$ where $0 \leq s_1 < ... < s_k \leq s$. Then for $\alpha > 0, f \in b\mathcal{E}_\Delta$

$$\tilde{P}^x[\int_0^{\tilde{\zeta}} e^{-\alpha t} f(\tilde{Z}_{s+t})dt;\ \tilde{\Lambda}] = \tilde{P}^x[\tilde{U}^\alpha f(\tilde{Z}_s);\ \tilde{\Lambda}] \ . \tag{2.1}$$

Proof. We need only to prove this for $x \in G_1 \cup G_2$. By the Markov property of $\tilde{Z}_t$ at T_1, the left-hand-side of (2.1) equals

$$P^x[\int_0^{\zeta} e^{-\alpha t} f(Z_{s+t})dt + e^{-\alpha(\zeta - s)}\tilde{U}^\alpha f(Z_\zeta);\ \Lambda] \tag{2.2}$$

where $\Lambda = \{Z_{s_j} \in B_j, 1 \leq j \leq k; s < \zeta\}$. The right-hand-side of (2.1) is $P^x[\tilde{U}^\alpha f(Z_s); \Lambda]$. If $x \in G_1$, applying the Markov property of Z_t^1 at the time s and lemma 2.1 we have that this last expression equals (2.2). If $x \in G_2 - G_1$, write this expression as

$$P_2^x[\tilde{U}^\alpha f(Z_s^2); \Lambda, Z_s^2 \in G_2 - G_1] + P_2^x[\tilde{U}^\alpha f(Z_s^2); \Lambda, Z_s^2 \in G_1 \cap G_2] \ .$$

Apply the Markov property of Z_t^2 at time s, and use lemma 2.1 on the first term above and lemma 2.2 on the second term, to obtain (2.2). ∎

Lemma 2.4. $(\tilde{Z}_t, \tilde{P}^x)$ is simple Markov.

Proof. Let $x \in G_1 \cup G_2, u \geq 0$ and $\tilde{\Gamma}$ be of the form $\tilde{\Gamma} = \{\tilde{Z}_{u_j} \in A_j,\ 1 \leq j \leq m\}$ where $0 \leq u_1 < ... < u_m \leq u$. We need to show tht for $\alpha > 0, f \in b\mathcal{E}_\Delta$

$$\tilde{P}^x[\int_0^{\tilde{\zeta}} e^{-\alpha t} f(\tilde{Z}_{u+t})dt;\ \tilde{\Gamma}] = \tilde{P}^x[\tilde{U}^\alpha f(\tilde{Z}_u);\ \Gamma] \ . \tag{2.3}$$

Let $\tilde{\Gamma}_{nl} = \tilde{\Gamma} \cap \{u_{l-1} < T_n \le u_l \le u < T_{n+1}\}$, where u_0 stands for -1. Then using the Markov property of $\tilde{Z}_t$ at T_n we have

$$\begin{aligned}
&\tilde{P}^x[\int_0^{\tilde{\zeta}} e^{-\alpha t} f(\tilde{Z}_{u+t})dt;\ \tilde{\Gamma}_{nl}] \\
&= \tilde{P}^x[\tilde{P}^{\tilde{Z}(T_n(\tilde{\omega}))}\{\int_0^{\tilde{\zeta}} e^{-\alpha t} f(\tilde{Z}_{u-T_n(\tilde{\omega})+t})dt\ ; \\
&\quad \tilde{Z}_{u_j - T_n(\tilde{\omega})} \in A_j, l \le j \le m, u - T_n(\tilde{\omega}) < T_1\}\ ; \\
&\quad \tilde{Z}_{u_j}(\tilde{\omega}) \in A_j, 1 \le j < l, u_{l-1} < T_n(\tilde{\omega}) \le u_l]
\end{aligned}$$

(where the inner integrand is a function of $\tilde{\omega}'$). Apply lemma 2.3 with $x = \tilde{Z}(T_n(\tilde{\omega}))$ to reduce the above to

$$\begin{aligned}
&\tilde{P}^x[\tilde{P}^{\tilde{Z}(T_n(\tilde{\omega}))}\{\tilde{U}^\alpha f(\tilde{Z}_{u-T_n(\tilde{\omega})});\ \tilde{Z}_{u_j - T_n(\tilde{\omega})} \in A_j, l \le j \le m, u - T_n(\tilde{\omega}) < T_1\}\ ; \\
&\quad \tilde{Z}_{u_j}(\tilde{\omega}) \in A_j, 1 \le j < l, u_{l-1} < T_n(\tilde{\omega}) \le u_l]\ ,
\end{aligned}$$

which by the Markov property of $\tilde{Z}_t$ at T_n equals

$$\tilde{P}^x[\tilde{U}^\alpha f(\tilde{Z}_u);\ \tilde{\Gamma}_{nl}]\ .$$

Summing over n, l we obtain (2.3). ∎

Proof of theorem 1.1. By lemma 2.4 and a standard theorem, to complete the proof that $(\tilde{Z}_t, \tilde{P}^x)$ is a right process it suffices to show that for $\alpha > 0, f \in b\mathcal{E}_\Delta$ with $f \ge 0, t \to \tilde{U}^\alpha f(\tilde{Z}_t)$ is right continuous a.s. $\tilde{P}^x$ for all x. Using the Markov property of $\tilde{Z}_t$ at times T_n, it suffices to show $t \to \tilde{U}^\alpha f(\tilde{Z}_t)$ is right continuous on $[0, T_1)$ a.s. $\tilde{P}^y$ for all y, i.e. $t \to \tilde{U}^\alpha f(Z_t)$ is right continuous on $[0, \zeta)$ a.s. P^y for all y. By lemmas 2.1 and 2.2

$$\tilde{U}^\alpha f(z) = U_i^\alpha f(z) + P_i^z e^{-\alpha\zeta_i} \tilde{U}^\alpha f(Z^i_{\zeta_i}), \quad z \in G_i$$

for both $i = 1, 2$. The right-hand-side is obviously α-exessive w.r.t. Z_t^i, and so a.s. $P_i^y, t \to \tilde{U}^\alpha f(Z_t^i)$ is right continuous on $[0, \zeta_i)$. Thus $t \to \tilde{U}^\alpha f(Z_t)$ is right continuous on $[0, \zeta)$ a.s. P^y for all y. Finally, it remains to show

$$\tilde{Z}_t|_{G_i} = Z_t^i$$

for both i. But this is immediate from construction and lemma 2.2. ∎

Proof of theorem 1.2. Denote $\dot{Z}_t = \tilde{Z}_t|_F$. We show that $\hat{Z}_t$ and $\dot{Z}_t$ have identical hitting distributions; thus by the Blumenthal-Getoor-McKean theorem one has $\hat{Z}_t \sim \dot{Z}_t$. (For a modern version of the B-G-M theorem, see [2].) Let D be a compact set in E and $T_D = \inf\{t \geq 0 \colon \hat{Z}_t \in D\}$ or $\inf\{t \geq 0 \colon \dot{Z}_t \in D\}$. We must show that for all $x, \hat{P}^x(\hat{Z}(T_D) \in \cdot) = \dot{P}^x(\dot{Z}(T_D) \in \cdot)$. Define stopping times S_n in $\hat{Z}_t$ by: $S_0 = 0$ and

$$\begin{aligned} S_{n+1} &= \inf\{t \geq S_n \colon \hat{Z}_t \in D \cup (G_1 \cap F)^c\} \text{ if } \hat{Z}_{S_n} \in G_1 \cap F \\ &= \inf\{t \geq S_n \colon \hat{Z}_t \in D \cup (G_2 \cap F)^c\} \text{ if } \hat{Z}_{S_n} \in G_3 \cap F \\ &= \inf\{t \geq S_n \colon \hat{Z}_t \in D\} \text{ otherwise.} \end{aligned}$$

The same stopping times in $\dot{Z}_t$ are also denoted S_n. Now using the fact

$$\hat{Z}_t|_{F \cap G_i} \sim \dot{Z}_t|_{F \cap G_i} \ , \ i = 1, 2$$

one has by induction $\hat{P}^x[\hat{Z}(S_n) \in \cdot] = \dot{P}^x[\dot{Z}(S_n) \in \cdot]$ for all n. The desired equality of hitting distributions will follow from this and the convergence

$$\hat{P}^x[\hat{Z}(S_n) \in D \cap B] \uparrow \hat{P}^x[\hat{Z}(T_D) \in B]$$

for $B \subset E$ and the same convergence in $\dot{Z}_t$. The reason for this convergence is that if $S_n < T_D$ for all n, then for infinitely many n we have $\hat{Z}(S_n) \in G_3 \cap F$ and $\hat{Z}(S_{n+1}) \in G_2^c \cap F$, and so $\hat{Z}(S_n)$ diverges (because $\text{dist}(G_3, G_2^c) > 0$), which implies $\lim_n S_n = T_\Delta$ (because the paths have left limits on $(0, T_\Delta)$) and so $T_D = \infty$; and the same is valid for $\dot{Z}_t$. ∎

REFERENCES

[1] PH. COURRÈGE et P. PRIOURET. Recollements de processus de Markov. *Publ. Inst. Statist. Univ. Paris* **14**(1965) 275-377.

[2] P.J. FITZSIMMONS, R.K. GETOOR and M.J. SHARPE. The Blumenthal-Getoor-McKean theorem revisited. *Seminar on Stochastic Processes*, 1989. Birkhauser, Boston (1990) 35-57.

[3] JOSEPH GLOVER and JOANNA MITRO. Symmetries and functions of Markov processes. *Annals of Probab.* **18**(1990) 655-668.

[4] P.A. MEYER. Renaissance, recollements, mélanges, ralentissement de processus de Markov *Ann. Inst. Fourier, Grenoble* **23**(1975) 465-491.

[5] C.T. SHIH. Construction of right processes from hitting distributions. *Seminar on Stochastic Processes*, 1983. Birkhauser, Boston (1984) 189-256.

C.T. SHIH
Department of Mathematics
University of Michigan
Ann Arbor, Michigan 48109-1003

Measurability of the Solution of a Semilinear Evolution Equation

BIJAN Z. ZANGENEH

1 Introduction

Let H be a real separable Hilbert space with an inner product and a norm denoted by $<,>$ and $\| \|$, respectively. Let $(\Omega, \mathcal{F}, \mathcal{F}_t, P)$ be a complete stochastic basis with a right continuous filtration. Let Z be an H-valued cadlag semimartingale. Consider the initial value problem of the semilinear stochastic evolution equation of the form:

$$\begin{cases} dX_t &= A(t)X_t\,dt + f_t(X_t)dt + dZ_t \\ X(0) &= X_0, \end{cases} \tag{1}$$

where

• $f_t(\cdot) = f(t,\omega,\cdot) : H \to H$ is of monotone type, and for each $x \in H$, $f_t(x)$ is a stochastic process which satisfies certain measurability conditions;

• $A(t)$ is an unbounded closed linear operator which generates an evolution operator $U(t,s)$.

We say X_t is a *mild solution* of (1) if it is a strong solution of the integral equation

$$X_t = U(t,0)X_0 + \int_0^t U(t,s)f_s(X_s)ds + \int_0^t U(t,s)dZ_s. \tag{2}$$

Since Z is a cadlag semimartingale the stochastic convolution integral $\int_0^t U(t,s)dZ_s$ is known to be a cadlag adapted process [see Kotelenez(1982)]. More generally, instead of (2) we are going to study

$$X_t = U(t,0)X_0 + \int_0^t U(t,s)f_s(X_s)ds + V_t, \tag{3}$$

where V_t is a cadlag adapted process.

The existence and uniqueness of the solution of equation (3) in the case in which f is independent of ω and $V \equiv 0$ is a well-known theorem of Browder (1964) and Kato (1964).

In Theorem 4 of this paper we will show the solution of (3) is measurable in the appropriate sense. In addition diverse examples which have arisen in applications are shown to satisfy the hypotheses of Theorem 4 and consequently the results can be applied to these examples. This solution will be shown to be a weak limit of solutions of (3) in the case when $A \equiv 0$, which in turn have been constructed by the Galerkin approximation of the finite-dimensional equation.

In Section 2 we prove that the solution of (3) in the case when $A \equiv 0$ is measurable and in Section 3 we generalize this to the case when A is non-trivial.

In Zangeneh (1990) measurablity of the solution of (3) is used to prove the existence of the solution of stochastic semilinear integral equation

$$X_t = U(t,0)X_0 + \int_0^t U(t,s)f_s(X_s)ds + \int_0^t U(t,s)g_s(X)dW_s + V_t, \tag{4}$$

where

• $g_s(\cdot)$ is a uniformly-Lipschitz predictable functional with values in the space of Hilbert-Schmidt operators on H.

• $\{W_t, t \in \mathbf{R}\}$ is an H-valued cylindrical Brownian motion with respect to $(\Omega, \mathcal{F}, \mathcal{F}_t, P)$.

1.1 Notation and Definitions

Let g be an H-valued function defined on a set $D(g) \subset H$. Recall that g is *monotone* if for each pair

$$x, y \in D(g), \quad < g(x) - g(y), x - y > \geq 0,$$

and g is *semi-monotone* with *parameter* M if, for each pair $x, y \in D(g)$,

$$< g(x) - g(y), x - y > \geq -M\|x - y\|^2.$$

On the real line we can represent any semi-monotone function with parameter M, by $f(x) - Mx$; where f is a non-decreasing function on $\mathbf{R}$.

We say g is *bounded* if there exists an increasing continuous function ψ on $[0, \infty)$ such that $\|g(x)\| \leq \psi(\|x\|), \forall x \in D(g)$. g is *demi-continuous* if, whenever (x_n) is a sequence in $D(g)$ which converges strongly to a point $x \in D(g)$, then $g(x_n)$ converges weakly to $g(x)$.

Let $(\Omega, \mathcal{F}, \mathcal{F}_t, P)$ be a complete stochastic basis with a right continuous filtration.

We follow Yor (1974) and define cylindrical Brownian motion as

Definition 1 *A family of random linear functionals $\{W_t, t \geq 0\}$ on H is called a cylindrical Brownian motion on H if it satisfies the following conditions:*
(i) $W_0 = 0$ *and* $W_t(x)$ *is* $\mathcal{F}_t$*-adapted for every* $x \in H$.
(ii) *For every* $x \in H$ *such that* $x \neq 0$, $W_t(x)/\|x\|$ *is a one-dimensional Brownian motion.*

Note that cylindrical Brownian motion is not H-valued because its covariance is not nuclear. For the properties of cylindrical Brownian motion and the definition of stochastic integrals with respect to the cylindrical Brownian motion see Yor (1974).

2 Measurability of the Solution

2.1 Integral Equations in Hilbert Space

Let $(G, \mathcal{G})$ be a measurable space, i.e., G is a set and $\mathcal{G}$ is a σ-field of subsets of G. Let $T > 0$ and let $S = [0, T]$. Let β be the Borel field of S. Let $L^2(S, H)$ be the set of all H-valued square integrable functions on S.

Consider the integral equation

$$u(t,y) = \int_0^t f(s,y,u(s,y))ds + V(t,y), \quad t \in S,\ y \in G, \tag{5}$$

where $f : S \times G \times H \to H$ and $V : S \times G \to H$. The variable y is a parameter, which in practice will be an element ω of a probability space.

Our aim in this section is to show that under proper hypotheses on f and V there exists a unique solution u to (5), and that this solution is a $\beta \times \mathcal{G}$-measurable function of t and the parameter y.

We say $X(\cdot,\cdot)$ is *measurable* if it is $\beta \times \mathcal{G}$-measurable.

We will study (5) in the case where $-f$ is demi-continuous and semi-monotone on H and V is right continuous and has left limits in t *(cadlag)*.

This has been well-studied in the case in which V is continuous and f is bounded by a polynomial and does not depend on the parameter y. See for example Bensoussan and Temam (1972).

Let $\mathcal{H}$ be the Borel field of H. Consider functions f and V

$$f : S \times G \times H \to H$$

$$V : S \times G \to H.$$

We impose the following conditions on f and V:

Hypothesis 1 **(a)** *f is $\beta \times \mathcal{G} \times \mathcal{H}$-measurable and V is $\mathcal{G} \times \mathcal{H}$-measurable.*
(b) *For each $t \in S$ and $y \in G$, $x \to f(t,y,x)$ is demi-continuous and uniformly bounded in t. (That is, there is a function $\varphi = \varphi(x,y)$ on $\mathbf{R}_+ \times G$ which is continuous and increasing in x and such that for all $t \in S$, $x \in H$, and $y \in G$, $\|f(t,y,x)\| \le \varphi(y,\|x\|)$.)*
(c) *There exists a non-negative $\mathcal{G}$-measurable function $M(y)$ such that for each $t \in S$ and $y \in G$, $x \to -f(t,y,x)$ is semi-monotone with parameter $M(y)$.*
(d) *For each $y \in G$, $t \to V(t,y)$ is cadlag.*

Theorem 1 *Suppose f and V satisfy Hypothesis 1. Then for each $y \in G$, (5) has a unique cadlag solution $u(\cdot,y)$, and $u(\cdot,\cdot)$ is $\beta \times \mathcal{G}$-measurable. Furthermore*

$$\|u(t,y)\| \le \|V(t,y)\| + 2\int_0^t e^{M(y)(t-s)}\|f(s,y,V(s,y))\|ds; \tag{6}$$

$$\|u(\cdot,y)\|_\infty \le \|V(\cdot,y)\|_\infty + 2C_T\varphi(y,\|V(\cdot,y)\|_\infty), \tag{7}$$

where $\|u\|_\infty = \sup_{0\le t\le T}\|u(t)\|$, and

$$C_T = \begin{cases} \frac{1}{M(y)}e^{M(y)T} & \textit{if } M(y) \neq 0 \\ 1 & \textit{otherwise.} \end{cases}$$

Let us reduce this theorem to the case when $M = 0$ and $V = 0$. Define the transformation

$$X(t,y) = e^{M(y)t}(u(t,y) - V(t,y)) \tag{8}$$

and set

$$g(t,y,x) = e^{M(y)t}f(t,y,V(t,y) + xe^{-M(y)t}) + M(y)x. \tag{9}$$

Lemma 1 *Suppose f and V satisfy Hypothesis 1. Let X and g be defined by (8) and (9). Then g is $\beta \times \mathcal{G} \times \mathcal{H}$-measurable and $-g$ is monotone, demi-continuous, and uniformly bounded in t. Moreover u satisfies (5) if and only if X satisfies*

$$X(t,y) = \int_0^t g(s,y,X(s,y))ds, \quad \forall t \in S,\ y \in G. \tag{10}$$

Proof: The verification of this is straightforward. Suppose that V and f satisfy Hypothesis 1. We claim g satisfies the above conditions.

• g is $\beta \times \mathcal{G} \times \mathcal{H}$-measurable.

Indeed, if $h \in H$ then $< f(t,y,\cdot), h >$ is continuous and $V(t,y) + xe^{-M(y)t}$ is $\beta \times \mathcal{G} \times \mathcal{H}$-measurable, so $< f(t,y,V(t,y)+xe^{-M(y)t}), h >$ is $\beta \times \mathcal{G} \times \mathcal{H}$-measurable. Since H is separable then $f\left(t,y,V(t,y) + xe^{-M(y)t}\right)$ is also $\beta \times \mathcal{G} \times \mathcal{H}$-measurable, and since $e^{M(y)t}$ and $M(y)x$ are $\beta \times \mathcal{G} \times \mathcal{H}$-measurable, then g is $\beta \times \mathcal{G} \times \mathcal{H}$-measurable.

• g is bounded, since $\sup_t \|V_t(y)\| < \infty$ and $\|g(t,y,x)\| \le \phi(y, \|x\|)$, where

$$\phi(y,\xi) = e^{M(y)T}\phi(y, \xi + \sup_t \|V_t\|) + M(y)\xi.$$

• g is demi-continous.

• $-g$ is monotone.

Furthermore, one can check directly that if X is measurable, so is u. Since X is continuous in t and V is cadlag, u must be cadlag. It is easy to see that different solutions of (9) correspond to different solutions of (5). Q.E.D.

By Lemma 1, Theorem 1 is a direct consequence of the following.

Theorem 2 *Let $g = g(t,y,x)$ be a $\beta \times \mathcal{G} \times \mathcal{H}$-measurable function on $S \times G \times H$ such that for each $t \in S$ and $y \in G$, $x \to -g(t,y,x)$ is demi-continous, monotone and bounded by φ. Then for each $y \in G$ the equation (10) has a unique continuous solution $X(\cdot,y)$, and $(t,y) \to X(t,y)$ is $\beta \times \mathcal{G}$-measurable.*

Furthermore X satisfies (7) with $M = 0$ and $V = 0$.

Remark that the transformation (8) $u \to X$ is bicontinuous and in particular, implies if X satisfies (6) and (7) for $M = 0$ and $V = 0$, then u satisfies (6) and (7).

Note that y serves only as a nuisance parameter in this theorem. It only enters in the measurability part of the conclusion. In fact, one could restate the theorem somewhat informally as: if g depends measurably on a parameter y in (10), so does the solution.

The proof of Theorem 2 in the case in which f is independent of y is a well-known theorem of Browder (1964) and Kato (1964). One proof of this theorem can be found in Vainberg (1973), Th (26.1), page 322. The proof of the uniqueness and existence are in Vainberg (1973). In this section we will prove the uniqueness of the solution and inequalities (6) and (7). In subsection 2.3 we will prove the measurability and outline the proof of the existence of the solution of equation (10).

Since y is a nuisance parameter, which serves mainly to clutter up our formulas, we will only indicate it explicitly in our notation when we need to do so.

Let us first prove a lemma which we will need for proof of the uniqueness and for the proof of inequalities (6) and (7).

Lemma 2 *If $a(\cdot)$ is an H-valued integrable function on S and if $X(t) := X_0 + \int_0^t a(s)ds$, then*

$$\|X(t)\|^2 = \|X_0\|^2 + 2\int_0^t < X(s), a(s) > ds.$$

Proof: Since $a(s)$ is integrable, then $X(t)$ is absolutely continuous and $X'(t) = a(t)$ a.e. on S. Then $\|X(t)\|$ is also absolutely continuous and

$$\frac{d}{dt}\|X(t)\|^2 = 2 < \frac{dX(t)}{dt}, X(t) >= 2 < a(t), X(t) > \quad \text{a.e.}$$

so that

$$\int_0^t \frac{d}{ds}\|X(s)\|^2 ds = \|X(t)\|^2 - \|X_0\|^2.$$

Thus

$$\|X(t)\|^2 - \|X_0\|^2 = 2\int_0^t < X(s), a(s) > ds.$$

Q.E.D.

Now we can prove inequalities (6) and (7) in case $M = 0$ and $V = 0$.

Lemma 3 *If $M = V = 0$, the solution of the integral equation (10) satisfies the inequality*

$$\|X(t)\| \leq 2\int_0^t \|g(s,0)\|ds \leq 2T\varphi(0).$$

Proof: Since $X(t)$ is a solution of the integral equation (10), then by Lemma 2 we have

$$\begin{aligned}
\|X(t)\|^2 &= 2\int_0^t < g(s, X(s)),\ X(s) > ds \\
&= 2\int_0^t < g(s, X(s)) - g(s,0), X(s) > ds \\
&\quad + 2\int_0^t < g(s,0), X(s) > ds \\
&\leq 2\int_0^t < g(s, X(s)) - g(s,0), X(s) > ds \\
&\quad + 2\int_0^t \|g(s,0)\|\ \|X(s)\|ds.
\end{aligned}$$

Since $-g$ is monotone, the first integral is negative. We can bound the second integral and rewrite the above inequality as

$$\begin{aligned}
\|X(t)\|^2 &\leq 2\int_0^t \|g(s,0)\|\|X(s)\|ds \\
&\leq 2\sup_{0\leq s\leq t}\|X(s)\| \int_0^t \|g(s,0)\|ds.
\end{aligned}$$

Thus $\sup_{0\leq s\leq t}\|X(s)\| \leq 2\int_0^t \|g(s,0)\|ds$. Since $\sup_{0\leq s\leq t}\|g(s,x)\| \leq \varphi(\|x\|)$, the proof is complete. Q.E.D.

Proof of Uniqueness

Let X and Y be two solutions of (10). Then we have

$$X(t,y) - Y(t,y) = \int_0^t [g(s,y,X(s,y)) - g(s,y,Y(s,y))]ds.$$

By Lemma 2 one has

$$\|X(t,y)-Y(t,y)\|^2 = \int_0^t < g(s,y,X(s,y))-g(s,y,Y(s,y)), X(s,y)-Y(s,y) > ds.$$

Since $-g$ is monotone, the right hand side of the above equation is negative, so

$$X(t,y) = Y(t,y).$$

Q.E.D.

2.2 Measurability of the Solution in Finite-dimensional Space

Consider the integral equation

$$X(t,y) = \int_0^t h(s,y,X(s,y))ds, \tag{11}$$

where $h(\cdot,\cdot)$ satisfies the following hypothesis.

Hypothesis 2 **(a)** *h satisfies Hypothesis 1 (a), (b).*
(b) *For each $t \in S$ and $y \in G$, $-h(t,y,\cdot)$ is continuous and monotone.*

Since h is measurable and uniformly bounded in t, then $h(\cdot,y,x)$ is integrable. As $h(t,y,\cdot)$ is continuous, the integral equation (11) is a classical deterministic integral equation in $\mathbf{R}^n$ and the existence of its solution is well known. In subsection 2.1 we proved that (11) has a unique bounded solution, so we only need to prove the measurability of the solution.

The existence, uniqueness and measurability of the solution of (11) is known (see Krylov and Rozovskii (1979) for a proof in a more general situation). Since the measurability result is easy to prove in our setting, we will include a proof in the following theorem for the sake of completeness.

Theorem 3 *The solution of the integral equation (11) is measurable.*

Proof: For the proof of measurability we are going to construct a sequence of solutions of other integral equations which converges uniformly to a solution of (11).
First: Let $\psi(\cdot)$ be a positive C^∞-function on $H_n \simeq \mathbf{R}^n$ with support $\{\|x\| \le T\varphi(0)+2\}$, which is identically equal to one on $\{\|x\| \le T\varphi(y,0)+1\}$. Now define $\tilde{h}(t,x) = h(t,x)\psi(x)$.

$-\tilde{h}$ is semi-monotone. This can be seen because if $\|X\| > T\varphi(0)+2$ and $\|Z\| > T\varphi(0)+2$,then $\tilde{h}(t,X) = \tilde{h}(t,Z) = 0$ and so

$$< \tilde{h}(t,X) - \tilde{h}(t,Z), X - Z >= 0.$$

Let $\|Z\| \le T\varphi(0)+2$. Then

$$\begin{aligned} < \tilde{h}(t,X) - \tilde{h}(t,Z), X - Z > &= < h(t,X)\psi(X) - h(t,Z)\psi(X), X - Z > \\ &\quad + < h(t,Z)\psi(X) - h(t,Z)\psi(Z), X - Z > . \end{aligned}$$

By the Schwarz inequality this is

$$\leq \psi(X) < h(t,X) - h(t,Z), X - Z > +\|h(t,Z)\| \, |\psi(X) - \psi(Z)| \, \|X - Z\|.$$

Since $-h$ is monotone and ψ is positive, the first term of the right hand side of the inequality is negative. Now as Z is bounded and ψ is C^∞ with compact support, the second term is $\leq M(y)\|X - Z\|^2$ for some $M(y)$.

Since by Lemma 3 the solution of (11) is bounded by $T\varphi(0)$, it never leaves the set $\{\|x\| \leq T\varphi(0) + 1\}$, so the unique solution of (11) is also the unique solution of the equation $X(t) = \int_0^t \tilde{h}(s, X(s))ds$. Thus without loss of generality we can assume $h(t,\cdot)$ has compact support.

Second: Define $k(x)$ to be equal to $C\exp\{\frac{1}{\|x\|^2-1}\}$ on $\{\|x\| < 1\}$ and equal to zero on $\{\|x\| \geq 1\}$. Then $k(x)$ is C^∞ with support in the unit ball $\{\|x\| \leq 1\}$. Choose C such that $\int_{\mathbf{R}^n} k(x)dx = 1$. Introduce, for $\varepsilon > 0$

$$I_\varepsilon u(x) = \varepsilon^n \int_{R^n} k(\frac{x-z}{\varepsilon})u(z)dz.$$

This is a C^∞-function called the mollifier of u.

Now define $h_\varepsilon(t,x) = I_\varepsilon h(t,\cdot)(x)$. Since for any ε the first derivatives with respect to x of $J_\varepsilon u(x)$ and also $J_\varepsilon u(x)$ itself are bounded in terms of the maximum of $\|u(x)\|$, then h_ε and $D_x h_\varepsilon$ are bounded in terms of the maximum of $\|h(t,x)\|$. Thus there exist $K_1(y)$ and $K_2(y)$ independent of ε such that

$$K_1(y) \geq \sup_{\|x\| \leq T\varphi(y,0)+2} \|D_x h_\varepsilon(x)\| \quad \text{and} \quad K_2(y) \geq \sup_{\|x\| \leq T\varphi(y,0)+2} \|h_\varepsilon(x)\|.$$

By the mean value theorem we have

$$\|h_\varepsilon(t,y,x_2) - h_\varepsilon(t,y,x_1)\| \leq K_1(y)\|x_2 - x_1\|. \tag{12}$$

Now consider the following integral equation:

$$X_\varepsilon(t) = \int_0^t h_\varepsilon(s, X_\varepsilon(s))ds. \tag{13}$$

Equation (13) can be solved by the Picard method. Since $y \to h(t,y,x)$ is measurable in (t,y), $y \to h_\varepsilon(t,y,x)$ is measurable in (t,y). Then the solution X_ε of equation (13) is measurable and so is $\lim_{\varepsilon\to 0} X_\varepsilon$. To complete the proof of Theorem 3 we need to prove the following lemma.

Lemma 4 *As $\varepsilon \to 0$, the solution X_ε of (13) converges uniformly to a solution X of (11).*

Proof: From (11) and (13) we have

$$X_\varepsilon(t) - X(t) = \int_0^t (h_\varepsilon(s, X_\varepsilon(s)) - h(s, X(s)))ds.$$

Then

$$\begin{aligned} \|X_\varepsilon(t) - X(t)\| \leq & \int_0^t \|h_\varepsilon(s, X_\varepsilon(s)) - h_\varepsilon(s, X(s))\|ds \\ & + \int_0^t \|h_\varepsilon(s, X(s) - h(s, X(s))\|ds. \end{aligned}$$

By (12) we see this is

$$\leq K_1(y) \int_0^t \|X_\varepsilon(s) - X(s)\| ds$$
$$+ \int_0^t \|h_\varepsilon(s, X(s)) - h(s, X(s))\| ds.$$

By Gronwall's inequality we have

$$\sup_{0\leq t\leq T}\|X_\varepsilon(t) - X(t)\| \leq \exp(TK_1) \int_0^T \|h_\varepsilon(s, X(s)) - h(s, X(s))\| ds.$$

But $h_\varepsilon(s, X(s)) \to h(s, X(s))$ pointwise and $\|h_\varepsilon(t, X(t))\| \leq K_2$ so by the dominated convergence theorem,

$$\sup_{0\leq t\leq T}\|X_\varepsilon(t) - X(t)\| \to 0.$$

Q.E.D.

2.3 The Proof of the Measurability in Theorem 2

Now we shall briefly outline the proof of the existence from Vainberg (1973), Th(26.1), page 322 and give a proof of the measurability of the solution of equation (10).

Vainberg constructs a solution of this equation by first solving the finite-dimensional projections of the equation, and then taking the limit. Since the solution of the infinite-dimensional case is constructed as a limit of finite-dimensional solutions, one merely needs to trace the proof and check that the measurability holds at each stage. There is one extra hypothesis in [Vainberg, Th(26.1)], namely that $t \to g(t, x)$ is demi-continuous, whereas in our case, we merely assume g is measurable and uniformly bounded in t [Hypothesis 1 (a) (b)]. However, the demi-continuity of g is not used in showing the existence of the solution of the integral equation (10). It is only used to show the inequality (6) for the finite-dimensional case. We have reproved (6) in Lemma 3.

Now let (H_n) be an increasing sequence of finite-dimensional subspaces of H such that $\cup_n H_n$ is dense in H, and let J_n be the orthogonal projection of H onto H_n, so that $J_n \to I$ strongly.
Consider the integral equation

$$X_n(t) = \int_0^t J_n\, g(s, X_n(s))\, ds. \qquad (14)$$

First let us show that $J_n\, g$ satisfies Hypothesis 2.

- $J_n\, g(t, y, \cdot)$ is continuous.

Since $g(t, y, \cdot)$ is demi-continuous, $g(t, x_k) \to g(t, x)$ weakly when $\|x_k - x\| \to 0$. But $J_n\, g$ takes its values in the finite-dimensional space H_n, where weak and strong convergence coincide, therefore

$$\|J_n\, g(t, x_k) - J_n\, g(t, x)\| \to 0,$$

and $J_n\, g(t, y, \cdot)$ is continuous.

- $J_n\, g(t, y, \cdot)$ is monotone from H_n to H_n.

Let $X, Z \in H_n$. Then

$$< J_n\, g(t, X) - J_n\, g(t, Z), X - Z > = < g(t, X) - g(t, Z), J_n X - J_n Z > \qquad (15)$$

since $J_n = J_n^*$. For $X, Z \in H_n$, $J_n(X - Z) = X - Z$ so the left hand side of (15) is negative, hence $J_n g(t, y, \cdot)$ is monotone.

- $J_n g(t, y)$ satisfies Hypothesis 1(a).
- $J_n g(t, y)$ is uniformly bounded by φ.

Now by Theorem 3, equation (14) has a unique continuous measurable solution which satisfies

$$\|X_n(t)\| \leq 2\int_0^t \|J_n g(s,0)\|ds \leq 2\int_0^t \|g(s,0)\|ds \leq 2T\varphi(y,0). \tag{16}$$

Now we are going to prove

Lemma 5 *For each y, $X_n(\cdot, y)$ converges weakly in $L^2(S, H)$ to a solution $X(\cdot, y)$ of (10). Furthermore $X(\cdot, y)$ is continuous for each y.*

Proof: Let (X_{n_k}) be an arbitrary subsequence of (X_n). By (16) and Hypothesis 1 (b) we have

$$\|g(t, y, X_{n_k}(y,t))\| \leq \varphi(y, \|X_{n_k}(y,t)\|) \leq \varphi(y, 2T\varphi(y,0))$$

so $g(\cdot, X_{n_k}(\cdot))$ is a bounded sequence in $L^2(S, H)$. Then there is a further subsequence (n_{k_l}) such that $g(\cdot, X_{n_{k_l}}(\cdot)) \to Z(\cdot)$ weakly in $L^2(S, H)$ as $l \to \infty$. Each X_n satisfies (14) and it can be proved that $X_{n_{k_l}}(\cdot) \to \int_0^{\cdot} Z(s)ds$ weakly [see Vainberg]. We define X to be the weak limit of $X_{n_{k_l}}$ in $L^2(S, H)$. Vainberg proved that $X(y, \cdot)$ is continuous and is a solution of (10) [see Vainberg, pp. 325-326].

Since the solution $X(\cdot, y)$ is unique, every subsequence of (X_n) has in turn a subsequence which converges to $X(y, \cdot)$ weakly, it follows that the whole sequence X_n converges weakly to X. Q.E.D.

To complete the proof of Theorem 2 we need to show the measurability of $X(\cdot, \cdot)$.

Fix $t \in S$, $h \in H$, since by Theorem 3 X_n is measurable in (t, y), then $\int_0^t < X_n(s, y), h > ds$ is measurable in (t, y). But $\int_0^t < X_n(s, y), h > ds$ converges to $\int_0^t < X(s, y), h > ds$ pointwise, so $\int_0^t < X(s, y), h > ds$ is measurable in (t, y).

As the integrand $< X(s, y), h >$ is continuous in s, then

$$\frac{d}{dt}\int_0^t < X(s,y), h > ds = < X(t,y), h >$$

and since the integral is measurable in (t, y), the function $< X(t, y), h >$ is measurable. By the separablity of H, $X(t, y)$ is measurable in (t, y). Q.E.D.

3 The Semilinear Evolution Equation

3.1 The Main theorem

Suppose $A = \{A(t), t \in S\}$ is a family of operators satisfying the following hypothesis.

Hypothesis 3 **(a)** *There exists $\lambda \in \mathbf{R}$ such that for all $s > 0$, $(A(s) - \lambda I)$ is the generator of a contraction semigroup;*

(b) *the operator-valued function $(-A(t) + \mu I)^{-1}$ is strongly continuously differentiable with respect to t for $t \geq 0$ and $\mu > \lambda$;*

(c) *there exists a fundamental solution $U(t,s)$ of the linear equation $\dot{u}(t) = A(t)u(t)$. Moreover, if $u_0 \in H$ and $f \in C(S,H)$, then the equation*

$$\begin{cases} \dot{u}(t) &= A(t)u(t) + f(t) \\ u(0) &= u_0 \end{cases} \tag{17}$$

has a strong solution u given by

$$u(t) = U(t,0)u_0 + \int_0^t U(t,s)f(s)ds. \tag{18}$$

If $u_0 \in D(A(0))$ and $f \in C^1(S,H)$, then (18) is also a strong solution of (17).

Remark 1 *Note that Hypothesis 3 holds, for example, if $\{A(t), t \in \mathbf{R}^+\}$ is a family of closed operators in H with domain D independent of t, satisfying the following conditions:*

(i) *considered as a mapping of D (with graph norm) into H, $A(t)$ is C^1 in t on $\mathbf{R}^+$ in the strong operator topology;*

(ii) *if $A(t)^*$ is the adjoint of $A(t)$, then $D(A(t)^*) \subset D$ for all t;*

(iii) *$\exists \lambda \in R$ such that*

$$< A(t)x, x > \leq \lambda \|x\|^2, \quad \forall x \in D(A(t)), \quad \forall t \in S.$$

Proof: See Browder (1964)

In the following theorem which is our main theorem we will study the integral equation (3) in a more abstract setting, where $V \equiv V(t,y)$ and $f \equiv f(t,y,x)$ satisfy the hypotheses of Theorem 1.

Theorem 4 *Let $X_0(\cdot)$ be $\mathcal{G}$-measurable. Suppose that f and V satisfy Hypothesis 1 and suppose that $A(t)$ and $U(t,s)$ satisfy Hypothesis 3. Then for each $y \in G$, (3) has a unique cadlag solution $X(\cdot, y)$, and $X(\cdot,\cdot)$ is $\beta \times \mathcal{G}$-measurable. Furthermore*

$$\|X(t)\| \leq \|X_0\| + \|V(t)\| + \int_0^t e^{(\lambda+M)(t-s)} \|f(s, U(s,0)X_0 + V(s))\| ds, \tag{19}$$

$$\|X\|_\infty \leq \|X_0\| + \|V\|_\infty + C_T \varphi(\|X_0\| + \|V\|_\infty), \tag{20}$$

where

$$C_T = \begin{cases} \frac{1}{M+\lambda} e^{(M+\lambda)T} & \text{if } M + \lambda \neq 0 \\ 1 & \text{otherwise.} \end{cases}$$

If X_1 and X_2 are solutions corresponding to different initial values X_{01} and X_{02}, then

$$\|X_2(t) - X_1(t)\| \leq e^{(\lambda+M)t} \|X_{01} - X_{02}\|, \quad t \in S. \tag{21}$$

Proof: By using the transformations (8), and (9) we can assume by Lemma 1 that $X_0 = 0$, $M = 0$ and $V = 0$ in (3). We can also suppose $\lambda \equiv 0$ in Hypothesis 3(a) [see Zangeneh (1990) Lemma 3.1, page 30]. Thus we consider

$$X(t,y) = \int_0^t U(t,s) f(s,y,X(s,y)) ds, \quad t \in S, \quad y \in G. \tag{22}$$

Here y serves only as nuisance parameter. It only enters in the measurability part of the conclusion. The proof of Theorem 4 in the case in which f is independent of y is a well-known theorem of Browder (1964) and Kato (1964).

The existence and uniqueness are therefore known. To establish the measurability and inequalities (19)–(21) we follow the proof of Vainberg (1973), Th (26.2) page 331. Let $A_n(t) := A(t)(I - n^{-1}A(t))^{-1}$, and consider the equation

$$X_n(t,y) = \int_0^t (A_n(s)X_n(s,y) + f(s,y,X_n(s,y))ds. \tag{23}$$

A_n is a bounded operator with $\|A_n(t)\|_L \leq 2n$ which converges strongly to $A(t)$. Vainberg shows that (23) has a unique solution X_n, and moreover that there is a subsequence (X_{n_k}) of X_n which converges weakly in $L^2(S,H)$ to a limit X, which is a solution of (22); and for each y, $X(\cdot,y)$ is continuous.

But now by Lemma 5 X_n converges weakly to X in $L^2(S,H)$. Moreover $f_n(x) := A_n x + f(x)$ satisfies the hypotheses of Theorem 2 so that $X_n(\cdot,\cdot)$ is $\beta\times\mathcal{G}$-measurable. It follows by the proof of Theorem 2 that $X(\cdot,\cdot)$ is $\beta\times\mathcal{G}$-measurable.

The proofs of the inequalities (19)–(21) in case $M = 0$, $\lambda = 0$ and $V \equiv 0$ are in Vainberg (1973), and the extension to the general case of Theorem 4 follows immediately from transformation (8) and (9). Q.E.D.

As an application of Theorem 4 we can show the existence and uniqueness of the solution of (3) when X_0 , f and V satisfy the following conditions.

Hypothesis 4 (a) $X_0 \in \mathcal{F}_0$.

(b) $f = f(t,\omega,x)$ *and* $V = V(t,\omega)$ *are optional;*

(c) *There exists a set* $G \subset \Omega$ *such that* $P(G) = 1$, *and if* $\omega \in G$, *then* f *and* V *satisfy Hypothesis 1.*

Corollary 1 *Suppose that* X_0, f *and* V *satisfy Hypothesis 4. Suppose* A *and* U *satisfy Hypothesis 3. Then (3) has a unique adapted cadlag (continuous, if* V_t *is continuous) solution.*

Proof: The existence and uniqueness of a cadlag solution is immediate from Theorem 4. We need only prove that it is adapted. To see this, fix $s < t$, take $S = [0,s]$, and take $\mathcal{G} = \mathcal{F}_t|_G$ in Theorem 4, where G is the set of Hypothesis 4. Now $\Omega - G$ has measure 0 so it is in $\mathcal{F}_0 \subset \mathcal{F}_t$.

Theorem 4 implies $X(s,\cdot)|_G$ is $\mathcal{G}$-measurable; as all subsets of $\Omega - G$ are in $\mathcal{F}_t$ by completeness, $X(s,\cdot)$ itself is $\mathcal{F}_t$-measurable. By right continuity of the filtration,

$$X(s,\cdot) \in \mathcal{F}_s = \cap_{t>s}\mathcal{F}_t.$$

Thus $\{X(t,\cdot), t \in S\}$ is adapted.

Note that any discontinuity of the solution in general comes from a discontinuity of V. Q.E.D.

3.2 Some Examples

Example (1)

Let A be a closed, self-adjoint, negative definite unbounded operator such that A^{-1} is nuclear. Let $U(t) \equiv e^{tA}$ be a semigroup generated by A. Since A is self-adjoint then U satisfies Hypotheses 3, so it satisfies all the conditions we impose on U.

Let $W(t)$ be a cylindrical Brownian motion on H. Consider the initial-value problem:

$$\begin{cases} dX_t &= AX_t\,dt + f_t(X_t)dt + dW(t), \\ X(0) &= X_0, \end{cases} \tag{24}$$

where X_0, and f satisfy Hypothesis 4.

Let X be a mild solution of (24), i.e. a solution of the integral equation:

$$X_t = U(t)X(0) + \int_0^t U(t-s)f_s(X_s)ds + \int_0^t U(t-s)dW(s). \tag{25}$$

Note that since A^{-1} is nuclear then $\int_0^t U(t-s)dW(s)$ is H-valued [see Dawson (1972)].

The existence and uniqueness of the solution of (25) have been studied in Marcus (1978). He assumed that f is independent of $\omega \in \Omega$ and $t \in S$ and that there are $M > 0$, and $p \geq 1$ for which

$$< f(u) - f(v), u - v > \leq -M\|u-v\|^p$$

and

$$\|f(u)\| \leq C(1 + \|u\|^{p-1}).$$

He proved that this integral equation has a unique solution in $L^p(\Omega, L^p(S,H))$.

As a consequence of Corollary 1 we can extend Marcus' result to more general f and we can show the existence of a strong solution of (25) which is continuous instead of merely being in $L^p(\Omega, L^p(S,H))$.

The Ornstein–Uhlenbeck process $V_t = \int_0^t U(t-s)dW(s)$ has been well-studied e.g. in [Iscoe et al. (1990)] where they show that V_t has a continuous version. We can rewrite (25) as

$$X_t = U(t)X(0) + \int_0^t U(t-s)f_s(X_s)ds + V_t,$$

where V_t is an adapted continuous process. Then by Corollary 1 the equation (25) has a unique continuous adapted solution.

Example (2) Let D be a bounded domain with a smooth boundary in $\mathbf{R}^d$. Let $-A$ be a uniformly strongly elliptic second order differential operator with smooth coefficients on D. Let B be the operator $B = d(x)D_N + e(x)$, where D_N is the normal derivative on ∂D, and d and e are in $C^\infty(\partial D)$. Let A (with the boundary condition $Bf \equiv 0$) be self-adjoint.

Consider the initial-boundary-value problem

$$\begin{cases} \frac{\partial u}{\partial t} + Au &= f_t(u) + \dot{W} \quad \text{on} \quad D \times [0,\infty) \\ Bu &= 0 \quad \text{on} \quad \partial D \times [0,\infty) \\ u(0,x) &= 0 \quad \text{on} \quad D, \end{cases} \tag{26}$$

where $\dot{W} = \dot{W}(t,x)$ is a white noise in space–time [for the definition and properties of white noise see J.B Walsh (1986)], and f_t is a non-linear function that will be defined below. Let $p > \frac{d}{2}$. W can be considered as a *Brownian motion* $\tilde{W}_t$ *on the Sobolov space* H_{-p} [see Walsh (1986), Chapter 4. Page 4.11]. There is a complete orthonormal basis $\{e_k\}$ for H_p.

The operator A (plus boundary conditions) has eigenvalues $\{\lambda_k\}$ with respect to $\{e_k\}$ i.e. $Ae_k = \lambda_k e_k$, $\forall k$. The eigenvalues satisfy $\Sigma_j(1+\lambda_j)^{-p} < \infty$ if $p > \frac{d}{2}$

[see Walsh (1986), Chapter 4, page 4.9]. Then $[A^{-1}]^p$ is nuclear and $-A$ generates a contraction semigroup $U(t) \equiv e^{-tA}$. This semigroup satisfies Hypotheses 3.

Now consider the initial-boundary-value problem (26) as a semilinear stochastic evolution equation

$$du_t + Au_t\,dt = f_t(u_t)dt + d\tilde{W}_t \tag{27}$$

with initial condition $u(0) = 0$, where $f : S \times \Omega \times H_{-p} \to H_{-p}$ satisfies Hypotheses 4(b) and 4(c) relative to the separable Hilbert space $H = H_{-p}$. Now we can define the mild solution of (27) (which is also a mild solution of (26)), as the solution of

$$u_t = \int_0^t U(t-s)f_s(u_s)ds + \int_0^t U(t-s)d\tilde{W}_s. \tag{28}$$

Since $\tilde{W}_t$ is a continuous local martingale on the separable Hilbert space H_{-p}, then $\int_0^t U(t-s)d\tilde{W}_s$ has an adapted continuous version [see Kotelenez (1982)]. If we define

$$V_t := \int_0^t U(t-s)d\tilde{W}_s,$$

then by Corollary 1, equation (28) has a unique continuous solution with values in H_{-p}.

3.3 A Second Order Equation

Let Z_t be a cadlag semimartingale with values in H. Let A satisfy the following:

Hypothesis 5 *A is a closed strictly positive definite self-adjoint operator on H with dense domain D(A), so that there is a* $K > 0$ *such that* $< Ax, x > \geq K\|x\|^2, \quad \forall x \in D(A)$.

Consider the Cauchy problem, written formally as

$$\begin{cases} \frac{\partial^2 x}{\partial t^2} + Ax &= \dot{Z} \\ x(0) &= x_0, \\ \frac{\partial x}{\partial t}(0) &= y_0. \end{cases} \tag{29}$$

Following Curtain and Pritchard (1978), we may write (29) formally as a first-order system

$$\begin{cases} dX(t) &= \mathcal{A}X(t)dt + d\tilde{Z}_t \\ X(0) &= X_0, \end{cases} \tag{30}$$

where $X(t) = \begin{pmatrix} x(t) \\ y(t) \end{pmatrix}$, $\tilde{Z}_t = \begin{pmatrix} 0 \\ Z_t \end{pmatrix}$, $X_0 = \begin{pmatrix} x_0 \\ y_0 \end{pmatrix}$, and $\mathcal{A} = \begin{pmatrix} 0 & I \\ -A & 0 \end{pmatrix}$.

Introduce a Hilbert space $\mathcal{K} = D(A^{1/2}) \times H$ with inner product

$$< X, \bar{X} >_{\mathcal{K}} = < A^{1/2}x, A^{1/2}\bar{x} > + < y, \bar{y} >,$$

and norm

$$\|X\|_{\mathcal{K}}^2 = \|A^{1/2}x\|^2 + \|y\|^2,$$

where $X = \begin{pmatrix} x \\ y \end{pmatrix}$, $\bar{X} = \begin{pmatrix} \bar{x} \\ \bar{y} \end{pmatrix}$ [see Chapter 4, page, 93, Vilenkin (1972)].

Now for $X \in D(\mathcal{A}) = D(A) \times D(A^{1/2})$, we have

$$< X, \mathcal{A}X >_{\mathcal{K}} = < Ax, y > + < y, -Ax > = 0$$

Thus

$$< (\mathcal{A}-\lambda I)X, X >_{\mathcal{K}} = < \mathcal{A}X, X >_{\mathcal{K}} -\lambda\|X\|^2_{\mathcal{K}} = -\lambda\|X\|^2_{\mathcal{K}}.$$

Since

$$| < (\mathcal{A}-\lambda I)X, X >_{\mathcal{K}} | \leq \|(\mathcal{A}-\lambda I)X\|_{\mathcal{K}}\|X\|_{\mathcal{K}},$$

we have

$$\|(\mathcal{A}-\lambda I)X\|_{\mathcal{K}} \geq \lambda\|X\|_{\mathcal{K}}.$$

The adjoint of $\mathcal{A}^*$ of $\mathcal{A}$ is easily shown to be $-\mathcal{A}$. With the same logic

$$\|(\mathcal{A}^*-\lambda I)X\|_{\mathcal{K}} \geq \lambda\|X\|_{\mathcal{K}}.$$

Then $\mathcal{A}$ generates a contraction semigroup $U(t) \equiv e^{t\mathcal{A}}$ on $\mathcal{K}$. [see Curtain and Pritchard (1978), Th (2.14), page 22]. Moreover $\mathcal{A}$ and $U(t)$ satisfy Hypothesis 3 with $\lambda = 0$.

Now consider the mild solution of (30):

$$V_t = U(t)X_0 + \int_0^t U(t-s)d\tilde{Z}_s. \tag{31}$$

Since $\tilde{Z}_t$ is a cadlag semimartingale on $\mathcal{K}$, the stochastic convolution integral $\int_0^t U(t-s)d\tilde{Z}_s$ has a cadlag version [see Kotelenez (1982)], so V_t is a cadlag adapted process on $\mathcal{K}$.

Now let us consider the semilinear Cauchy problem, written formally as

$$\begin{cases} \frac{\partial^2 x(t)}{\partial t^2} + Ax(t) & = & f(x(t), \frac{\partial x(t)}{\partial t}) + \dot{Z}_t \\ x(0) & = & x_0, \\ \frac{\partial x}{\partial t}|_{t=0} & = & y_0, \end{cases} \tag{32}$$

where $f : D(A^{1/2}) \times H \rightarrow H$ satisfies the following conditions:

Hypothesis 6 **(a)** $-f(x,\cdot) : H \rightarrow H$ *is semi-monotone i.e.* $\exists M > 0$ *such that for all* $x \in D(A^{1/2})$ *and all* $y_1, y_2 \in H$

$$< f(x,y_2) - f(x,y_1), y_2 - y_1 > \leq M\|y_2 - y_1\|^2;$$

(b) *for all* $x \in D(A^{1/2})$, $f(x,\cdot)$ *is demi-continuous and there is a continuous increasing function* $\varphi : \mathbf{R}^+ \rightarrow \mathbf{R}^+$ *such that* $\|f(0,y)\| \leq \varphi(\|y\|)$;
(c) $f(\cdot,y) : D(A^{1/2}) \rightarrow H$ *is uniformly Lipschitz , i.e.* $\exists M > 0$ *such that* $\forall y \in H$

$$\|f(x_2,y) - f(x_1,y)\| \leq M\|A^{1/2}(x_2 - x_1)\|.$$

[The completeness of $D(A^{1/2})$ under the norm $\|A^{1/2}x\|$ follows from the strict positivity of $A^{1/2}$.]

Note that any uniformly Lipschitz function $f : D(A^{1/2}) \times H \rightarrow H$ satisfies Hypothesis 6.

Proposition 1 *If f satisfies Hypothesis 6, then the Cauchy problem (32) has a unique mild adapted cadlag solution $x(t)$ with values in $D(A^{1/2})$. Moreover $\frac{dx(t)}{dt}$ is an H-valued cadlag process. If Z_t is continuous, $(x, \frac{dx}{dt})$ is continuous in $\mathcal{K}$.*

Proof: Define a mapping F from $\mathcal{K}$ to $\mathcal{K}$ by $F(x,y) = \begin{pmatrix} 0 \\ f(x,y) \end{pmatrix}$. We are going to show that F satisfies the hypotheses of Corollary 1.

- F is semi-monotone.

Let $X_1 = \begin{pmatrix} x_1 \\ y_1 \end{pmatrix}$ and $X_2 = \begin{pmatrix} x_2 \\ y_2 \end{pmatrix}$. Then

$$\begin{aligned} < F(X_2) - F(X_1), X_2 - X_1 >_{\mathcal{K}} &= < f(x_2,y_2) - f(x_1,y_1), y_2 - y_1 > \\ &= < f(x_2,y_2) - f(x_2,y_1), y_2 - y_1 > \\ &\quad + < f(x_2,y_1) - f(x_1,y_1), y_2 - y_1 > . \end{aligned}$$

By Hypothesis 6(a) and the Schwartz inequality this is

$$\leq M\|y_2 - y_1\|^2 + \|f(x_2,y_1) - f(x_1,y_1)\|\|y_2 - y_1\|.$$

By Hypothesis 6(c) this is

$$\begin{aligned} &\leq M\|y_2 - y_1\|^2 + M\|A^{1/2}(x_2 - x_1)\|\|y_2 - y_1\| \\ &\leq M\|y_2 - y_1\|^2 + M/2\|A^{1/2}(x_2 - x_1)\|^2 + M/2\|y_2 - y_1\|^2 \\ &\leq 3M/2(\|A^{1/2}(x_2 - x_1)\|^2 + \|y_2 - y_1\|^2) \\ &= 3M/2\|X_2 - X_1\|^2_{\mathcal{K}}. \end{aligned}$$

Thus $-F : \mathcal{K} \to \mathcal{K}$ is semi-monotone.

- F is demi-continuous in the pair (x, y) because it is demi-continuous in y and uniformly continuous in x.
- F is bounded since

$$\|F(x)\|_{\mathcal{K}} = \|f(x,y)\| \leq \|f(x,y) - f(0,y)\| + \|f(0,y)\|;$$

by Hypotheses 6(b) and 6(c) this is

$$\leq M\|A^{1/2}x\| + \varphi(\|y\|),$$

and since $\|A^{1/2}x\| \leq \|X\|_{\mathcal{K}}$ and $\|y\| \leq \|X\|_{\mathcal{K}}$ then

$$\|F(X)\|_{\mathcal{K}} \leq M\|X\|_{\mathcal{K}} + \varphi(\|X\|_{\mathcal{K}}).$$

Thus F is bounded by the function $\psi(r) = Mr + \varphi(r)$. Then F satisfies the hypotheses of Corollary 1 on $\mathcal{K}$.

Now as in the linear case we may write (32) as a first order initial value problem:

$$\begin{cases} dX_t &= \mathcal{A}(t)X_t dt + F(X(t))dt + d\tilde{Z}_t, \\ X(0) &= X_0. \end{cases}$$

Since A generates a contraction semigroup $U(t)$ we can write the above initial value problem as

$$X(t) = U(t)X(0) + \int_0^t U(t-s)F(X(s))ds + \int_0^t U(t-s)d\tilde{Z}_t.$$

By (31) we can rewrite this as

$$X(t) = \int_0^t U(t-s)F(X_s)ds + V_t.$$

Since V_t is cadlag and adapted then F, U and V satisfy all the conditions of Corollary 1. Then there is an adapted cadlag solution on $\mathcal{K}$. If Z_t is continuous, V_t is continuous too and X_t is a continuous solution of (32) on $\mathcal{K}$. Q.E.D.

Remark 2 *We assume $f : D(A^{1/2}) \times H \to H$. We could let f depend on $\omega \in \Omega$ and $t \in S$ as well. This would not involve any essential modification of the proof.*

Example (3): Let D, A, B, and W be as in Example (2). Let $p > d/2$ and consider a mixed problem of the form:

$$\begin{cases} \frac{\partial^2 u}{\partial t^2} + Au &= f(u, \frac{\partial u}{\partial t}) + \dot{W} \quad \text{on } D \times [0, \infty) \\ Bu &= 0 \quad \text{on } \partial D \times [0, \infty) \\ u(x, 0) &= 0 \quad \text{on } D \\ \frac{\partial u}{\partial t}(x, 0) &= 0 \quad \text{on } D, \end{cases} \tag{33}$$

where $f : H_{-p+1} \times H_{-p} \to H_{-p}$.

As in Example (2) we consider W as a Brownian motion $\tilde{W}_t$ on the Sobolev space H_{-p}. Now A is a strictly positive definite self-adjoint operator on H_{-p}, and $[A^{-1}]^p$ is nuclear. Since all of the eigenvalues of A are strictly positive, then

$$< Ax, x >_{H_{-p}} \geq \lambda_0 \|x\|^2_{H_{-p}}, \tag{34}$$

for all $X \in D(A) = H_{-p+2}$.

Then we can write (33) as the following Cauchy problem on the Sobolev space H_{-p}:

$$\begin{cases} du_t &= \dot{u}_t \, dt \\ d\dot{u}_t &= -Au_t dt + f(u_t, \dot{u}_t) dt + d\tilde{W}_t \\ u(0) &= 0 \\ \dot{u}(0) &= 0. \end{cases} \tag{35}$$

Now A satisfies (34) and it is a positive definite self-adjoint operator on H_{-p}. Note that if $f \in H_n$, then $A^{1/2} f \in H_{n-1}$ [see, Walsh (1986), Example 3, Page 4.10]. Then $D(A^{1/2}) = H_{-p+1}$. Since $\tilde{W}_t$ is continuous then by Proposition 1, (35) has a continuous mild solution $u_t \in C(S, H_{-p+1})$ and, moreover, $u_t \in C^1(S, H_{-p})$ i.e., the mild solution of (33) is continuous process in H_{-p} for any $p > d/2 - 1$, and it is a differentiable process in H_{-p} for any $p > d/2$.

Acknowledgement

This work has been part of the author's Ph.D. dissertation. He wishes to express his gratitude to his supervisor Professor J.B. Walsh for his guidence and encouragement.

References

[1] **Bensoussan, A. & Temam, R.E.T. (1972)**. Equations aux derivées partielles stochastiques non lineaires (1). *Israel Journal of Mathematics*, 11, 95-129.

[2] **Browder, F.E. (1964)**. Non-linear equations of evolution. *Ann. of Math.*, 80, 485-523

[3] **Curtain, R.F. & Pritchard, A.J. (1978)**. Infinite dimensional linear system theory. *LN in control and information sciences.* 8, Springer-Verlag, Berlin-Heidelberg, New York.

[4] **Dawson, D.A. (1972)**. Stochastic evolution equations. *Mathematical Biosci.*, 15, 287-316.

[5] **Iscoe, I., Marcus, M.B., McDonald, D., Talagrand, M. & Zinn, J. (1990)**. Continuity of l^2-valued Ornstein-Uhlenbeck Process. *The Annals of Probability,* 18(1), 68-84.

[6] **Kato, T. (1964)**. Nonlinear evolution equations in Banach spaces. *Proc. Symp. Appl. Math.*, 17, 50-67.

[7] **Kotelenz, P. (1982)**. A submartingale type inequality with applications to stochastic evolution equations. *Stochastics*, 8, 139-151.

[8] **Krylov, N.V. & Rozovskii, B.L. (1981)**. Stochastic evolution equations. *J. Soviet Math.*, 16, 1233-1277.

[9] **Marcus, R. (1978)**. Parabolic Ito equations with monotone non-linearities. *J. Functional Analysis*, 29, 275-286.

[10] **Vainberg, M.M. (1973)**. *Variational method and method of monotone operators in the theory of nonlinear equations.* John Wiley & Sons, Ltd.

[11] **Vilenkin, N.Ya. (1972)**. *Functional analysis*, Wolters-Noordhoff Publishing, Groninge, The Netherlands.

[12] **Walsh, J.B. (1986)**. An introduction to stochastic partial differential equations. *Lecture Notes in Math.*, 1180, 266-439.

[13] **Yor, M. (1974)**. Existence et unicité de diffusions à valeurs dans un espace de Hilbert. *Ann. Inst. H.Poincaré*, 10, 55-88.

[14] **Z. Zangeneh, B. (1990)**. Semilinear Stochastic Evolution Equations, Ph.D thesis, *University of British Columbia.*

BIJAN Z. ZANGENEH
Department of Mathematics
University of British Columbia
Vancouver, B.C. V6T1Y4
CANADA